K&E Series on Knots and Everything – Vol. 4

GAUGE FIELDS,
KNOTS AND GRAVITY

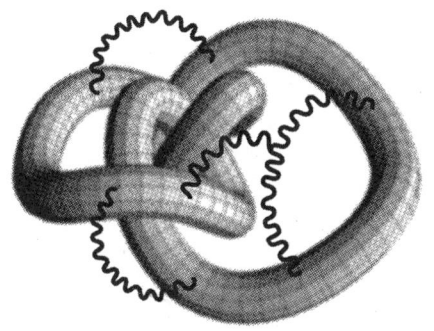

$$R_{\mu\nu} - \tfrac{1}{2} g_{\mu\nu} R = 8\pi\kappa\, T_{\mu\nu}$$

K&E Series on Knots and Everything – Vol. 4

John Baez & Javier P. Muniain
University of California, Riverside

GAUGE FIELDS,
KNOTS AND GRAVITY

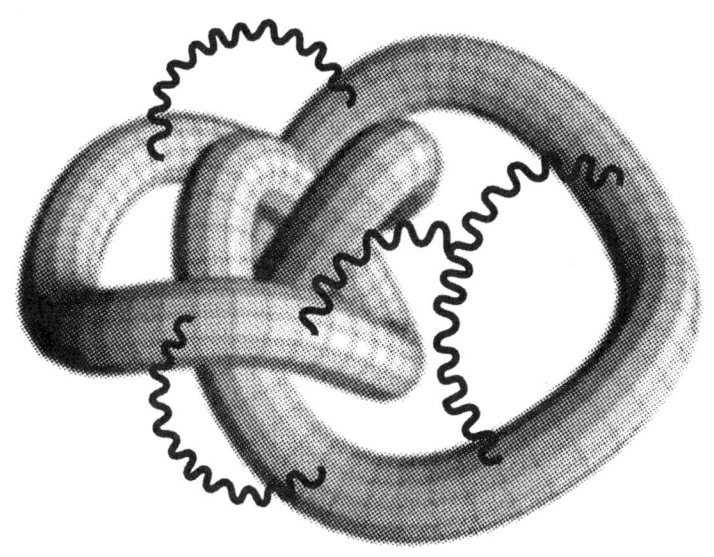

$$R_{\mu\nu} - \tfrac{1}{2} g_{\mu\nu} R = 8\pi \kappa\, T_{\mu\nu}$$

NEW JERSEY · LONDON · SINGAPORE · BEIJING · SHANGHAI · HONG KONG · TAIPEI · CHENNAI

Published by

World Scientific Publishing Co. Pte. Ltd.
5 Toh Tuck Link, Singapore 569224
USA office: 27 Warren Street, Suite 401-402, Hackensack, NJ 07601
UK office: 57 Shelton Street, Covent Garden, London WC2H 9HE

Library of Congress Cataloging-in-Publication Data
Baez, John C., 1961–
  Gauge fields, knots, and gravity / John C. Baez and Javier P. Muniain.
    p. cm. -- (Series on knots and everything ; vol. 4)
  Includes index.
  ISBN-13 978-981-02-1729-7 -- ISBN-10 981-02-1729-3
  ISBN-13 978-981-02-2034-1 (pbk) -- ISBN-10 981-02-2034-0 (pbk)
  1. Gauge fields (Physics) 2. Quantum gravity. 3. Knot theory.
  4. General relativity (Physics) 5. Electromagnetism. I. Muniain, Javier P. II. Title. III. Series: K & E series on knots and everything ; vol. 4.
  QC793.3.F5B33 1994
  530.1'4--dc20   94-3438
                                                                          CIP

British Library Cataloguing-in-Publication Data
A catalogue record for this book is available from the British Library.

First published 1994
Reprinted 2001, 2003, 2006, 2008, 2011, 2013, 2015

Copyright © 1994 by World Scientific Publishing Co. Pte. Ltd.

All rights reserved. This book, or parts thereof, may not be reproduced in any form or by any means, electronic or mechanical, including photocopying, recording or any information storage and retrieval system now known or to be invented, without written permission from the Publisher.

For photocopying of material in this volume, please pay a copying fee through the Copyright Clearance Center, Inc., 222 Rosewood Drive, Danvers, MA 01923, USA. In this case permission to photocopy is not required from the publisher.

Printed by Fulsland Offset Printing (S) Pte Ltd Singapore

To Our Parents

Preface

Two of the most exciting developments of 20th century physics were general relativity and quantum theory, the latter culminating in the 'standard model' of particle interactions. General relativity treats gravity, while the standard model treats the rest of the forces of nature. Unfortunately, the two theories have not yet been assembled into a single coherent picture of the world. In particular, we do not have a working theory of gravity that takes quantum theory into account. Attempting to 'quantize gravity' has led to many fascinating developments in mathematics and physics, but it remains a challenge for the 21st century.

The early 1980s were a time of tremendous optimism concerning string theory. This theory was very ambitious, taking as its guiding philosophy the idea that gravity could be quantized only by unifying it with all the other forces. As the theory became immersed in ever more complicated technical issues without any sign of an immediate payoff in testable experimental predictions, some of this enthusiasm diminished among physicists. Ironically, at the same time, mathematicians found string theory an ever more fertile source of new ideas. A particularly appealing development to mathematicians was the discovery by Edward Witten in the late 1980s that Chern-Simons theory — a quantum field theory in 3 dimensions that arose as a spin-off of string theory — was intimately related to the invariants of knots and links that had recently been discovered by Vaughan Jones and others. Quantum field theory and 3-dimensional topology have become firmly bound together ever since, although there is much that remains mysterious about the relationship.

While less popular than string theory, a seemingly very different approach to quantum gravity also made dramatic progress in the 1980s. Abhay Ashtekar, Carlo Rovelli, Lee Smolin and others discovered how to rewrite general relativity in terms of 'new variables' so that it more closely resembled the other forces of nature, allowing them to apply a new set of techniques to the problem of quantizing gravity. The philosophy of these researchers was far more conservative than that of the string theorists. Instead of attempting a 'theory of everything' describing all forces and all particles, they attempted to understand quantum gravity *on its own*, following as closely as possible the traditional guiding principles of both general relativity and quantum theory. Interestingly, they too were led to the study of knots and links. Indeed, their approach is often known as the 'loop representation' of quantum gravity. Furthermore, quantum gravity in 4 dimensions turned out to be closely related to Chern-Simons theory in 3 dimensions. Again, there is much that remains mysterious about this. For example, one wonders why Chern-Simons theory shows up so prominently both in string theory and the loop representation of quantum gravity. Perhaps these two approaches are not as different as they seem!

It is the goal of this text to provide an *elementary* introduction to some of these developments. We hope that both physicists who wish to learn more differential geometry and topology, and mathematicians who wish to learn more gauge theory and general relativity, will find this book a useful place to begin. The main prerequisites are some familiarity with electromagnetism, special relativity, linear algebra, and vector calculus, together with some of that undefinable commodity known as 'mathematical sophistication'.

The book is divided into three parts that treat electromagnetism, gauge theory, and general relativity, respectively. Part I of this book introduces the language of modern differential geometry, and shows how Maxwell's equations can be drastically simplified using this language. We stress the coordinate-free approach and the relevance of global topological considerations in understanding such things as the Bohm-Aharonov effect, wormholes, and magnetic monopoles. Part II introduces the mathematics of gauge theory — fiber bundles, connections and curvature — and then introduces the Yang-Mills equation, Chern classes, and Chern-Simons classes. It also includes a brief introduction to knot theory and its relation to Chern-Simons theory. Part

Preface ix

III introduces the basic concepts of Riemannian and semi-Riemannian geometry and then concentrates on topics in general relativity of special importance to quantum gravity: the Einstein-Hilbert and Palatini formulations of the action principle for gravity, the ADM formalism, and canonical quantization. Here we emphasize tensor computations written in the notation used in general relativity. We conclude this part with a sketch of Ashtekar's 'new variables' and the way Chern-Simons theory provides a solution to the Wheeler-DeWitt equation (the basic equation of canonical quantum gravity).

While we attempt to explain everything 'from scratch' in a self-contained manner, we really hope to lure the reader into further study of differential geometry, topology, gauge theory, general relativity and quantum gravity. For this reason, we provide copious notes at the end of each part, listing our favorite reading material on all these subjects. Indeed, the reader who wishes to understand any of these subjects in depth may find it useful to read some of these references in parallel with our book. This is especially true because we have left out many relevant topics in order to keep the book coherent, elementary, and reasonable in size. For example, we have not discussed fermions (or mathematically speaking, spinors) in any detail. Nor have we treated principal bundles. Also, we have not done justice to the experimental aspects of particle physics and general relativity, focusing instead upon their common conceptual foundation in gauge theory. The reader will thus have to turn to other texts to learn about such matters.

One really cannot learn physics or mathematics except by doing it. For this reason, this text contains over 300 exercises. Of course, far more exercises are assigned in texts than are actually done by the readers. At the very least, we urge the reader to read and ponder the exercises, the results of which are often used later on. The text also includes 130 illustrations, since we wish to emphasize the geometrical and topological aspects of modern physics. Terms appear in boldface when they are defined, and all such definitions are referred to in the index.

This book is based on the notes of a seminar on knot theory and quantum gravity taught by J.B. at U. C. Riverside during the school year 1992-1993. The seminar concluded with a conference on the subject, the proceedings of which will appear in a volume entitled *Knots* and Quantum Gravity.

We would like to thank Louis Kauffman for inviting us to write this book, and also Chris Lee and Ms. H. M. Ho of World Scientific for helping us at every stage of the writing and publication process. We also wish to express our thanks to Edward Heflin and Dardo D. Píriz for reading parts of the manuscript and to Carl Yao for helping us with some LaTeXcomplications. Scott Singer of the Academic Computing Graphics and Visual Imaging Lab of the U. C. Riverside deserves special thanks and recognition for helping us to create the book cover. Some of the graphics used for the design of the cover were generated with *Mathematica*, by Wolfram Research, Inc.; these were kindly given to us by Joe Grohens of WRI.

J.B. is indebted to many mathematicians and physicists for useful discussions and correspondence, but he would particularly like to thank Abhay Ashtekar, whose work has done so much to unify the study of gauge fields, knots and gravity. He would also like to thank the readers of the USENET physics and mathematics newsgroups, who helped in many ways with the preparation of this book. He dedicates this book to his parents, Peter and Phyllis Baez, with profound thanks for their love. He also gives thanks and love to his mathematical muse, Lisa Raphals.

J.M. dedicates this book to his parents, Luis and Crescencia Pérez de Muniáin y Mohedano, for their many years of continued love and support. He is grateful to Eleanor Anderson for being a patient and inspiring companion during the long and hard hours taken to complete this book. He also acknowledges José Wudka for many discussions on quantum field theory.

Contents

	Preface	vii
I	**Electromagnetism**	**1**
1	Maxwell's Equations	3
2	Manifolds	15
3	Vector Fields	23
4	Differential Forms	39
5	Rewriting Maxwell's Equations	69
6	DeRham Theory in Electromagnetism	103
	Notes to Part I	153
II	**Gauge Fields**	**159**
1	Symmetry	161
2	Bundles and Connections	199
3	Curvature and the Yang-Mills Equation	243
4	Chern-Simons Theory	267

5	Link Invariants from Gauge Theory	291
	Notes to Part II	353

III Gravity 363

1	Semi-Riemannian Geometry	365
2	Einstein's Equation	387
3	Lagrangians for General Relativity	397
4	The ADM Formalism	413
5	The New Variables	437
	Notes to Part III	451
	Index	457

Part I

Electromagnetism

Chapter 1

Maxwell's Equations

Our whole progress up to this point may be described as a gradual development of the doctrine of relativity of all physical phenomena. Position we must evidently acknowledge to be relative, for we cannot describe the position of a body in any terms which do not express relation. The ordinary language about motion and rest does not so completely exclude the notion of their being measured absolutely, but the reason of this is, that in our ordinary language we tacitly assume that the earth is at rest.... There are no landmarks in space; one portion of space is exactly like every other portion, so that we cannot tell where we are. We are, as it were, on an unruffled sea, without stars, compass, sounding, wind or tide, and we cannot tell in what direction we are going. We have no log which we can cast out to take a dead reckoning by; we may compute our rate of motion with respect to the neighboring bodies, but we do not know how these bodies may be moving in space. – James Clerk Maxwell, 1876.

Starting with Maxwell's beautiful theory of electromagnetism, and inspired by it, physicists have made tremendous progress in understanding the basic forces and particles constituting the physical world. Maxwell showed that two seemingly very different forces, the electric and magnetic forces, were simply two aspects of the 'electromagnetic field'. In so doing, he was also able to explain *light* as a phenomenon in which ripples in the electric field create ripples in the magnetic field, which in turn create new ripples in the electric field, and so on. Shockingly, however, Maxwell's theory also predicted that light emitted by a moving body would travel no faster than light from a stationary body.

Eventually this led Lorentz, Poincaré and especially Einstein to realize that our ideas about space and time had to be radically revised. That the motion of a body can only be measured relative to another body had been understood to some extent since Galileo. Taken in conjunction with Maxwell's theory, however, this principle forced the recognition that in addition to the rotational symmetries of space there must be symmetries that mingle the space and time coordinates. These new symmetries also mix the electric and magnetic fields, charge and current, energy and momentum, and so on, revealing the world to be much more coherent and tightly-knit than had previously been suspected.

There are, of course, forces in nature besides electromagnetism, the most obvious of which is gravity. Indeed, it was the simplicity of gravity that gave rise the first conquests of modern physics: Kepler's laws of planetary motion, and then Newton's laws unifying celestial mechanics with the mechanics of falling bodies. However, reconciling the simplicity of gravity with relativity theory was no easy task! In seeking equations for gravity consistent with his theory of special relativity, Einstein naturally sought to copy the model of Maxwell's equations. However, the result was not merely a theory in which ripples of some field propagate through spacetime, but a theory in which the geometry of spacetime itself ripples and bends. Einstein's equations say, roughly, that energy and momentum affect the *metric* of spacetime (whereby we measure time and distance) much as charges and currents affect the electromagnetic field. This served to heighten hopes that much or perhaps even all of physics is fundamentally *geometrical* in character.

There were, however, severe challenges to these hopes. Attempts by Einstein, Weyl, Kaluza and Klein to further unify our description of the forces of nature using ideas from geometry were largely unsuccessful. The reason is that the careful study of atoms, nuclei and subatomic particles revealed a wealth of phenomena that do not fit easily into any simple scheme. Each time technology permitted the study of smaller distance scales (or equivalently, higher energies), new puzzles arose. In part, the reason is that physics at small distance scales is completely dominated by the principles of *quantum theory*. The naive notion that a particle is a point tracing out a path in spacetime, or that a field assigns a number or vector to each point of spacetime, proved to be wholly inadequate, for one cannot measure the position and velocity

Maxwell's Equations

of a particle simultaneously with arbitrary accuracy, nor the value of a field and its time derivative. Indeed, it turned out that the distinction between a particle and field was somewhat arbitrary. Much of 20th century physics has centered around the task of making sense of microworld and developing a framework with which one can understand subatomic particles and the forces between them in the light of quantum theory.

Our current picture, called the standard model, involves three forces: electromagnetism and the weak and strong nuclear forces. These are all 'gauge fields', meaning that they are described by equations closely modelled after Maxwell's equations. These equations describe *quantum* fields, so the forces can be regarded as carried by particles: the electromagnetic force is carried by the photon, the weak force is carried by the W and Z particles, and the strong force is carried by gluons. There are also charged particles that interact with these force-carrying particles. By 'charge' here we mean not only the electric charge but also its analogs for the other forces. There are two main kinds of charged particles, quarks (which feel the strong force) and leptons (which do not). All of these charged particles have corresponding antiparticles of the same mass and opposite charge.

Somewhat mysteriously, the charged particles come in three families or 'generations'. The first generation consists of two leptons, the electron e and the electron neutrino ν_e, and two quarks, the up and down, or u and d. Most of the matter we see everyday is made out of these first-generation particles. For example, according to the standard model the proton is a composite of two up quarks and one down, while the neutron is two downs and an up. There is a second generation of quarks and leptons, the muon μ and muon neutrino ν_μ, and the charmed and strange quarks c, s. For the most part these are heavier than the corresponding particles in the first generation, although all the neutrinos appear to be massless or nearly so. For example, the muon is about 207 times as massive as the electron, but almost identical in every other respect. Then there is a third, still more massive generation, containing the tau τ and tau neutrino ν_τ, and the top and bottom quarks t and b. For many years the top quark was merely conjectured to exist, but just as this book went to press, experimentalists announced that it may finally have been found.

Finally, there is a very odd charged particle in the standard model, the Higgs particle, which is neither a quark nor a lepton. This has not been observed either, and is hypothesized to exist primarily to explain the relation between the elecromagnetic and weak forces.

Even more puzzling than all the complexities of the standard model, however, is the question of where *gravity* fits into the picture! Einstein's equations describing gravity do *not* take quantum theory into account, and it has proved very difficult to 'quantize' them. We thus have not one picture of the world, but two: the standard model, in which all forces except gravity are described in accordance with quantum theory, and general relativity, in which gravity alone is described, not in accordance with quantum theory. Unfortunately it seems difficult to obtain guidance from experiment; simple considerations of dimensional analysis suggest that quantum gravity effects may become significant at distance scales comparable to the **Planck length**,

$$\ell_p = (\hbar\kappa/c^3)^{1/2},$$

where $\hbar$ is Planck's constant, κ is Newton's gravitational constant, and c is the speed of light. The Planck length is about $1.616 \cdot 10^{-35}$ meters, far below the length scales we can probe with particle accelerators.

Recent developments, however, hint that gravity may be closer to the gauge theories of the standard model than had been thought. Fascinatingly, the relationship also involves the study of *knots* in 3-dimensional space. While this work is in its early stages, and may not succeed as a theory of physics, the new mathematics involved is so beautiful that it is difficult to resist becoming excited. Unfortunately, understanding these new ideas depends on a thorough mastery of quantum field theory, general relativity, geometry, topology, and algebra. Indeed, it is almost certain that *nobody* is sufficiently prepared to understand these ideas fully! The reader should therefore not expect to understand them when done with this book. Our goal in this book is simply to start fairly near the beginning of the story and bring the reader far enough along to see the frontiers of current research in dim outline.

We must begin by reviewing some geometry. These days, when mathematicians speak of geometry they are usually referring not to Euclidean geometry but to the many modern generalizations that fall

Maxwell's Equations

under the heading of 'differential geometry'. The first theory of physics to *explicitly* use differential geometry was Einstein's general relativity, in which gravity is explained as the curvature of spacetime. The gauge theories of the standard model are of a very similar geometrical character (although quantized). But there is also a lot of differential geometry lurking in Maxwell's equations, which after all were the inspiration for both general relativity and gauge theory. So, just as a good way to master auto repair is to take apart an old car and put in a new engine so that it runs better, we will begin by taking apart Maxwell's equations and putting them back together using modern differential geometry.

In their classic form, Maxwell's equations describe the behavior of two vector fields, the **electric field** $\vec{E}$ and the **magnetic field** $\vec{B}$. These fields are defined throughout space, which is taken to be $\mathbb{R}^3$. However, they are also functions of time, a real-valued parameter t. The electric and magnetic fields depend on the electric **charge density** ρ, which is a time-dependent function on space, and also on the electric **current density** $\vec{j}$, which is time-dependent vector field on space. (For the mathematicians, let us note that unless otherwise specified, functions are assumed to be real-valued, and functions and vector fields on $\mathbb{R}^n$ are assumed to be **smooth**, that is, infinitely differentiable.)

In units where the speed of light is equal to 1, **Maxwell's equations** are:

$$\nabla \cdot \vec{B} = 0$$
$$\nabla \times \vec{E} + \frac{\partial \vec{B}}{\partial t} = 0$$
$$\nabla \cdot \vec{E} = \rho$$
$$\nabla \times \vec{B} - \frac{\partial \vec{E}}{\partial t} = \vec{j}.$$

There are a number of interesting things about these equations that are worth understanding. First, there is the little fact that we can only determine the direction of the magnetic field experimentally if we know the difference between right and left. This is easiest to see from the **Lorentz force law**, which says that the force on a charged particle with charge q and velocity $\vec{v}$ is

$$\vec{F} = q\,(\vec{E} + \vec{v} \times \vec{B}).$$

To measure $\vec{E}$, we need only measure the force $\vec{F}$ on a static particle and divide by q. To figure out $\vec{B}$, we can measure the force on charged particles with a variety of velocities. However, recall that the definition of the cross product involves a completely arbitrary right-hand rule! We typically define

$$\vec{v} \times \vec{B} = (v_y B_z - v_z B_y, v_z B_x - v_x B_z, v_x B_y - v_y B_x).$$

However, this is just a convention; we could have set

$$\vec{v} \times \vec{B} = (v_z B_y - v_y B_z, v_x B_z - v_z B_x v_y B_x - v_x B_y),$$

and all the mathematics of cross products would work just as well. If we used this 'left-handed cross product' when figuring out $\vec{B}$ from measurements of $\vec{F}$ for various velocities $\vec{v}$, we would get an answer for $\vec{B}$ with the opposite of the usual sign! It may seem odd that $\vec{B}$ depends on an arbitrary convention this way. In fact, this turns out to be an important clue as to the mathematical structure of Maxwell's equations.

Secondly, Maxwell's equations naturally come in two pairs. The pair that does not involve the electric charge and current densities

$$\nabla \cdot \vec{B} = 0 \qquad \nabla \times \vec{E} + \frac{\partial \vec{B}}{\partial t} = 0,$$

looks very much like the pair that *does*:

$$\nabla \cdot \vec{E} = \rho \qquad \nabla \times \vec{B} - \frac{\partial \vec{E}}{\partial t} = \vec{j}.$$

Note the funny minus sign in the second pair. The symmetry is clearest in the **vacuum** Maxwell equations, where the charge and current densities vanish:

$$\nabla \cdot \vec{B} = 0 \qquad \nabla \times \vec{E} + \frac{\partial \vec{B}}{\partial t} = 0,$$

$$\nabla \cdot \vec{E} = 0 \qquad \nabla \times \vec{B} - \frac{\partial \vec{E}}{\partial t} = 0.$$

Maxwell's Equations

Then the transformation

$$\vec{B} \mapsto \vec{E}, \qquad \vec{E} \mapsto -\vec{B}$$

takes the first pair of equations to the second and vice versa! This symmetry is called **duality** and is a clue that the electric and magnetic fields are part of a unified whole, the electromagnetic field. Indeed, if we introduce a complex-valued vector field

$$\vec{\mathcal{E}} = \vec{E} + i\vec{B},$$

duality amounts to the transformation

$$\vec{\mathcal{E}} \mapsto i\vec{\mathcal{E}},$$

and the vacuum Maxwell equations boil down to two equations for $\vec{\mathcal{E}}$:

$$\nabla \cdot \vec{\mathcal{E}} = 0, \qquad \nabla \times \vec{\mathcal{E}} = i\frac{\partial \vec{\mathcal{E}}}{\partial t}$$

This trick has very practical applications. For example, one can use it to find solutions that correspond to plane waves moving along at the speed of light, which in the units we are using equals 1.

Exercise 1. *Let $\vec{k}$ be a vector in $\mathbb{R}^3$ and let $\omega = |\vec{k}|$. Fix $\vec{\mathbf{E}} \in \mathbb{C}^3$ with $\vec{k} \cdot \vec{\mathbf{E}} = 0$ and $i\vec{k} \times \vec{\mathbf{E}} = \omega \vec{\mathbf{E}}$. Show that*

$$\vec{\mathcal{E}}(t, \vec{x}) = \vec{\mathbf{E}}\, e^{-i(\omega t - \vec{k} \cdot \vec{x})}$$

satisfies the vacuum Maxwell equations.

The symmetry between $\vec{E}$ and $\vec{B}$ does not, however, extend to the non-vacuum Maxwell equations. We can consider making ρ and $\vec{\jmath}$ complex, and writing down:

$$\nabla \cdot \vec{\mathcal{E}} = \rho, \qquad \nabla \times \vec{\mathcal{E}} = i(\frac{\partial \vec{\mathcal{E}}}{\partial t} + \vec{\jmath}).$$

However, this amounts to introducing magnetic charge and current density, since if we split ρ and $\vec{\jmath}$ into real and imaginary parts, we see that the imaginary parts play the role of magnetic charge and current densities:

$$\rho = \rho_e + i\rho_m,$$
$$\vec{j} = \vec{j}_e + i\vec{j}_m.$$

We get

$$\nabla \cdot \vec{B} = \rho_m \qquad \nabla \times \vec{E} + \frac{\partial \vec{B}}{\partial t} = -\vec{j}_m,$$

$$\nabla \cdot \vec{E} = \rho_e \qquad \nabla \times \vec{B} - \frac{\partial \vec{E}}{\partial t} = \vec{j}_e.$$

These equations are quite charming, but unfortunately no magnetic charges — so called **magnetic monopoles** — have been observed! (We will have a bit more to say about this in Chapter 6.) We could simply keep these equations and say that ρ and $\vec{j}$ are real-valued on the basis of experimental evidence. But it is a mathematical as well as a physical challenge to find a better way of understanding this phenomenon. It turns out that the formalism of gauge theory makes it seem quite natural.

Finally, there is the connection between Maxwell's equations and special relativity. The main idea of special relativity is that in addition to the symmetries of space (translations and rotations) and time (translations) there are equally important symmetries mixing space and time, the Lorentz transformations. The idea is that if you and I are both unaccelerated, so that my velocity with respect to you is constant, the coordinates I will naturally use, in which I am at rest, will differ from yours, in which you are at rest. If your coordinate system is (t, x, y, z) and I am moving with velocity v in the x direction with respect to you, for example, the coordinates in which I am at rest are given by

$$\begin{aligned} t' &= (\cosh \phi)t - (\sinh \phi)x \\ x' &= -(\sinh \phi)t + (\cosh \phi)x \\ y' &= y \\ z' &= z, \end{aligned}$$

where ϕ is a convenient quantity called the **rapidity**, defined so that $\tanh \phi = v$. Note the close resemblance to the formula for rotations in

Maxwell's Equations

space. The idea is that just as the x, y, and z components of position are all just aspects of something more important, the position itself, space and time are just aspects of a unitary whole, *spacetime*.

Maxwell's equations are invariant under these Lorentz transformations — indeed, this was the main fact that led Einstein to special relativity! He realized that Maxwell's equations predict that *any* unaccelerated observer will measure light moving in *any* direction in the vacuum to have the *same* speed. Mathematically speaking, the point is that if we have a solution of Maxwell's equations and we do a Lorentz transformation on the coordinates together with a certain transformation of $\vec{E}$, $\vec{B}$, ρ and $\vec{j}$, we again have a solution.

For example, suppose that we do a Lorentz transformation of velocity v in the x direction, as above. The precise recipe for transforming the charge and current densities is

$$\begin{aligned}
\rho' &= (\cosh\phi)\rho - (\sinh\phi)j_x \\
j'_x &= -(\sinh\phi)\rho + (\cosh\phi)j_x \\
j'_y &= j_y \\
j'_z &= j_z.
\end{aligned}$$

Note that ρ and $\vec{j}$ get mixed up together. In fact, we shall see that they are really just two aspects of a single thing called the 'current', which has ρ as its component in the time direction and j_x, j_y, j_z as its components in the space directions.

The formula for transforming the electric and magnetic fields under the same Lorentz transformation is somewhat more complicated:

$$\begin{aligned}
E'_x &= E_x \\
E'_y &= (\cosh\phi)E_y - (\sinh\phi)B_z \\
E'_z &= (\sinh\phi)B_y + (\cosh\phi)E_z,
\end{aligned}$$

$$\begin{aligned}
B'_x &= B_x \\
B'_y &= (\cosh\phi)B_y + (\sinh\phi)E_z \\
B'_z &= -(\sinh\phi)E_y + (\cosh\phi)B_z.
\end{aligned}$$

The most important message here is that the electric and magnetic fields are two aspects of a unified 'electromagnetic field'. Also, we see that the electromagnetic field is more complicated in character than the current, since it has six independent components that transform in a more subtle manner. It turns out to be a '2-form'.

When we have rewritten Maxwell's equations using the language of differential geometry, all the things we have just discussed will be much clearer — at least if we succeed in explaining things well. The key step, which is somewhat shocking to the uninitiated, is to work as much as possible in a manner that does not require a choice of coordinates. After all, as far as we can tell, the world was *not* drawn on graph paper. Coordinates are merely something *we* introduce for our own convenience, and the laws of physics should not care which coordinates we happen to use. If we postpone introducing coordinates until it is actually necessary, we will not have to do anything to show that Maxwell's equations are invariant under Lorentz transformations; it will be *manifest*.

Just for fun, let us write down the new version of Maxwell's equations right away. We will explain what they *mean* quite a bit later, so do not worry if they are fairly cryptic. They are:

$$dF = 0$$
$$\star d \star F = J.$$

Here F is the 'electromagnetic field' and J is the 'current', while the d and $\star$ operators are slick ways of summarizing all the curls, divergences and time derivatives that appear in the old-fashioned version. The equation $dF = 0$ is equivalent to the first pair of Maxwell's equations, while the equation $\star d \star F = J$ is equivalent to the second pair. The 'funny minus sign' in the second pair will turn out to be a natural consequence of how the $\star$ operator works.

If the reader is too pragmatic to get excited by the terse beauty of this new-fangled version of Maxwell's equations, let us emphasize that this way of writing them is a warm-up for understanding gauge theory, and allows us to study Maxwell's equations and gauge theory on curved spacetimes, as one needs to in general relativity. Indeed, we will start by developing enough differential geometry to do a fair amount of physics on general spacetimes. Then we will come back to Maxwell's equations. We warn the reader that the next few sections are not really

Maxwell's Equations

a solid course in differential geometry. Whenever something is at all tricky to prove we will skip it! The easygoing reader can take some facts on faith; the careful reader may want to get ahold of a good book on differential geometry to help fill in these details. Some suggestions on books appear in the notes at the end of Part I.

Chapter 2

Manifolds

We therefore reach this result: In the general theory of relativity, space and time cannot be defined in such a way that differences of the spatial co-ordinates can be directly measured by the unit measuring-rod, or differences in the time co-ordinate by a standard clock.

The method hitherto employed for laying co-ordinates into the space-time continuum in a definite manner thus breaks down, and there seems to be no other way which would allow us to adapt systems of co-ordinates to the four-dimensional universe so that we might expect from their application a particularly simple formulation of the laws of nature. So there is nothing for it but to regard all imaginable systems of co-ordinates, on principle, as equally suitable for the description of nature. This comes to requiring that:

The general laws of nature are to be expressed by equations which hold good for all systems of co-ordinates, that is, are co-variant with respect to any substitutions whatever (generally covariant). — Albert Einstein

In order to do modern physics we need to be able to handle spaces and spacetimes that are more general than good old $\mathbb{R}^n$. The kinds of spaces we will be concerned with are those that look *locally* like $\mathbb{R}^n$, but perhaps not *globally*. Such a space is called an n-dimensional 'manifold'. For example, the sphere

$$x^2 + y^2 + z^2 = 1,$$

looks locally like the plane $\mathbb{R}^2$, which is why some people thought the Earth was flat. These days we call this sphere S^2 — the **2-sphere** — to indicate that it is a 2-dimensional manifold. Similarly, while the space we live in looks locally like $\mathbb{R}^3$, we have no way yet of ruling out the possibility that it is really S^3, the **3-sphere**:

$$w^2 + x^2 + y^2 + z^2 = 1,$$

and indeed, in many models of cosmology space is a 3-sphere. In such a universe one could, if one had time, sail around the cosmos in a spaceship just as Magellan circumnavigated the globe. More generally, it is even possible that spacetime has more than 4 dimensions, as is assumed in so-called 'Kaluza-Klein theories'. For a while, string theorists seemed quite sure that the universe must either be 10 or 26-dimensional! More pragmatically, there is a lot of interest in low-dimensional physics, such as the behavior of electrons on thin films and wires. Also, classical mechanics uses 'phase spaces' that may have very many dimensions.

These are some of the physical reasons why it is good to generalize vector calculus so that it works nicely on any manifold. On the other hand, mathematicians have many reasons of their own for dealing with manifolds. For example, the set of solutions of an equation is often a manifold (see the equation for the 3-sphere above).

We now head towards a precise definition of a manifold. First of all, we remind the reader that a **topological space** is a set X together with a family of subsets of X, called the **open sets**, required to satisfy the conditions:

1) The empty set and X itself are open,
2) If $U, V \subseteq X$ are open, so is $U \cap V$,
3) If the sets $U_\alpha \subseteq X$ are open, so is the union $\bigcup U_\alpha$.

The collection of sets taken to be open is called the **topology** of X. An open set containing a point $x \in X$ is called a **neighborhood** of x. The complement of an open set is called **closed**.

A basic example is $\mathbb{R}^n$, where a set U is taken to be open if for every $x \in U$, all points sufficiently close to x are also in U:

Manifolds

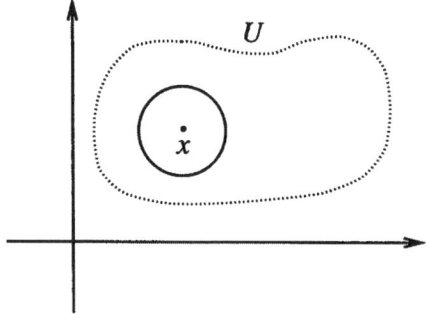

Fig. 1. An open set in $\mathbb{R}^2$

The use of a topology is that it allows us to define continuous functions. Roughly speaking, a function is continuous if it sends nearby points to nearby points. The trick is making the notion of 'nearby' precise using open sets. A function $f: X \to Y$ from one topological space to another is defined to be **continuous** if, given any open set $U \subseteq Y$, the inverse image $f^{-1}U \subseteq X$ is open.

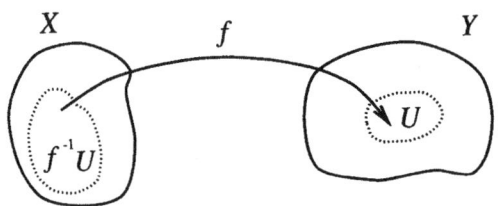

Fig. 2. A continuous function from X to Y

If one has not yet, one should do the following exercise.

Exercise 2. *Show that a function $f: \mathbb{R}^n \to \mathbb{R}^m$ is continuous according to the above definition if and only if it is according to the epsilon-delta definition: for all $x \in \mathbb{R}^n$ and all $\epsilon > 0$, there exists $\delta > 0$ such that $\|y - x\| < \delta$ implies $\|f(y) - f(x)\| < \epsilon$.*

The idea of a manifold is that, like the globe, we can cover it with patches that look just like $\mathbb{R}^n$. More precisely, we say that a collection

U_α of open sets **covers** a topological space X if their union is all of X. Given a topological space X and an open set $U \subseteq X$, we define a **chart** to be a continuous function $\varphi: U \to \mathbb{R}^n$ with a continuous inverse (the inverse being defined on the set $\varphi(U)$).

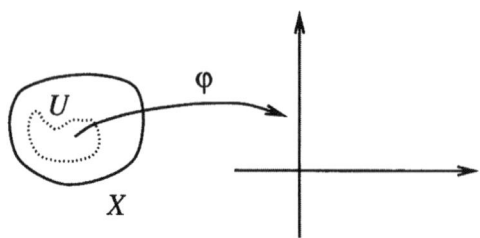

Fig. 3. A chart on X

As long as we work 'in the chart φ' we can pretend we are working in $\mathbb{R}^n$, just as the Europeans could pretend they lived on $\mathbb{R}^2$ as long as they did not go too far from home. For example, if we have a function $f: U \to \mathbb{R}$, we can turn it into a function on $\mathbb{R}^n$ by using $f \circ \varphi^{-1}: \mathbb{R}^n \to \mathbb{R}$.

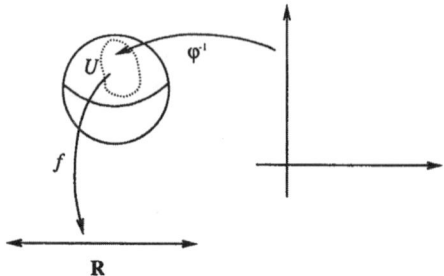

Fig. 4. Turning a function on U into a function on $\mathbb{R}^n$

Finally, we say that an n-**dimensional manifold**, or n-**manifold**, is a topological space M equipped with charts $\varphi_\alpha: U_\alpha \to \mathbb{R}^n$, where U_α are open sets covering M, such that the **transition function** $\varphi_\alpha \circ \varphi_\beta^{-1}$ is smooth where it is defined. Such a collection of charts is called an

Manifolds

atlas.

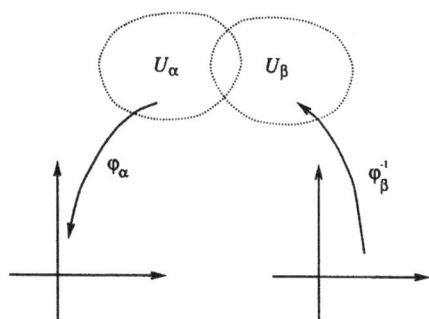

Fig. 5. Two charts and the transition function

What does this definition mean? First, every point of M lives in some open subset U_α that looks like $\mathbb{R}^n$, or in other words, we can 'patch together' the whole manifold out of bits that look like $\mathbb{R}^n$. Second, it means that we can tell using charts if a function on M is smooth, without any ambiguity, because the transition functions between charts are smooth. To be precise, we say a function $f: M \to \mathbb{R}$ is **smooth** if for all α, $f \circ \varphi_\alpha^{-1}: \mathbb{R}^n \to \mathbb{R}$ is smooth. Suppose you are using the chart $\varphi_\alpha: U_\alpha \to \mathbb{R}^n$ and I am using the chart $\varphi_\beta: U_\beta \to \mathbb{R}^n$, and let $V = U_\alpha \cap U_\beta$ be the overlap of our two charts. Suppose that you think the function f is smooth on V, that is, suppose $f \circ \varphi_\alpha^{-1}$ is smooth on $\varphi_\alpha V$, as below:

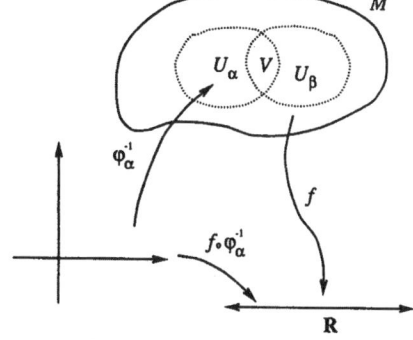

Fig. 6. Your picture

Then I will agree that f is smooth on V, that is, $f \circ \varphi_\beta^{-1}$ will be smooth on $\varphi_\beta V$ too:

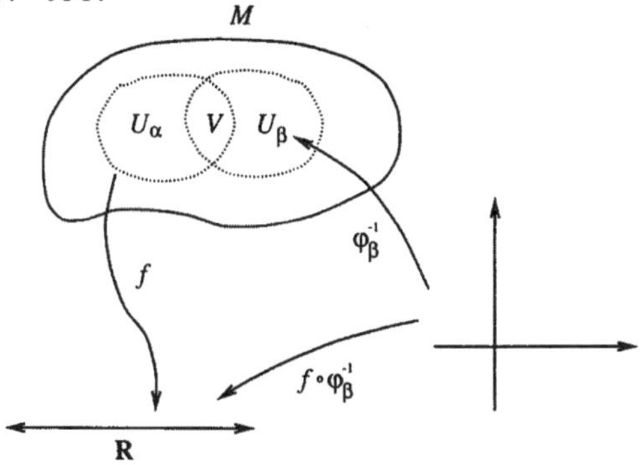

Fig. 7. My picture

Why? Because we can express my function in terms of your function and the transition function:

$$f \circ \varphi_\beta^{-1} = (f \circ \varphi_\alpha^{-1}) \circ (\varphi_\alpha \circ \varphi_\beta^{-1}).$$

Strictly speaking, the sort of manifold we have defined here is called a **smooth manifold**. There are also, for example, **topological manifolds**, where the transition functions are only required to be continuous. For us, 'manifold' will always mean 'smooth manifold'. Also, we will always assume our manifolds are 'Hausdorff' and 'paracompact'. These are topological properties that we prefer to avoid explaining here, which are satisfied by all but the most bizarre and useless examples.

In the following exercises we describe some examples of manifolds, leaving the reader to check that they really are manifolds.

Exercise 3. *Given a topological space X and a subset $S \subseteq X$, define the* **induced topology** *on S to be the topology in which the open sets are of the form $U \cap S$, where U is open in X. Let S^n, the* **n-sphere**, *be the unit sphere in $\mathbb{R}^{n+1}$:*

$$S^n = \{\vec{x} \in \mathbb{R}^{n+1} | \sum_{i=1}^{n+1}(x^i)^2 = 1\}.$$

Show that $S^n \subset \mathbb{R}^{n+1}$ with its induced topology is a manifold.

Exercise 4. *Show that if M is a manifold and U is an open subset of M, then U with its induced topology is a manifold.*

Manifolds

Exercise 5. *Given topological spaces X and Y, we give $X \times Y$ the* **product topology** *in which a set is open if and only if it is a union of sets of the form $U \times V$, where U is open in X and V is open in Y. Show that if M is an m-dimensional manifold and N is an n-dimensional manifold, $M \times N$ is an $(m + n)$-dimensional manifold.*

Exercise 6. *Given topological spaces X and Y, we give $X \cup Y$ the* **disjoint union topology** *in which a set is open if and only if it is the union of an open subset of X and an open subset of Y. Show that if M and N are n-dimensional manifolds the disjoint union $M \cup N$ is an n-dimensional manifold.*

There are many different questions one can ask about a manifold, but one of the most basic is whether it extends indefinitely in all directions like $\mathbb{R}^3$ or is 'compact' like S^3. There is a way to make this precise which proves to be very important in mathematics. Namely, a topological space X is said to be **compact** if for every cover of X by open sets U_α there is a finite collection $U_{\alpha_1}, \ldots, U_{\alpha_n}$ that covers X. For manifolds, there is an equivalent definition: a manifold M is compact if and only if every sequence in M has a convergent subsequence. A basic theorem says that a subset of $\mathbb{R}^n$ is compact if and only if it is closed and fits inside a ball of sufficiently large radius.

The study of manifolds is a fascinating business in its own right. However, since our goal is to do physics on manifolds, let us turn to the basic types of fields that live on manifolds: vector fields and differential forms.

Chapter 3
Vector Fields

And it is a noteworthy fact that ignorant men have long been in advance of the learned about vectors. Ignorant people, like Faraday, naturally think in vectors. They may know nothing of their formal manipulation, but if they think about vectors, they think of them as vectors, that is, directed magnitudes. No ignorant man could or would think about the three components of a vector separately, and disconnected from one another. That is a device of learned mathematicians, to enable them to evade vectors. The device is often useful, especially for calculating purposes, but for general purposes of reasoning the manipulation of the scalar components instead of the vector itself is entirely wrong. — Oliver Heaviside

Heaviside was one of the first advocates of modern vector analysis, as well as a very sarcastic fellow. In the quote above, he was making the point that the great physicist Faraday did not need to worry about coordinates, because Faraday had a direct physical understanding of vectors. Pictorially, a vector field on a manifold can be visualized as a field of arrows. For example, a vector field on S^2 is basically just a field of arrows tangent to the sphere:

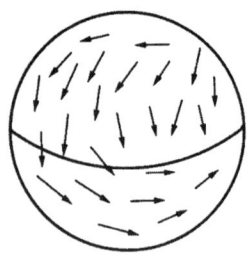

Fig. 1. Vector field on S^2

To do calculations with vector fields, however, it is nice to define them in an algebraic sort of way. The key to defining vector fields on manifolds is to note that given a field of arrows, one can differentiate a function in the direction of the arrows. In particular, given a function f and a vector field v on $\mathbb{R}^n$, we can form the directional derivative of f in the direction v, which we will write simply as vf.

Let us write a formula for vf in this case. The formula for a directional derivative should not be news to the readers of this book, but we will rewrite it using some slick physics notation. We will write $x^1, \ldots, x^n$ for the coordinates on $\mathbb{R}^n$, and write just ∂_μ for the partial derivative $\partial/\partial x^\mu$. (When we are dealing with three or fewer dimensions we will sometimes write x, y, z instead of x^1, x^2, x^3, and write $\partial_x, \partial_y, \partial_z$ for $\partial_1, \partial_2, \partial_3$.) Also, we will use the **Einstein summation convention** and always sum over repeated indices that appear once as a subscript and once as a superscript. Then if v has components $(v^1, \ldots, v^n)$, we have the formula

$$vf = v^\mu \partial_\mu f.$$

If this seems enigmatic, remember that it is just short for

$$vf = v^1 \frac{\partial f}{\partial x^1} + \cdots + v^n \frac{\partial f}{\partial x^n}.$$

In fact, since the formula $vf = v^\mu \partial_\mu f$ holds for all f, we can be even more slick and write

$$v = v^\mu \partial_\mu.$$

Vector Fields

What does this mean, though? The sight of the partial derivatives ∂_μ sitting there with nothing to differentiate is only slightly unnerving; we can always put a function f to the right of them whenever we want. Much odder is that we are saying the vector field v *is* the linear combination of these partial derivatives. What we are doing might be regarded as rather sloppy, since we are identifying two different, although related, things: the vector field v, and the operator $v^\mu \partial_\mu$ that takes a directional derivative in the direction of v. In fact, this 'sloppy' attitude turns out to be extremely convenient, and next we will go even further and use it to *define* vector fields on manifolds. It is important to realize that in mathematics it is often crucial to think about familiar objects in a new way in order to generalize them to a new situation.

Now let us define vector fields on a manifold M. Following the philosophy outlined above, these will be entities whose sole ambition in life is to differentiate functions. First a bit of jargon. The set of smooth (real-valued) functions on a manifold M is written $C^\infty(M)$, where the C^∞ is short for 'having infinitely many continuous derivatives'. Note that $C^\infty(M)$ is an **algebra** over the real numbers, meaning that it is closed under (pointwise) addition and multiplication, as well as multiplication by real numbers, and the following batch of rules holds:

$$\begin{aligned} f + g &= g + f \\ f + (g + h) &= (f + g) + h \\ f(gh) &= (fg)h \\ f(g + h) &= fg + fh \\ (f + g)h &= fh + gh \\ 1f &= f \\ \alpha(\beta f) &= (\alpha \beta) f \\ \alpha(f + g) &= \alpha f + \alpha g \\ (\alpha + \beta) f &= \alpha f + \beta f, \end{aligned}$$

where $f, g, h \in C^\infty(M)$ and $\alpha, \beta \in \mathbb{R}$. Of course it is a **commutative** algebra, that is, $fg = gf$.

Now, a **vector field** v on M is defined to be a function from $C^\infty(M)$ to $C^\infty(M)$ satisfying the following properties:

$$v(f+g) = v(f) + v(g)$$
$$v(\alpha f) = \alpha v(f)$$
$$v(fg) = v(f)g + fv(g),$$

for all $f, g \in C^\infty(M)$ and $\alpha \in \mathbb{R}$. Here we have isolated all the basic rules a directional derivative operator should satisfy. The first two simply amount to linearity, and it is the third one, the product rule or **Leibniz law**, that really captures the essence of differentiation.

This definition may seem painfully abstract. We will see in a bit that it really is just a way of talking about a field of arrows on M. For now, note the main good feature of this definition: *it does not rely on any choice of coordinates on M!* A basic philosophy of modern physics is that the universe does not come equipped with a coordinate system. While coordinate systems are necessary for doing specific concrete calculations, the choice of the coordinate system to use is a matter of convenience, and there is often no 'best' coordinate system. One should strive to write the laws of physics in a *manifestly* coordinate-independent manner, so one can see what they are really saying and not get distracted by things that might depend on the coordinates.

Let $\mathrm{Vect}(M)$ denote the set of all vector fields on M. We leave it to the reader to check that one can add vector fields and multiply them by functions on M as follows. Given $v, w \in \mathrm{Vect}(M)$, we define $v + w$ by

$$(v + w)(f) = v(f) + w(f),$$

and given $v \in \mathrm{Vect}(M)$ and $g \in C^\infty(M)$, we define gv by

$$(gv)(f) = gv(f).$$

Exercise 7. *Show that $v + w$ and $gw \in \mathrm{Vect}(M)$.*

Exercise 8. *Show that the following rules for all $v, w \in \mathrm{Vect}(M)$ and $f, g \in C^\infty(M)$:*

$$f(v + w) = fv + fw$$
$$(f + g)v = fv + gv$$
$$(fg)v = f(gv)$$
$$1v = v.$$

(Here '1' denotes the constant function equal to 1 on all of M.) Mathematically, we summarize these rules by saying that Vect(M) is a **module** over $C^\infty(M)$.

It turns out that the vector fields $\{\partial_\mu\}$ on $\mathbb{R}^n$ **span** Vect($\mathbb{R}^n$) as a module over $C^\infty(M)$. In other words, every vector field on $\mathbb{R}^n$ is a linear combination of the form

$$v^\mu \partial_\mu = v^1 \partial_1 + \cdots + v^n \partial_n,$$

for some functions $v^\mu \in C^\infty(\mathbb{R}^n)$. It is also true that the vector fields $\{\partial_\mu\}$ on $\mathbb{R}^n$ are **linearly independent**

Exercise 9. *Show that if $v^\mu \partial_\mu = 0$, that is, $v^\mu \partial_\mu f = 0$ for all $f \in C^\infty(\mathbb{R}^n)$, we must have $v^\mu = 0$ for all μ.*

This implies that every vector field v on $\mathbb{R}^n$ has a unique representation as a linear combination $v^\mu \partial_\mu$; we say that the vector fields $\{\partial_\mu\}$ form a **basis** of Vect($\mathbb{R}^n$). The functions v^μ are called the **components** of the vector field v.

Tangent Vectors

Often it is nice to think of a vector field on M as really assigning an 'arrow' to each point of M. This kind of arrow is called a tangent vector. For example, we may think of a tangent vector at a point $p \in S^2$ as a vector in the plane tangent to p:

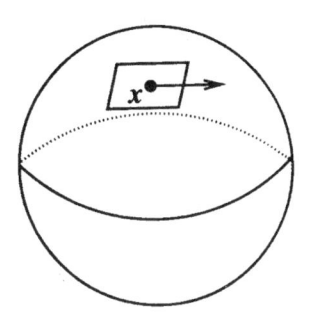

Fig. 2. Tangent vector

To get a precise definition of a tangent vector at $p \in M$, note that a tangent vector should let us take directional derivatives at the point p. For example, given a vector field v on M, we can take the derivative $v(f)$ of any function $f \in C^\infty(M)$, and then evaluate the function $v(f)$ at p. We can think of the result, $v(f)(p)$, as being the result of differentiating f in the direction 'v_p' at the point p. In other words, we can *define*

$$v_p \colon C^\infty(M) \to \mathbb{R}$$

by

$$v_p(f) = v(f)(p),$$

and think of v_p as a tangent vector at p. We call v_p the value of v at p.

Note that v_p has three basic properties, which follow from the definition of a vector field:

$$v_p(f + g) = v_p(f) + v_p(g)$$
$$v_p(\alpha f) = \alpha v_p(f)$$
$$v_p(fg) = v_p(f)g(p) + f(p)v_p(g).$$

Henceforth, we will simply *define* a **tangent vector** at $p \in M$ to be a function from $C^\infty(M)$ to $\mathbb{R}$ satisfying these three properties. Let T_pM, the **tangent space** at p, denote the set of all tangent vectors at $p \in M$.

It now follows rigorously from our definitions that for each $p \in M$, a vector field $v \in \mathrm{Vect}(M)$ determines a tangent vector $v_p \in T_pM$. One can also show, though it takes a bit of work, that *every* tangent vector at p is of the form v_p for some vector field or other. A related fact, which is much easier to show, is the following:

Exercise 10. *Let $v, w \in \mathrm{Vect}(M)$. Show that $v = w$ if and only if $v_p = w_p$ for all $p \in M$.*

Why do tangent vectors as we have defined them 'look like arrows'? First of all, we can add two tangent vectors $v, w \in T_pM$ by

$$(v + w)(f) = v(f) + w(f),$$

and multiply tangent vectors by real numbers:

$$(\alpha v)(f) = \alpha v(f).$$

Tangent Vectors

(Now we are using the letters v, w to denote tangent vectors, not vector fields!) With addition and multiplication defined this way, the tangent space is really a vector space. For example, in Figure 2 we have drawn a tangent space to look like a little plane. The tangent vectors can be thought of as arrows living in this vector space.

Exercise 11. *Show that T_pM is a vector space over the real numbers.*

Another reason why tangent vectors really look like arrows is that curves have tangent vectors:

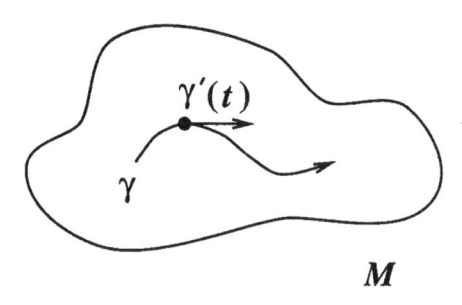

Fig. 3. The tangent vector to a curve in M

By a **curve** we will always mean a function from $\mathbb{R}$ or some interval to M that is smooth, i.e., such that for any $f \in C^\infty(M)$, $f(\gamma(t))$ depends smoothly on t. Given a curve $\gamma: \mathbb{R} \to M$ and any $t \in \mathbb{R}$, the tangent vector $\gamma'(t)$ should be a vector in the tangent space $T_{\gamma(t)}M$. We define $\gamma'(t)$ in the only sensible way possible: it is the function from $C^\infty(M)$ to $\mathbb{R}$ that sends any function $f \in C^\infty(M)$ to the derivative

$$\frac{d}{dt}f(\gamma(t)).$$

In other words, the tangent vector $\gamma'(t)$ differentiates functions in the direction that γ is moving in at time t.

Exercise 12. *Check that $\gamma'(t) \in T_{\gamma(t)}M$ using the definitions.*

If the curve γ describes the motion of a particle through space, the tangent vector $\gamma'(t)$ represents its velocity. For this reason, we will sometimes write

$$\frac{d\gamma}{dt}$$

for $\gamma'(t)$, especially when we are not particularly concerned with which value of t we are talking about.

Note that for manifolds it generally makes no sense to say that a tangent vector $v \in T_pM$ is 'the same' as another one, $w \in T_qM$, unless the points p and q are the same. For example, there is no 'best' way to compare tangent vectors at the north pole of S^2 to tangent vectors at the equator. It also makes no sense to add tangent vectors at different points!

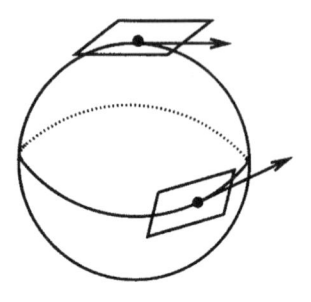

Fig. 4. Tangent vectors at different points of S^2

We mention this because the reader may be used to $\mathbb{R}^n$, where one often says the following two vectors are 'the same', even though they are at different points in $\mathbb{R}^n$:

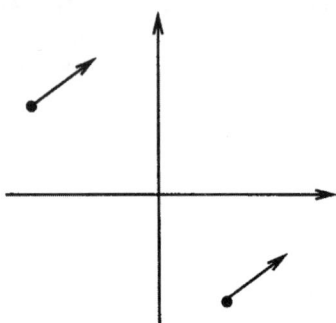

Fig. 5. Tangent vectors at different points of $\mathbb{R}^n$

The reason why one can get away with this is that for any point p in

$\mathbb{R}^n$, the tangent vectors
$$(\partial_\mu)_p \in T_p\mathbb{R}^n,$$
form a basis. This allows one to relate tangent vectors at different points of $\mathbb{R}^n$ — one can sloppily say that the vector
$$v = v^\mu(\partial_\mu)_p \in T_p\mathbb{R}^n$$
and the vector
$$w = w^\mu(\partial_\mu)_q \in T_q\mathbb{R}^n$$
are 'the same' if $v^\mu = w^\mu$, even though v and w are not literally equal. Later we will get a deeper understanding of this issue, which requires a theory of 'parallel transport', the process of dragging a vector at one point of a manifold over to another point. This turns out to be a crucial idea in physics, and in fact the root of gauge theory!

Covariant Versus Contravariant

A lot of modern mathematics and physics requires keeping track of which things in life are covariant, and which things are contravariant. Let us begin to explain these ideas by comparing functions and tangent vectors. Say we have a function $\phi: M \to N$ from one manifold to another. If we have a real-valued function on N, say $f: N \to \mathbb{R}$, we can get a real-valued function on M by composing it with f.

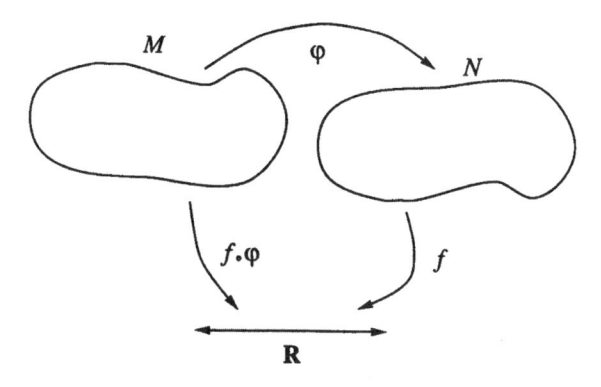

Fig. 6. Pulling back f from N to M

We call this process **pulling back** f from N to M by ϕ. We define

$$\phi^* f = f \circ \phi,$$

and call $\phi^* f$ the **pullback** of f by ϕ. The point is that while ϕ goes 'forwards' from M to N, the pullback operation ϕ^* goes 'backwards', taking functions on N to functions on M. We say that real-valued functions on a manifold are **contravariant** because of this perverse backwards behavior.

Exercise 13. *Let $\phi: \mathbb{R} \to \mathbb{R}$ be given by $\phi(t) = e^t$. Let x be the usual coordinate function on $\mathbb{R}$. Show that $\phi^* x = e^x$.*

Exercise 14. *Let $\phi: \mathbb{R}^2 \to \mathbb{R}^2$ be rotation counterclockwise by an angle θ. Let x, y be the usual coordinate functions on $\mathbb{R}^2$. Show that*

$$\phi^* x = (\cos\theta)x - (\sin\theta)y$$
$$\phi^* y = (\sin\theta)x + (\cos\theta)y.$$

By the way, we say that $\phi: M \to N$ is **smooth** if $f \in C^\infty(N)$ implies that $\phi^* f \in C^\infty(M)$. Henceforth we will assume functions from manifolds to manifolds are smooth unless otherwise stated, and we will often call such functions **maps**.

Exercise 15. *Show that this definition of smoothness is consistent with the previous definitions of smooth functions $f: M \to \mathbb{R}$ and smooth curves $\gamma: \mathbb{R} \to M$.*

Using our new jargon, we have: given any map

$$\phi: M \to N,$$

pulling back by ϕ is an operation

$$\phi^*: C^\infty(N) \to C^\infty(M).$$

Tangent vectors, on the other hand, are **covariant**: a tangent vector $v \in T_p M$ and a smooth function $\phi: M \to N$ gives a tangent vector $\phi_* v \in T_{\phi(p)} N$, called the **pushforward** of v by ϕ. This is defined by

$$(\phi_* v)(f) = v(\phi^* f).$$

Covariant Versus Contravariant

We say we are **pushing forward** v by ϕ. Note that we use a subscript asterisk for pushforwards and a superscript for pullbacks! One way to think of the pushforward is that if γ is a curve in M with tangent vector $\gamma'(t) \in T_pM$, the curve $\phi \circ \gamma$ is a curve with tangent vector

$$(\phi \circ \gamma)'(t) = \phi_*(\gamma'(t)) \in T_{\phi(p)}N.$$

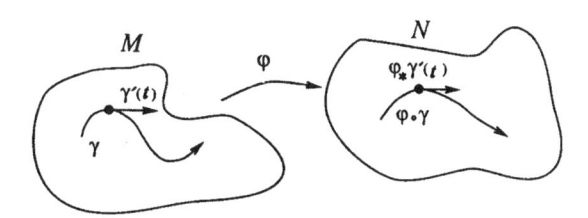

Fig. 7. Pushing forward the tangent vector of a curve from M to N

Exercise 16. *Prove that $(\phi \circ \gamma)'(t) = \phi_*(\gamma'(t))$.*

Exercise 17. *Show that the pushfoward operation*

$$\phi_*: T_pM \to T_{\phi(p)}N$$

is linear.

Exercise 18. *Show that if $\phi: M \to N$ is a diffeomorphism, we can push forward a vector field v on M to obtain a vector field ϕ_*v on N satisfying*

$$(\phi_*v)_q = \phi_*(v_p)$$

whenever $\phi(p) = q$.

Exercise 19. *Let $\phi: \mathbb{R}^2 \to \mathbb{R}^2$ be rotation counterclockwise by an angle θ. Let ∂_x, ∂_y be the coordinate vector fields on $\mathbb{R}^2$. Show that at any point of $\mathbb{R}^2$*

$$\phi_*\partial_x = (\cos\theta)\partial_x + (\sin\theta)\partial_y$$
$$\phi_*\partial_y = -(\sin\theta)\partial_x + (\cos\theta)\partial_y.$$

It is traditional in mathematics, by the way, to write pushforwards and other covariant things with lowered asterisks, and to write pullbacks and other contravariant things with raised asterisks. It might help as a mnemonic to remember that the tangent vectors ∂_μ are written with the μ downstairs, and are covariant. In the next chapter we will discuss things similar to tangent vectors, but which are contravariant! These things will have their indices upstairs. We warn the reader, however, that while the vector field ∂_μ is covariant and has its indices downstairs, physicists often think of a vector field v as *being* its components v^μ. These have their indices upstairs, so physicists say that the v^μ are contravariant! This is one of those little differences that makes communication between the two subjects a bit more difficult.

Flows and the Lie Bracket

One sort of vector field that comes up in physics is the velocity vector field of a fluid, such as water. Imagine that the velocity vector field v is constant as a function of time, so that each molecule of water traces out a curve $\gamma(t)$ as time passes, with the tangent vector of γ equal to the value of v at the point $\gamma(t)$:

$$\gamma'(t) = v_{\gamma(t)}$$

for all t. If the curve starts at some point $p \in M$, that is $\gamma(0) = p$, we call γ the **integral curve through** p of the vector field v:

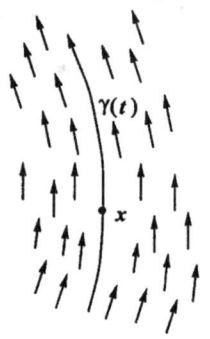

Fig. 8. Integral curve through p of the vector field v

Flows and the Lie Bracket 35

Calculating the integral curves of a vector field amounts to solving a first-order differential equation. One has to be careful, because the solution might 'shoot off to infinity' in a finite amount of time:

Exercise 20. *Let v be the vector field $x^2 \partial_x + y \partial_y$ on $\mathbb{R}^2$. Calculate the integral curves $\gamma(t)$ and see which ones are defined for all t.*

We say that the vector field v is **integrable** if all the integral curves are defined for all t.

Suppose v is an integrable vector field on M, which we think of as the velocity vector field of some water. If we keep track of how *all* the molecules of water are moving along, we have something called a 'flow'. Let $\phi_t(p)$ be the integral curve of v through the point $p \in M$. For each time t, the map

$$\phi_t \colon M \to M$$

turns out to be smooth, by a result on the smooth dependence of solutions of differential equations on the initial conditions. Water that was at p at time zero will be at $\phi_t(p)$ by time t, so we call the family of maps $\{\phi_t\}$ the **flow generated by** v. The defining equation for the flow is (rewriting our equation for γ):

$$\frac{d}{dt} \phi_t(p) = v_{\phi_t(p)}.$$

Exercise 21. *Show that ϕ_0 is the identity map $\mathrm{id} \colon X \to X$, and that for all $s, t \in \mathbb{R}$ we have $\phi_t \circ \phi_s = \phi_{t+s}$.*

There is an important way to get new vector fields from old ones that is related to the concept of flows. This is called the **Lie bracket** or **commutator** of vector fields. Given $v, w \in \mathrm{Vect}(M)$, the Lie bracket $[v, w]$ is defined by

$$[v, w](f) = v(w(f)) - w(v(f)),$$

for all $f \in C^\infty(M)$, or, for short,

$$[v, w] = vw - wv.$$

Let us show that the Lie bracket defined in this way actually is a vector field on the manifold M. It is easy to prove linearity, so the crucial thing is the Leibniz rule: if $u = [v, w]$, we have

$$\begin{aligned} u(fg) &= (vw - wv)(fg) \\ &= v[w(f)g + fw(g)] - w[v(f)g + fv(g)] \\ &= vw(f)g + f\,vw(g) - wv(f)g - f\,wv(g) \\ &= u(f)g + fu(g). \end{aligned}$$

Here we used the Leibniz law twice and then used the definition of the Lie brackets.

The Lie bracket measures the failure of 'mixed directional derivatives' to commute. Of course, ordinary mixed partial derivatives *do* commute:

$$[\partial_\mu, \partial_\nu] = 0.$$

We can think of this pictorially, as follows: flowing a little bit first in the ∂_μ direction and then in the ∂_ν direction gets us to the same place as if we had done it in the other order:

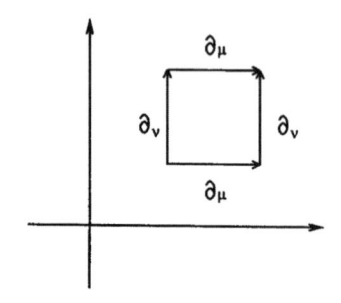

Fig. 9. $[\partial_\mu, \partial_\nu] = 0$

However, if we take some other vector fields, this does not usually work:

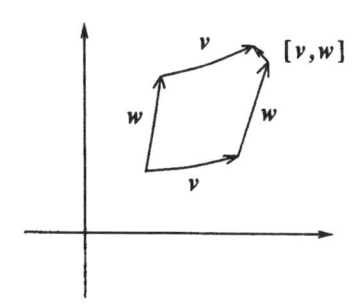

Fig. 10. $[v, w] \neq 0$

We say in this case that the vector fields do not **commute**.

Exercise 22. *Consider the normalized vector fields in the r and θ directions on the plane in polar coordinates (not defined at the origin):*

$$v = \frac{x\partial_x + y\partial_y}{\sqrt{x^2 + y^2}}$$

$$w = \frac{x\partial_y - y\partial_x}{\sqrt{x^2 + y^2}}.$$

Calculate $[v, w]$.

To make the relationship with flows precise, suppose that v generates the flow ϕ_t, and w generates the flow ψ_t. Then for any $f \in C^\infty(M)$

$$(vf)(p) = \frac{d}{dt}f(\phi_t(p))\Big|_{t=0},$$

and similarly

$$(wf)(p) = \frac{d}{ds}f(\psi_s(p))\Big|_{s=0},$$

so one can check that

$$[v, w](f)(p) = \frac{\partial^2}{\partial t \partial s}\Big(f(\psi_s(\phi_t(p))) - f(\phi_t(\psi_s(p)))\Big)\Big|_{s=t=0}.$$

If you think about it, this is related to what we said above. In $f(\phi_t(\psi_s(p)))$ we are starting at p, flowing along w a little bit, then along v a little bit, and then evaluating f, while in $f(\psi_s(\phi_t(p)))$ we are flowing first along v and then w. The Lie bracket measures (infinitesimally, as it were) how these flows fail to commute!

Exercise 23. *Check the equation above.*

The Lie bracket of vector fields satisfies some identities which we will come back to in Part II. For now, we simply let the reader prove them:

Exercise 24. *Show that for all vector fields u, v, w on a manifold, and all real numbers α and β, we have:*

1) $[v, w] = -[w, v]$.
2) $[u, \alpha v + \beta w] = \alpha[u, v] + \beta[u, w]$.
*3) The **Jacobi identity**:* $[u, [v, w]] + [v, [w, u]] + [w, [u, v]] = 0$.

Chapter 4

Differential Forms

> *As a herald it's my duty*
> *to explain those forms of beauty.* — *Goethe*, Faust.

1-forms

The electric field, the magnetic field, the electromagnetic field on spacetime, the current — all these are examples of differential forms. The gradient, the curl, and the divergence can all be thought of as different aspects of single operator d that acts on differential forms. The fundamental theorem of calculus, Stokes' theorem, and Gauss' theorem are all special cases of a single theorem about differential forms. So while they are somewhat abstract, differential forms are a powerful unifying notion.

We begin with 1-forms. Our goal is to generalize the concept of the gradient of a function to functions on arbitrary manifolds. What we will do is to make up, for each smooth function f on M, an object called df that is supposed to be like the usual gradient ∇f defined on $\mathbb{R}^n$. Remember that the directional derivative of a function f on $\mathbb{R}^n$ in the direction v is just the dot product of ∇f with v:

$$\nabla f \cdot v = vf.$$

In other words, the gradient of f is a thing that keeps track of the directional derivatives of f in all directions. We want our 'df' to do the same job on any manifold M.

The gradient of a function on $\mathbb{R}^n$ is a vector field, so one might want to say that df should be a vector field. The problem is the dot product in the formula above. On $\mathbb{R}^n$ there is a well-established way to take the dot product of tangent vectors, but manifolds do not come pre-equipped with a way to do this. Geometers call a way of taking dot products of tangent vectors a 'metric'. In fact, we will see that in general relativity the gravitational field is described by the metric on spacetime. Far from there being a single 'best' metric on a manifold, there are typically *lots* that satisfy Einstein's equations of general relativity. This makes it nice to avoid using a particular metric unless we actually *need* to. Therefore we will not think of df as a vector field, but as something else, a '1-form'.

The trick is to realize what ∇f is doing in the formula $\nabla f \cdot v = vf$. For each vector field v that we choose, this formula spits out a function vf, the directional derivative of f in the direction v. In other words, what really matters is the *operator*

$$v \mapsto \nabla f \cdot v,$$

or, what is the same thing,

$$v \mapsto vf.$$

Let us isolate the essential properties of this map. There is really only one: linearity! This means that

$$\nabla f \cdot (v+w) = \nabla f \cdot v + \nabla f \cdot w$$

for any vector fields v and w, and

$$\nabla f \cdot (gv) = g(\nabla f \cdot v)$$

where g is any smooth function on $\mathbb{R}^n$. Since we can pull out any function $g \in C^\infty(\mathbb{R}^n)$ in the above formula, not just constants, mathematicians say that

$$v \mapsto \nabla f \cdot v$$

is **linear over** $C^\infty(\mathbb{R}^n)$ — not just linear over the real numbers.

So, abstracting a bit, we define a **1-form** on any manifold M to be a map from Vect(M) to $C^\infty(M)$ that is linear over $C^\infty(M)$. In other

1-forms

words, if we feed a vector field v to a 1-form ω, it spits out a function $\omega(v)$ in a way satisfying

$$\omega(v+w) = \omega(v) + \omega(w),$$

$$\omega(gv) = g\omega(v).$$

We use $\Omega^1(M)$ to denote the space of all 1-forms on a manifold M. Later on we will talk about 2-forms, 3-forms, and so on.

The basic example of a 1-form is this: for any smooth function f on M there is a 1-form df defined by

$$df(v) = vf.$$

(Think of this as a slick way to write $\nabla f \cdot v = vf$.) To show that df is really a 1-form, we just need to check linearity:

$$df(v+w) = (v+w)f = vf + wf = df(v) + df(w),$$

and

$$df(gv) = (gv)(f) = g\,v(f) = g\,df(v).$$

We call the 1-form df the **differential** of f, or the **exterior derivative** of f.

Just as we can add vector fields or multiply them by functions, we can do the same for 1-forms. We can add two 1-forms ω and μ and get a 1-form $\omega + \mu$ by defining

$$(\omega + \mu)(v) = \omega(v) + \mu(v),$$

and we can multiply a 1-form ω by a smooth function f and get a 1-form $f\omega$ by defining

$$(f\omega)(v) = f\omega(v).$$

Exercise 25. *Show that $\omega + \mu$ and $f\omega$ are really 1-forms, i.e., show linearity over $C^\infty(M)$.*

Exercise 26. *Show that $\Omega^1(M)$ is a module over $C^\infty(M)$ (see the definition in Exercise 8.)*

The map $d: C^\infty(M) \to \Omega^1(M)$ that sends each function f to its differential df is also called the differential, or exterior derivative. It is interesting in its own right, and has the following nice properties:

Exercise 27. *Show that*

$$d(f+g) = df + dg$$
$$d(\alpha f) = \alpha\, df$$
$$(f+g)dh = f\, dh + g\, dh$$
$$d(fg) = f\, dg + g\, df,$$

for any $f, g, h \in C^\infty(M)$ *and any* $\alpha \in \mathbb{R}$.

The first three properties in the exercise above are just forms of linearity, but the last one is a version of the product rule, or Leibniz law:

$$d(fg) = f\, dg + g\, df.$$

It is the Leibniz law that makes the exterior derivative really act like a derivative, so if you only want to do *part* of Exercise 27 check that the Leibniz law holds! It is worth mentioning, by the way, that when Leibniz was inventing calculus he first guessed that $d(fg) = df\, dg$, and only got it right the next day.

In fact, the reader has seen differentials before, in calculus. They start out as part of the expressions for differentiation

$$\frac{dy}{dx}$$

and integration

$$\int f(x)\, dx$$

but soon take on a mysterious life of their own, as in

$$d\sin x = \cos x\, dx!$$

We bet you remember wondering what the heck these differentials *really are!* In physics one thinks of dx as an 'infinitesimal change in position', and so on — but this is mystifying in its own right. Early in the history of calculus, the philosopher Berkeley complained about these

1-forms

infinitesimals, writing "They are neither finite quantities, nor quantities infinitely small, nor yet nothing. May we not call them ghosts of departed quantities?" More recently, people have worked out an alternative approach to the real numbers, called 'nonstandard analysis', that includes a logically satisfactory theory of infinitesimals — puny numbers that are greater than zero but less than any 'standard' real number. Most people these days, however, prefer to think of differentials as 1-forms.

Let us show that $d\sin x = \cos x\, dx$ is really true as an equation concerning 1-forms on the real line. We need to show that no matter what vector field we feed these two 1-forms, they spit out the same thing. This is not hard. Any vector field v on $\mathbb{R}$ is of the form $v = f(x)\partial_x$, so on one hand we have

$$(d\sin x)(v) = v\sin x = f(x)\partial_x \sin x = f(x)\cos x,$$

and on the other hand:

$$(\cos x\, dx)(v) = (\cos x)\, v(x) = f(x)\cos x\, \partial_x x = f(x)\cos x.$$

This is in fact just a special case of the following:

Exercise 28. *Suppose $f(x^1, \ldots, x^n)$ is a function on $\mathbb{R}^n$. Show that*

$$df = \partial_\mu f\, dx^\mu.$$

This means that on $\mathbb{R}^n$ the exterior derivative of a function is really just a different way of thinking about its gradient, since in old-fashioned language we had

$$\nabla f = (\partial_1 f, \ldots, \partial_n f).$$

To do the exercise above one needs to use the fact that the vector fields $\{\partial_\mu\}$ form a basis of vector fields on $\mathbb{R}^n$. In fact, this implies that the 1-forms $\{dx^\mu\}$ form a basis of 1-forms on $\mathbb{R}^n$. The key is that

$$dx^\mu(\partial_\nu) = \partial_\nu x^\mu = \delta^\mu_\nu$$

where the **Kronecker delta** δ^μ_ν equals 1 if $\mu = \nu$ and 0 otherwise. Now suppose we have a 1-form ω on $\mathbb{R}^n$. Then we can define some functions

$$\omega_\mu = \omega(\partial_\mu),$$

and we claim that
$$\omega = \omega_\mu dx^\mu.$$
This will imply that the 1-forms $\{dx^\mu\}$ span the 1-forms on $\mathbb{R}^n$. To show that ω equals $\omega_\mu dx^\mu$, we just need to feed both of them a vector field and show that they spit out the same function! Feed them $v = v^\nu \partial_\nu$, for example. Then on the one hand
$$\omega(v) = \omega(v^\nu \partial_\nu) = v^\nu \omega(\partial_\nu) = v^\nu \omega_\nu,$$
while on the other hand,
$$(\omega_\mu dx^\mu)(v) = (\omega_\mu dx^\mu)(v^\nu \partial_\nu) = \omega_\mu v^\nu dx^\mu(\partial_\nu) = \omega_\nu v^\nu$$
using the fact that $dx^\mu(\partial_\nu) = \delta^\mu_\nu$.

We leave it to the reader to finish the proof that the 1-forms $\{dx^\mu\}$ form a basis of $\Omega^1(\mathbb{R}^n)$:

Exercise 29. *Show that the 1-forms $\{dx^\mu\}$ are linearly independent, i.e., if*
$$\omega = \omega_\mu dx^\mu = 0$$
then all the functions ω_μ are zero.

Cotangent Vectors

Just as a vector field on M gives a tangent vector at each point of M, a 1-form on M gives a kind of vector at each point of M called a 'cotangent vector'. Given a manifold M and a point $p \in M$, a **cotangent vector** ω at p is defined to be a linear map from the tangent space $T_p M$ to $\mathbb{R}$. Let $T_p^* M$ denote the space of all cotangent vectors at p.

For example, if we have a 1-form ω on M, we can define a cotangent vector $\omega_p \in T_p^* M$ by saying that for any vector field v on M,
$$\omega_p(v_p) = \omega(v)(p).$$
Here the right-hand side stands for the function $\omega(v)$ evaluated at the point p.

Cotangent Vectors

Exercise 30. *For the mathematically inclined: show that the ω_p is really well-defined by the formula above. That is, show that $\omega(v)(p)$ really depends only on v_p, not on the values of v at other points. Also, show that a 1-form is determined by its values at points. In other words, if ω, ν are two 1-forms on M with $\omega_p = \nu_p$ for every point $p \in M$, then $\omega = \nu$.*

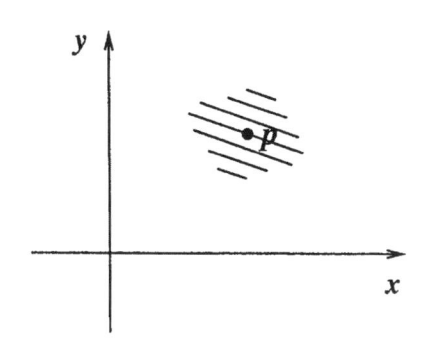

Fig. 1. A picture of the cotangent vector $(df)_p$

How can we visualize a cotangent vector? A tangent vector is like a little arrow; it points somewhere. A cotangent vector does not. A nice heuristic way to visualize a cotangent vector is as a little stack of parallel hyperplanes. For example, if we have a function f on a manifold M, we can visualize df at a point $p \in \mathbb{R}^2$ by drawing the level curves of f right near p, which look like a little stack of parallel lines. The picture in Figure 1 is two-dimensional, so level surfaces are just contour lines, and hyperplanes are just lines.

The bigger df is, the more tightly packed the hyperplanes are. When we take a tangent vector $v \in T_p M$, the number $df(v)$ basically just counts how many little hyperplanes in the stack df the vector v crosses. In Figure 2 we show a situation where $df(v) = 3$. By definition, of course, the number $df(v)$ is just the directional derivative $v(f)$!

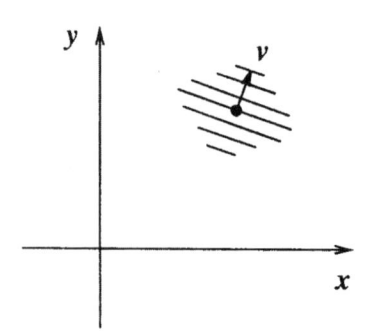

Fig. 2. $df(v) = 3$

Actually we must be a bit careful about thinking about $df(v)$ in terms of pictures, because it could be negative! If we think of the little stack of hyperplanes as 'contour lines', we should really count the number of them v crosses with a plus sign if v is pointing 'uphill' and a minus sign if it is pointing 'downhill'.

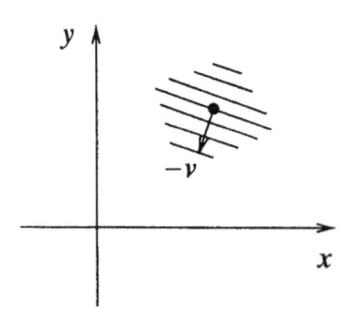

Fig. 3. $df(-v) = -3$

If this way of thinking of 1-forms is confusing, feel free to ignore it — but people with a strong taste for visualization may find it very handy.

Now let us explain precisely what we mean by 1-forms being dual to vector fields. First of all, given any vector space V, the **dual** vector space V^* is defined to be the space of all linear functionals $\omega: V \to \mathbb{R}$. In particular, the cotangent space T_p^*M is the dual of the tangent space T_pM. More generally, if we have a linear map from one vector space to

Cotangent Vectors

another,
$$f: V \to W,$$
we automatically get a map from W^* to V^*, the **dual** of f, written
$$f^*: W^* \to V^*$$
and defined by
$$(f^*\omega)(v) = \omega(f(v)).$$
Thus the dual of a vector space is a contravariant sort of beast: linear maps between vector spaces give rise to maps between their duals that go 'backwards'.

Exercise 31. *Show that the dual of the identity map on a vector space V is the identity map on V^*. Suppose that we have linear maps $f: V \to W$ and $g: W \to X$. Show that $(gf)^* = f^*g^*$.*

This means that cotangent vectors are contravariant. In other words, suppose we have a map $\phi: M \to N$ from one manifold to another with $\phi(p) = q$. We saw in the last section that there is a linear map
$$\phi_*: T_p M \to T_q N.$$
This gives a dual map, which we write as ϕ^*, going the other way:
$$\phi^*: T_q^* N \to T_p^* M.$$
If ω is a cotangent vector at $\phi(x)$, we call $\phi^*\omega$ the **pullback** of ω by ϕ. Explicitly, if $v \in T_p M$ and $\omega \in T_q N$, we have
$$(\phi^*\omega)(v) = \omega(\phi_* v).$$
We can also do this 'pulling back' globally. That is, given a 1-form ω on N, we get a 1-form $\phi^*\omega$ on M defined by
$$(\phi^*\omega)_p = \phi^*(\omega_q)$$
where $\phi(p) = q$.

Exercise 32. *Show that the pullback of 1-forms defined by the formula above really exists and is unique.*

Recall from the previous section that we can also pull back functions on N to functions on M when we have a map $\phi: M \to N$. There is a marvelous formula saying that the exterior derivative is compatible with pullbacks. Namely, given a function f on N and a map $\phi: M \to N$, we have
$$\phi^*(df) = d(\phi^* f).$$
Mathematicians summarize this by saying that the exterior derivative is **natural**. For example, if $\phi: \mathbb{R}^n \to \mathbb{R}^n$ is a diffeomorphism representing some change of coordinates, the above formula implies that we can compute d of a function on $\mathbb{R}^n$ either before or after changing coordinates, and get the same answer. (We discuss this a bit more in the next section.) So naturality can be regarded as a grand generalization of coordinate-independence.

To prove the above equation we just need to show that both sides, which are 1-forms on M, give the same cotangent vector at every point p in M:
$$(\phi^*(df))_p = (d(\phi^* f))_p.$$
This, in turn, means that
$$(\phi^*(df))_p(v) = (d(\phi^* f))_p(v)$$
for all $v \in T_p M$. To prove this, work out the left hand side using all the definitions and show it equals the right hand side:
$$\begin{aligned}(\phi^*(df))_p(v) &= (df)_q(\phi_* v) \\ &= ((\phi_* v)f)(q) \\ &= v(\phi^* f)(p) \\ &= (d(\phi^* f))_p(v)\end{aligned}$$
To make this more concrete it might be good to work out some examples:

Exercise 33. *Let $\phi: \mathbb{R} \to \mathbb{R}$ be given by $\phi(t) = \sin t$. Let dx be the usual 1-form on $\mathbb{R}$. Show that $\phi^* dx = \cos t\, dt$.*

Exercise 34. *Let $\phi: \mathbb{R}^2 \to \mathbb{R}^2$ denote rotation counterclockwise by the angle θ. Let dx, dy be the usual basis of 1-forms on $\mathbb{R}^2$. Show that*
$$\begin{aligned}\phi^* dx &= \cos\theta\, dx - \sin\theta\, dy \\ \phi^* dy &= \sin\theta\, dx + \cos\theta\, dy.\end{aligned}$$

The formula
$$\phi^*(df) = d(\phi^* f)$$
is a very good reason why the differential of a function has to be a 1-form instead of a vector field. Both functions and 1-forms are contravariant, so if $\phi: M \to N$ and $f \in C^\infty(N)$, both sides above are 1-forms on N. If one tried to make the differential of a function be a vector field, there would be no way to write down a sensible formula like this, since vector fields are covariant. (Try it!)

Change of Coordinates

Indeed, from childhood we have become familiar with the appearance of physical equations in non-Cartesian systems, such as polar coordinates, and in non-inertial systems, such as rotating coordinates. — Steven Weinberg

The introduction of numbers as coordinates [...] is an act of violence whose only practical vindication is the special calculatory manageability of the ordinary number continuum with its four basic operations. — Hermann Weyl

So far we have been avoiding coordinates as much as possible. The reason, of course, is that the world does not come equipped with coordinates! As far as we can tell, coordinates are something *we* impose upon the world when we want to talk about where things are. They are extremely useful, and in many applications quite essential. Unfortunately, different people might pick different coordinates! So it is good to know how the components of a vector field or 1-form depend on the coordinates used.

First let us describe how one can use coordinates locally on any manifold to work with vector fields and differential forms. We described the basic idea back in Chapter 2: given an n-dimensional manifold M, a chart is a diffeomorphism φ from an open set U in M to $\mathbb{R}^n$. This allows us to turn calculations on U into calculations on $\mathbb{R}^n$.

For example, we can use φ to pull back the coordinate functions x^μ from $\mathbb{R}^n$ to U. Instead of calling these functions $\varphi^* x^\mu$ as one really should, we usually call them simply x^μ. This is not too confusing as long as we know we are 'working in the chart' $\varphi: U \to \mathbb{R}^n$. The functions

x^μ on U are known as **local coordinates** on U. Any function on U can be written as a function $f(x^1, \ldots, x^n)$ of these local coordinates.

Similarly, the coordinate vector fields ∂_μ are a basis of vector fields on $\mathbb{R}^n$, and we may push these forwards by φ^{-1} to a basis of vector fields on U. As with the local coordinates, people usually denote these vector fields simply as ∂_μ. These are called the **coordinate vector fields** associated to the local coordinates x^μ on U. One thus writes any vector field v on U as

$$v = v^\mu \partial_\mu.$$

In the same way, the coordinate 1-forms dx^μ are a basis of 1-forms on $\mathbb{R}^n$, which we may pull back to U by φ, obtaining a basis of 1-forms on U. These are called the **coordinate 1-forms** associated to the local coordinates x^μ. These are written simply as dx^μ. Note that our use of x^μ and dx^μ to denote functions and 1-forms on U, while sloppy, is consistent:

Exercise 35. *Show that the coordinate 1-forms dx^μ really are the differentials of the local coordinates x^μ on U.*

We can write any 1-form ω on U as

$$\omega = \omega_\mu dx^\mu.$$

We should emphasize that it is bad to think of vector fields v or 1-forms ω as *being* their components v^μ or ω_μ. Instead, we should think of them as *having* components, which depend on the basis used. For example, the usual coordinate functions $x^1, \ldots, x^n$ on $\mathbb{R}^n$ give a basis $\{\partial_\mu\}$ for $\text{Vect}(\mathbb{R}^n)$. Given any vector field v on $\mathbb{R}^n$, I can write it uniquely as

$$v = v^\mu \partial_\mu,$$

where the v^μ are functions on $\mathbb{R}^n$. But suppose you chose some other coordinates on $\mathbb{R}^n$ — that is, some functions $x'^1, \ldots, x'^n$ on $\mathbb{R}^n$ such that $\{\partial'_\nu\}$ was another basis for $\text{Vect}(\mathbb{R}^n)$. Then you would write

$$v = v'^\nu \partial'_\nu.$$

The vector field v is the same in both cases — it is blissfully unaware of which coordinates we mere mortals are using. But its components

Change of Coordinates

depend on a coordinate system, and for us to talk to each other, we need to know how your components, v'^ν, are expressed in terms of mine, v^μ.

First, since your vector fields form a basis, we can express mine as linear combinations of yours:

$$\partial_\mu = T^\nu_\mu \partial'_\nu,$$

where the T^ν_μ are a matrix of functions on $\mathbb{R}^n$. It is not too hard to figure out these functions. Just apply both sides of the equation, which are vector fields, to the coordinate function x'^λ:

$$\partial_\mu x'^\lambda = T^\nu_\mu \partial'_\nu x'^\lambda.$$

The partial derivative $\partial'_\nu x'^\lambda$ is just the Kronecker delta δ^λ_ν, so actually we just have

$$\partial_\mu x'^\lambda = T^\lambda_\mu.$$

We can write this out somewhat more impressively as follows:

$$T^\lambda_\mu = \frac{\partial x'^\lambda}{\partial x^\mu}.$$

This implies that

$$\partial_\mu = \frac{\partial x'^\nu}{\partial x^\mu} \partial'_\nu.$$

Then, to express the components v'^μ in terms of the components v^μ, start with the fact that $v'^\mu \partial'_\mu = v^\mu \partial_\mu$, and use the equation above to get

$$v'^\nu \partial'_\nu = v^\mu \frac{\partial x'^\nu}{\partial x^\mu} \partial'_\nu.$$

Equating coefficients, we get

$$v'^\nu = \frac{\partial x'^\nu}{\partial x^\mu} v^\mu.$$

Now we can talk to each other! In short, to translate from my components to yours, I simply multiply by a matrix of partial derivatives corresponding to the change of coordinates.

1-forms work the same way, and we leave them as an important exercise for the reader:

Exercise 36. *In the situation above, show that*

$$dx'^\nu = \frac{\partial x'^\nu}{\partial x^\mu} dx^\mu.$$

Show that for any 1-form ω on $\mathbb{R}^n$, writing

$$\omega = \omega_\mu dx^\mu = \omega'_\nu dx'^\nu,$$

your components ω'_ν are related to my components ω_μ by

$$\omega'_\nu = \frac{\partial x^\mu}{\partial x'^\nu} \omega_\mu.$$

There is an interesting distinction between 'active' or 'passive' coordinate transformations. A **passive** coordinate transformation is a change of coordinate functions (on $\mathbb{R}^n$, or on a chart), which is what we have just been considering. We are not moving points of our space around, just changing the functions we use to describe them. An **active** coordinate transformation is just another name for a diffeomorphism

$$\phi: M \to M;$$

it moves the points of M around. We can push vector fields forwards by a diffeomorphism, and pull functions and 1-forms back. It is nice to know how these look in the special case of $\mathbb{R}^n$ (or a chart). Not surprisingly, the formulas look similar to the formulas for passive coordinate transformations that we have just derived!

There is, however, something a bit tricky about this business. The simplest example of this trickiness occurs when people in certain places switch from standard time to daylight saving time in the spring. The mnemonic formula is 'spring forward, fall back'. This is supposed to remind you to set your clock forward in the spring and back in the fall. The hard part is remembering what setting a clock 'forward' means! Is one supposed to move the hour hand to a *later* time, so one has to wake up *earlier* than one otherwise would? Or is one supposed to move the hour hand to an *earlier* time, so one can stay in bed *later*? Note that it takes a clock and a point in time to give a number that we call the 'time' t. More generally, it takes a coordinate system *together* with a point in spacetime to give a number. Changing the coordinate system

Change of Coordinates

one way has a similar effect to moving points of spacetime around the opposite way.

Let us now consider the effect a map $\phi\colon \mathbb{R}^m \to \mathbb{R}^n$ has on coordinate vector fields and 1-forms. If $n = m$ and ϕ is a diffeomorphism, this is an 'active coordinate transformation', but it is actually easier to keep things straight if we work in the general case. Write $x^1, \ldots, x^m$ for the coordinates on $\mathbb{R}^m$, and $x'^1, \ldots, x'^n$ for the coordinates on $\mathbb{R}^n$. First note that we can pull *back* the coordinate functions x'^ν on $\mathbb{R}^n$ to functions $\phi^* x'^\nu$ on $\mathbb{R}^m$ using ϕ. The definition is that

$$(\phi^* x'^\nu)(p) = x'^\nu(\phi(p))$$

for any point p in $\mathbb{R}^m$. In what follows, we will be sloppy and write

$$\frac{\partial x'^\nu}{\partial x^\mu}$$

when we really mean

$$\frac{\partial}{\partial x^\mu} \phi^* x'^\nu.$$

The reason we do this is simply that everyone does it, and the reader will have to get used to it.

Now consider the coordinate vector field ∂_μ on $\mathbb{R}^m$. We can push ∂_μ *forward* by ϕ, and we claim that

$$\phi_* \partial_\mu = \frac{\partial x'^\nu}{\partial x^\mu} \partial'_\nu.$$

To see this, just apply both sides to any coordinate function x'^λ on $\mathbb{R}^n$ and show that we get the same answer. The left hand side gives

$$\begin{aligned}(\phi_* \partial_\mu)(x'^\lambda) &= \partial_\mu(\phi^* x'^\lambda) \\ &= \frac{\partial x'^\lambda}{\partial x^\mu},\end{aligned}$$

where in the last step we are being sloppy in the way described above. The right hand side gives

$$\begin{aligned}\frac{\partial x'^\nu}{\partial x^\mu} \partial'_\nu x'^\lambda &= \frac{\partial x'^\nu}{\partial x^\mu} \delta^\lambda_\nu \\ &= \frac{\partial x'^\lambda}{\partial x^\mu},\end{aligned}$$

which is the same.

Finally, consider a coordinate 1-form dx'^ν. We can pull this *back* by ϕ. We claim that
$$\phi^*(dx'^\nu) = \frac{\partial x'^\nu}{\partial x^\mu} dx^\mu.$$

Exercise 37. *Show this.*

With these basic formulas in hand, you should be able to transform between coordinates both actively and passively!

To conclude, we should note that sometimes it is nice to be more general and work with a basis e_μ of vector fields on a chart that are not the coordinate vector fields. These are easy to come by:

Exercise 38. *Let*
$$e_\mu = T_\mu^\nu \partial_\nu,$$
where ∂_ν are the coordinate vector fields associated to local coordinates on an open set U, and T_μ^ν are functions on U. Show that the vector fields e_μ are a basis of vector fields on U if and only if for each $p \in U$ the matrix $T_\mu^\nu(p)$ is invertible.

If we have such a basis, we automatically get a **dual basis** of 1-forms f^μ on U such that
$$f^\mu(e_\nu) = \delta_\nu^\mu,$$
the Kronecker delta.

Exercise 39. *Use the previous exercise to show that the dual basis exists and is unique.*

We can write any vector field v on U as a linear combination
$$v = v^\mu e_\mu$$
where $v^1, \ldots, v^n$ are functions on U, called the **components** of v in the basis e_μ. Similarly, we can write any 1-form ω on U as a linear combination
$$\omega = \omega_\mu f^\mu.$$

We will use these more general bases quite a bit in the next chapter, when we discuss the notion of a 'metric'. This is like an inner product, and it will be handy to work with 'orthonormal' bases of vector fields and 1-forms on a chart. We leave it to the reader to work out how the components of a vector field or 1-form change when we perform an arbitrary change of basis:

Exercise 40. *Let e_μ be a basis of vector fields on U and let f^μ be the dual basis of 1-forms. Let*
$$e'_\mu = T_\mu^\nu e_\nu$$
be another basis of vector fields, and let f'^μ be the corresponding dual basis of 1-forms. Show that
$$f'^\mu = (T^{-1})_\nu^\mu f^\nu.$$
Show that if $v = v^\mu e_\mu = v'^\mu e'_\mu$, then
$$v'^\mu = (T^{-1})_\nu^\mu v^\nu,$$
and that if $\omega = \omega_\mu f^\mu = \omega'_\mu f'^\mu$ then
$$\omega'_\mu = T_\mu^\nu \omega_\nu.$$

p-forms

By the geometrical product of two vectors, we mean the surface content of the parallelogram determined by these vectors; we however fix the position of the plane in which the parallelogram lies. We refer to two surface areas as geometrically equal only when they are equal in content and lie in parallel planes. By the geometrical product of three vectors we mean the solid (a parallelepiped) formed from them. — Hermann Grassmann

If you ever seriously wondered how to take cross products in 4 dimensions, you were well on your way to reinventing differential forms. In fact, if you ever wondered why the definition of cross products requires a 'right-hand rule', you were getting close. (This rule is especially irksome to those who happen to be left-handed.) Differential forms allow one to generalize cross products to any number of dimensions, and it turns out that if one does things correctly, no right-hand rule is necessary! Interestingly, though, it turns out to be better to define the cross product not for tangent vectors (or vector fields) but for cotangent vectors (or 1-forms). If we do this, we get an extra bonus. Namely, we can show that the gradient, curl, and divergence are all different versions of the same thing, and see how to define them on arbitrary manifolds.

Let us plunge right in. Let V be a vector space. We want to be able to multiply two vectors in V somehow, and we want the basic property of the cross product, the antisymmetry,

$$\vec{v} \times \vec{w} = -\vec{w} \times \vec{v},$$

to hold. But we will call this generalized sort of cross product the 'wedge product' (or 'exterior product') and write it with a $\wedge$. We proceed in an abstract, algebraic sort of way. Namely, we will define a bigger vector space ΛV, in fact an algebra, so that the wedge product of any number of vectors in V will lie in this algebra. First we will give the definition as a mathematician would: the **exterior algebra** over V, denoted ΛV, is the algebra generated by V with the relations

$$v \wedge w = -w \wedge v$$

for all $v, w \in V$. What does this mean? Roughly, it means that we start with the vectors in V together with an element 1, and then form an algebra by taking all linear combinations of formal products of the form $v_1 \wedge \cdots \wedge v_p$, where $v_i \in V$; the only relations we impose upon these linear combinations are those in the definition of an algebra (as defined above in Chapter 3) together with the 'anticommutative' rule $v \wedge w = -w \wedge v$.

For example, say V is 3-dimensional. Then everything in ΛV is a linear combination of wedge products of elements of V. Suppose V has a basis dx, dy, dz. (We write the basis this way because in a bit we will want V to be a space of cotangent vectors.) Then for starters we have

$$1 \in \Lambda V$$

and

$$dx, dy, dz \in \Lambda V,$$

along with all linear combinations of these. But we can also take the wedge product of any two elements $v, w \in V$ and get an element of ΛV. If

$$\begin{aligned} v &= v_x dx + v_y dy + v_z dz \\ w &= w_x dy + w_y dy + w_z dz \end{aligned}$$

then we have

$$v \wedge w = (v_x dx + v_y dy + v_z dz) \wedge (w_x dx + w_y dy + w_z dz)$$
$$= (v_x w_y - v_y w_x) dx \wedge dy + (v_y w_z - v_z w_y) dy \wedge dz + (v_z w_x - v_x w_z) dz \wedge dx,$$

where all we did is use the definition of an algebra together with the 'anticommutative' rule. Notice that this looks a whole lot like the formula for the cross product! If we have a third element of V, say

$$u = u_x dx + u_y dy + u_z dz,$$

we can get another element of ΛV, namely $u \wedge v \wedge w$. This triple wedge product is closely related to the 'triple product' of three vectors in $\mathbb{R}^3$, $\vec{u} \cdot (\vec{v} \times \vec{w})$. We can also take wedge products of four or more vectors, but if V is 3-dimensional, this is always zero:

Exercise 41. *Show that*

$$u \wedge v \wedge w = \det \begin{pmatrix} u_x & u_y & u_z \\ v_x & v_y & v_z \\ w_x & w_y & w_z \end{pmatrix} dx \wedge dy \wedge dz.$$

Compare this to $\vec{u} \cdot (\vec{v} \times \vec{w})$.

Exercise 42. *Show that if a, b, c, d are four vectors in a 3-dimensional space then $a \wedge b \wedge c \wedge d = 0$.*

Exercise 43. *Describe ΛV if V is 1-dimensional, 2-dimensional, or 4-dimensional.*

In general, for any vector space V, we define $\Lambda^p V$ to be the subspace of ΛV consisting of linear combinations of p-fold products of vectors in V, e.g.

$$v_1 \wedge \cdots \wedge v_p.$$

Elements of ΛV that lie in $\Lambda^p V$ are said to have **degree** p. For example, $\Lambda^1 V$ is just V itself, while $\Lambda^0 V$ is by convention defined to be $\mathbb{R}$, since numbers can be regarded as wedge products of *no* vectors. Copying the example above, one can show the following:

Exercise 44. *Let V be an n-dimensional vector space. Show that $\Lambda^p V$ is empty for $p > n$, and that for $0 \leq p \leq n$ the dimension of $\Lambda^p V$ is $n!/p!(n-p)!$*

Recall that a vector space V is a **direct sum** of subspaces $L_1, \ldots, L_n$ if every vector $v \in V$ can be uniquely expressed as $v_1 + \cdots + v_n$, where $v_i \in L_i$. In this situation, we may think of vectors in L as n-tuples $(v_1, \ldots, v_n)$ where $v_i \in L_i$. Alternatively, given vector spaces $V_1, \ldots, V_n$, the **direct sum** $V_1 \oplus \cdots \oplus V_n$, sometimes written

$$\bigoplus_{i=1}^{n} V_i,$$

is defined as the vector space of all n-tuples $(v_1, \ldots, v_n)$ with $v_i \in V_i$, where addition and scalar multiplication are defined componentwise. The exterior algebra is an example of such a direct sum:

Exercise 45. *Show that ΛV is the direct sum of the subspaces $\Lambda^p V$:*

$$\Lambda V = \bigoplus \Lambda^p V,$$

and that the dimension of ΛV is 2^n if V is n-dimensional.

There is something very special about the exterior algebra in 3 dimensions! The wedge product of two vectors in V lies in $\Lambda^2 V$. Only in dimension 3 is the dimension of $\Lambda^2 V$ equal to that of V itself. So only in 3 dimensions can we pretend, if we so desire, that the wedge product of two vectors is again a vector! The way to do this (as we will see in Chapter 5) is to define a linear map called the 'star operator' that turns elements of $\Lambda^2 V$ into elements of ΛV. When V has the basis dx, dy, dz, the star operator is given by

$$\star: dx \wedge dy \mapsto dz$$
$$\star: dy \wedge dz \mapsto dx$$
$$\star: dz \wedge dx \mapsto dy.$$

The cross product really amounts to taking the wedge product and then applying the star operator. Note, however, that our definition of the star operator incorporates a right-hand rule. We could just as well have defined $\star\colon \Lambda^2 V \to V$ by

$$\star\colon dy \wedge dx \mapsto dz$$
$$\star\colon dz \wedge dy \mapsto dx$$
$$\star\colon dx \wedge dz \mapsto dy$$

which would amount to a left-hand rule. In short, the 'right-hand rule' nonsense enters when we unnaturally try to make the product of two elements of V to come out to an element of V, instead of $\Lambda^2 V$. This is noted in some physics books, where they say that the cross product of two vectors is a 'pseudovector' or 'axial vector', rather than a true vector. We prefer to say that the wedge product of 2 vectors lies in $\Lambda^2 V$ — this is true in all dimensions.

Exterior algebra is an interesting subject in itself, but we do not just want to generalize the cross product of vectors; we want to generalize the cross product of vector fields. Actually, as already mentioned, it is much better to take products of 1-forms! We will do this by copying our construction of ΛV, with the smooth functions $C^\infty(M)$ on some manifold M taking the place of the real numbers, and the 1-forms $\Omega^1(M)$ taking the place of the vector space V. Namely, we define the **differential forms** on M, denoted $\Omega(M)$, to be the algebra generated by $\Omega^1(M)$ with the relations

$$\omega \wedge \mu = -\mu \wedge \omega$$

for all $\omega, \mu \in \Omega^1(M)$. To be precise, we should emphasize that we form $\Omega(M)$ as an algebra 'over $C^\infty(M)$'. This means, first of all, that $\Omega(M)$ consists of linear combinations of wedge products of 1-forms with *functions* as coefficients. We allow all **locally finite** linear combinations, that is, those for which every point p in M has a neighborhood where only finitely many terms are nonzero. Secondly, it means that $\Omega(M)$ satisfies the rules of an algebra with *functions* taking the place of numbers. Maybe we should say again what all these rules are. We have, for all $\omega, \mu, \nu \in \Omega(M)$ and $f, g \in C^\infty(M)$,

$$\omega + \mu = \mu + \omega, \quad \omega + (\mu + \nu) = (\omega + \mu) + \nu, \quad \omega \wedge (\mu \wedge \nu) = (\omega \wedge \mu) \wedge \nu,$$

$$\omega \wedge (\mu + \nu) = \omega \wedge \mu + \omega \wedge \nu, \quad (\omega + \mu) \wedge \nu = \omega \wedge \nu + \mu \wedge \nu,$$

$$1\omega = \omega, \quad f(g\omega) = (fg)\omega, \quad f(\mu+\nu) = f\mu + f\nu, \quad (f+g)\omega = f\omega + g\omega.$$

We define the 0-forms, $\Omega^0(M)$, to be the functions themselves, and define the wedge product of a function with a differential form to be the ordinary product: $f \wedge \omega = f\omega$. We define the product of a number c and a differential form ω to be the product of the constant function $c \in \Omega^0(M)$ and ω. Elements that are linear combinations of products of p 1-forms are called p-**forms**, and we write the space of p-forms on M as $\Omega^p(M)$. We have

$$\Omega(M) = \bigoplus_p \Omega^p(M).$$

For example, suppose $M = \mathbb{R}^n$. The 0-forms on $\mathbb{R}^n$ are just functions, like

$$f.$$

The 1-forms all look like

$$\omega_\mu dx^\mu$$

where the coefficients ω_μ are functions. It is easy to check that the 2-forms all look like

$$\frac{1}{2} \omega_{\mu\nu} dx^\mu \wedge dx^\nu$$

where we have put in a factor of $\frac{1}{2}$ because $dx^\mu \wedge dx^\nu = -dx^\nu \wedge dx^\mu$. Also for this reason, we may as well assume that $\omega_{\mu\nu} = -\omega_{\nu\mu}$. Then on $\mathbb{R}^3$, for example, we have

$$\omega = \omega_{12} dx^1 \wedge dx^2 + \omega_{23} dx^2 \wedge dx^3 + \omega_{31} dx^3 \wedge dx^1.$$

Similarly, the 3-forms look like

$$\frac{1}{3!} \omega_{\mu\nu\lambda} dx^\mu \wedge dx^\nu \wedge dx^\lambda,$$

and we may as well assume that $\omega_{\mu\nu\lambda}$ is totally antisymmetric (that is, switches sign when we switch any two indices). On $\mathbb{R}^3$ we get

$$\omega = \omega_{123} dx^1 \wedge dx^2 \wedge dx^3.$$

p-forms

There are no nonzero 4-forms, 5-forms, etc., on $\mathbb{R}^3$. In general, there are no nonzero p-forms on an n-dimensional manifold if $p > n$.

We leave it for the reader to show some important facts about differential forms in the following exercises.

Exercise 46. *Given a vector space V, show that ΛV is a **graded commutative** or **supercommutative** algebra, that is, if $\omega \in \Lambda^p V$ and $\mu \in \Lambda^q V$, then*

$$\omega \wedge \mu = (-1)^{pq} \mu \wedge \omega.$$

Show that for any manifold M, $\Omega(M)$ is graded commutative.

Exercise 47. *Show that differential forms are contravariant. That is, show that if $\phi: M \to N$ is a map from the manifold M to the manifold N, there is a unique **pullback** map*

$$\phi^*: \Omega(N) \to \Omega(M)$$

agreeing with the usual pullback on 0-forms (functions) and 1-forms, and satisfying

$$\begin{aligned}
\phi^*(\alpha\omega) &= \alpha\phi^*\omega \\
\phi^*(\omega + \mu) &= \phi^*\omega + \phi^*\mu \\
\phi^*(\omega \wedge \mu) &= \phi^*\omega \wedge \phi^*\mu
\end{aligned}$$

for all $\omega, \mu \in \Omega(N)$ and $\alpha \in \mathbb{R}$.

Exercise 48. *Compare how 1-forms and 2-forms on $\mathbb{R}^3$ transform under parity. That is, let $P: \mathbb{R}^3 \to \mathbb{R}^3$ be the map*

$$P(x, y, z) = (-x, -y, -z),$$

known as the 'parity transformation'. Note that P maps right-handed bases to left-handed bases and vice versa. Compute $\phi^*(\omega)$ when ω is the 1-form $\omega_\mu dx^\mu$, and when it is the 2-form $\frac{1}{2}\omega_{\mu\nu} dx^\mu \wedge dx^\nu$.

In physics, the electric field $\vec{E}$ is called a vector, while the magnetic field $\vec{B}$ is called an axial vector, because $\vec{E}$ changes sign under parity transformation, while $\vec{B}$ does not. In Chapter 5 we will see that it is best to think of the electric field as a 1-form on space, and the magnetic field as a 2-form. In other words, while we may be used to thinking of

$\vec{E} = (E_x, E_y, E_z)$ and $\vec{B} = (B_x, B_y, B_z)$ as vector fields, it is better to use
$$E = E_x dx + E_y dy + E_z dz$$
and
$$B = B_x dy \wedge dz + B_y dz \wedge dx + B_z dx \wedge dy.$$
By the above exercise, this means that they transform differently under parity.

If the reader is frustrated because exterior algebras and differential forms seem difficult to *visualize*, we suggest taking a peek ahead to Figures 3 and 4 of Chapter 5. Grassmann, the inventor of the exterior algebra, visualized a wedge product $v_1 \wedge \cdots \wedge v_p$ as an oriented parallelepiped with sides given by the vectors $v_1, \ldots, v_p$. One must be careful, however, because the wedge product of 1-forms corresponds to a parallelepiped in the *cotangent* space.

The Exterior Derivative

We know from the first section of this chapter that the differential is a nice way to generalize the good old 'gradient' to manifolds. As we saw, the differential of a function, or 0-form, is a 1-form. Now we will show how to take the differential of a p-form and get a $(p+1)$-form:
$$d: \Omega^p(M) \to \Omega^{p+1}(M).$$
This will let us generalize the gradient, the curl and the divergence in one fell swoop, and see that they are secretly all the same thing. The big clue is that the curl of a gradient is zero:
$$\nabla \times (\nabla f) = 0$$
This suggests that we make d satisfy $d(df) = 0$ for any function f. Another clue is that the various product rules
$$\begin{aligned}
\nabla(fg) &= (\nabla f)g + f \nabla g \\
\nabla \times (fv) &= \nabla f \times v + f \nabla \times v \\
\nabla \cdot (fv) &= \nabla f \cdot v + f \nabla \cdot v \\
\nabla \cdot (v \times w) &= (\nabla \cdot v)w - v \nabla \cdot w
\end{aligned}$$

The Exterior Derivative

should all be special cases of some sort of Leibniz law for differential forms. Since the differential forms are graded commutative, it turns out that we need a graded version of the Leibniz law.

After scratching our head for a while, we define the **exterior derivative**, or **differential**, to be the unique set of maps

$$d\colon \Omega^p(M) \to \Omega^{p+1}(M)$$

such that the following properties hold:

1) $d\colon \Omega^0(M) \to \Omega^1(M)$ agrees with our previous definition.
2) $d(\omega + \mu) = d\omega + d\mu$ and $d(c\omega) = c\,d\omega$ for all $\omega, \mu \in \Omega(M)$ and $c \in \mathbb{R}$.
3) $d(\omega \wedge \mu) = d\omega \wedge \mu + (-1)^p \omega \wedge d\mu$ for all $\omega \in \Omega^p(M)$ and $\mu \in \Omega(M)$.
4) $d(d\omega) = 0$ for all $\omega \in \Omega(M)$.

To show that these properties uniquely determine the exterior derivative, one just needs the fact that any 1-form is a locally finite linear combination of those of the form df (with functions as coefficients). This fact is easy to see on $\mathbb{R}^n$, and can be shown in general using charts. Then to calculate d of any differential form, say

$$f\,dg \wedge dh,$$

we just use rules 1) - 4):

$$\begin{aligned}
d(f\,dg \wedge dh) &= df \wedge (dg \wedge dh) + f \wedge d(dg \wedge dh) \\
&= df \wedge dg \wedge dh + f\,d(dg) \wedge dh - f\,dg \wedge d(dh) \\
&= df \wedge dg \wedge dh.
\end{aligned}$$

To show that d with these properties is actually well-defined, it suffices (by the black magic of algebra) to show that this way of calculating d is compatible with the relations in the definition of differential forms. The most important one of these is the anticommutative law

$$\omega \wedge \mu = -\mu \wedge \omega$$

for 1-forms. For d to be well-defined, it had better be true that calculating $d(\omega \wedge \mu)$ gives the same answer as $d(-\mu \wedge \omega)$. This is where the graded Leibniz law is necessary: when ω and μ are 1-forms, we have

$$\begin{aligned} d(-\mu \wedge \omega) &= -d(\mu \wedge \omega) \\ &= -d\mu \wedge \omega + \mu \wedge d\omega, \\ &= -\omega \wedge d\mu + d\omega \wedge \mu \\ &= d(\omega \wedge \mu). \end{aligned}$$

Let us calculate the exterior derivative of 1-forms and 2-forms on $\mathbb{R}^3$. Taking any 1-form

$$\omega = \omega_x dx + \omega_y dy + \omega_z dz,$$

we get

$$d\omega = d\omega_x \wedge dx + d\omega_y \wedge dy + d\omega_z \wedge dz,$$

hence by the rule for d of a function and a little extra work

$$d\omega = (\partial_y \omega_z - \partial_z \omega_y) dy \wedge dz + (\partial_z \omega_x - \partial_x \omega_z) dz \wedge dx + (\partial_x \omega_y - \partial_y \omega_x) dx \wedge dy.$$

In other words, the exterior derivative of a 1-form on $\mathbb{R}^3$ is essentially just the curl! We need the right-hand rule to define the curl, however, while the exterior derivative involves no right-hand rule. This is because d of a 1-form is a 2-form; the right-hand rule only comes in when one tries to pretend that this 2-form is a 1-form, using the star operator as follows:

$$\star d\omega = (\partial_y \omega_z - \partial_z \omega_y) dx + (\partial_z \omega_x - \partial_x \omega_z) dy + (\partial_x \omega_y - \partial_y \omega_x) dz.$$

And, as noted, this pretense is only possible in 3 dimensions, while we can take d of a 1-form in any dimension:

Exercise 49. *Show that on $\mathbb{R}^n$ the exterior derivative of any 1-form is given by*

$$d(\omega_\mu dx^\mu) = \partial_\nu \omega_\mu dx^\nu \wedge dx^\mu.$$

Next, taking a 2-form on $\mathbb{R}^3$:

$$\omega = \omega_{xy} dx \wedge dy + \omega_{yz} dy \wedge dz + \omega_{zx} dz \wedge dx$$

The Exterior Derivative

we get

$$\begin{aligned} d\omega &= d\omega_{xy} \wedge dx \wedge dy + d\omega_{yz} \wedge dy \wedge dz + d\omega_{zx} \wedge dz \wedge dx \\ &= \partial_z \omega_{xy} dz \wedge dx \wedge dy + \partial_x \omega_{yz} dx \wedge dy \wedge dz + \partial_y \omega_{zx} dy \wedge dz \wedge dx \\ &= (\partial_z \omega_{xy} + \partial_x \omega_{yz} + \partial_y \omega_{zx}) dx \wedge dy \wedge dz. \end{aligned}$$

Thus the exterior derivative of a 2-form on $\mathbb{R}^3$ is just the divergence in disguise. In short, the exterior derivative has as special cases the following familiar operators:

◇ *Gradient* $\quad d \colon \Omega^0(\mathbb{R}^3) \to \Omega^1(\mathbb{R}^3)$

◇ *Curl* $\quad d \colon \Omega^1(\mathbb{R}^3) \to \Omega^2(\mathbb{R}^3)$

◇ *Divergence* $\quad d \colon \Omega^2(\mathbb{R}^3) \to \Omega^3(\mathbb{R}^3)$

In fact, there is a simple formula for the exterior derivative of any differential form on $\mathbb{R}^n$. Let I stand for a **multi-index**, that is, a p-tuple $(i_1, \ldots, i_p)$ of distinct integers between 1 and n. Let dx^I stand for the p-form

$$dx^{i_1} \wedge \cdots \wedge dx^{i_p}$$

on $\mathbb{R}^n$. Then any p-form on $\mathbb{R}^n$ can be expressed as

$$\omega = \omega_I dx^I$$

where following the Einstein summation convention we sum over all multi-indices I. We have

$$d\omega = d\omega_I \wedge dx^I$$

by the Leibniz law, since $d(dx^I) = 0$ (as can easily be checked). More concretely, using the formula for d of a function, we have

$$d\omega = (\partial_\mu \omega_I) \, dx^\mu \wedge dx^I.$$

Using this formula it is easy to derive an amazing identity:

$$d(d\omega) = 0$$

for any differential form on $\mathbb{R}^n$. Just compute:

$$\begin{aligned} d(d\omega) &= d(\partial_\mu \omega_I dx^\mu \wedge dx^I) \\ &= \partial_\nu \partial_\mu \omega_I dx^\nu \wedge dx^\mu \wedge dx^I \end{aligned}$$

and note that on the one hand

$$\partial_\nu \partial_\mu \omega_I = \partial_\mu \partial_\nu \omega_I$$

by the equality of mixed partials, but on the other hand

$$dx^\nu \wedge dx^\mu = -dx^\mu \wedge dx^\nu$$

by the anticommutative law. With a little thought one can see this means that $d(d\omega)$ is equal to the negative of itself, so it is zero. This rule is so important that people often write it as

$$d^2 \omega = 0$$

or even just

$$d^2 = 0.$$

On $\mathbb{R}^3$, d acts like the gradient on 0-forms, the curl on 1-forms and the divergence on 2-forms, so the identity $d^2 = 0$ contains within it the identities

$$\nabla \times (\nabla f) = 0$$

and

$$\nabla \cdot (\nabla \times v) = 0.$$

But this identity is better, since it applies to differential forms in any dimension. In fact it applies to any manifold! Here is an easy proof that does not use coordinates. By definition, any p-form on a manifold is a linear combination — with *constant* coefficients — of p-forms like

$$\omega = f_0 df_1 \wedge \cdots \wedge df_p.$$

So it suffices to prove the identity for p-forms of this sort. We have

$$d\omega = df_0 \wedge df_1 \wedge \cdots \wedge df_p$$

The Exterior Derivative

by the Leibniz law and the fact that $d(df) = 0$ for any function. Using the Leibniz law and $d(df) = 0$ again, we obtain

$$d(d\omega) = 0.$$

It turns out that the identity $d^2 = 0$ and its generalizations have profound consequences for physics, starting with Maxwell's equations. It is also the basis of a very important connection between geometry and topology, called deRham theory. We will explore these in Chapter 6. When we do, it is important to remember that this identity is just a way of saying that partial derivatives commute! As so often the case, the simplest facts in mathematics lie at the root of some of the most sophisticated developments.

We will wrap up this section by showing that the exterior derivative is **natural**. We already discussed this for functions in Section 4; it simply meant that d commutes with pullbacks. In fact, this is true for differential forms of any degree. In other words, for any map $\phi \colon M \to N$ between manifolds, and any differential form $\omega \in \Omega^p(N)$, we have

$$\phi^*(d\omega) = d(\phi^*\omega).$$

The proof is easy. By Exercise 47, ϕ^* is real-linear, so it suffices to treat the case where

$$\omega = f_0 df_1 \wedge \cdots \wedge df_p.$$

We then have, using Exercise 47 again together with the naturality of d on functions,

$$\begin{aligned}
\phi^*(d\omega) &= \phi^*(df_0 \wedge df_1 \wedge \cdots \wedge df_p) \\
&= \phi^* df_0 \wedge \cdots \wedge \phi^* df_p \\
&= d\phi^* f_0 \wedge \cdots \wedge d\phi^* f_p \\
&= d(\phi^* f_0 \wedge d\phi^* f_1 \wedge \cdots \wedge d\phi^* f_p) \\
&= d(\phi^* f_0 \wedge \phi^* df_1 \wedge \cdots \wedge \phi^* df_p) \\
&= d(\phi^*(f_0 \wedge df_1 \wedge \cdots \wedge df_p)) \\
&= d(\phi^*\omega)
\end{aligned}$$

as desired.

Chapter 5

Rewriting Maxwell's Equations

Hence space of itself, and time of itself, will sink into mere shadows, and only a union of the two shall survive. — *Hermann Minkowski*

The First Pair of Equations

We now have developed enough differential geometry to generalize the first pair of Maxwell equations,

$$\nabla \cdot \vec{B} = 0$$
$$\nabla \times \vec{E} + \frac{\partial \vec{B}}{\partial t} = 0,$$

to any manifold. We claim that they have a very beautiful form as a *single* equation in terms of differential forms.

Before giving away the answer, let us consider a special case: the static case. Then we just have two equations for vector fields on space, $\mathbb{R}^3$:

$$\nabla \cdot \vec{B} = 0, \qquad \nabla \times \vec{E} = 0$$

In the language of differential forms, the divergence becomes the exterior derivative on 2-forms on $\mathbb{R}^3$. Thus, instead of treating the magnetic field as a vector field $\vec{B} = (B_x, B_y, B_z)$ we will treat it as the 2-form

$$B = B_x dy \wedge dz + B_y dz \wedge dx + B_z dx \wedge dy.$$

Similarly, the curl becomes the exterior derivative on 1-forms on $\mathbb{R}^3$, so instead of treating the electric field as a vector field $\vec{E} = (E_x, E_y, E_z)$ we will treat it as the 1-form

$$E = E_x dx + E_y dy + E_z dz.$$

The first pair of static Maxwell's equations then become

$$dE = 0, \quad dB = 0.$$

Next consider the general, time-dependent case. Now we must think of the electric and magnetic fields as living on spacetime. We begin by working on Minkowski spacetime, $\mathbb{R}^4$, using the standard coordinate system, which we will number as (x^0, x^1, x^2, x^3). We will often write t instead of x^0 for the time coordinate, and x, y, z for the space coordinates (x^1, x^2, x^3). The electric and magnetic fields are 1-forms and 2-forms on $\mathbb{R}^4$, namely

$$E = E_x dx + E_y dy + E_z dz$$

and

$$B = B_x dy \wedge dz + B_y dz \wedge dx + B_z dx \wedge dy$$

We can combine both fields into a unified electromagnetic field F, a 2-form on $\mathbb{R}^4$, as follows:

$$F = B + E \wedge dt.$$

If we want to look at all the components,

$$F = \frac{1}{2} F_{\mu\nu} dx^\mu \wedge dx^\nu,$$

we can write them out as a matrix:

$$F_{\mu\nu} = \begin{pmatrix} 0 & -E_x & -E_y & -E_z \\ E_x & 0 & B_z & -B_y \\ E_y & -B_z & 0 & B_x \\ E_z & B_y & -B_x & 0 \end{pmatrix}.$$

The First Pair of Equations

The beauty of this way of unifying the electric and magnetic fields is that the first pair of Maxwell equations become simply

$$dF = 0.$$

To see this, first note that

$$dF = d(B + E \wedge dt) = dB + dE \wedge dt$$

Then split up the exterior derivative operator into a spacelike part and a timelike part. Recall that for any differential form ω we have

$$d\omega = \partial_\mu \omega_I \, dx^\mu \wedge dx^I,$$

where I ranges over all multi-indices and $\mu = 0, 1, 2, 3$. We can thus write $d\omega$ as a sum of the spacelike part

$$d_S \omega = \partial_i \omega_I \, dx^i \wedge dx^I$$

where i ranges over the 'spacelike' indices $1, 2, 3$, and the timelike part

$$dt \wedge \partial_t \omega = \partial_0 \omega_I \, dx^0 \wedge dx^I.$$

Then we have

$$\begin{aligned} dF &= dB + dE \wedge dt \\ &= d_S B + dt \wedge \partial_t B + (d_S E + dt \wedge \partial_t E) \wedge dt \\ &= d_S B + (\partial_t B + d_S E) \wedge dt \end{aligned}$$

Note that the first term has no dt in it while the second one does. Also note that the second one vanishes only if the expression in parentheses does. It follows that $dF = 0$ is equivalent to the pair of equations

$$\begin{aligned} d_S B &= 0, \\ \partial_t B + d_S E &= 0. \end{aligned}$$

These are just the first pair of Maxwell equations in slightly newfangled notation!

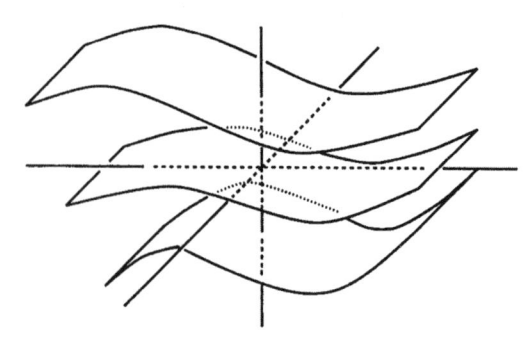

Fig. 1. Splitting spacetime into space and time

One advantage of the differential form language is its generality. We can take our spacetime to be any manifold M, of any dimension, and define the electromagnetic field to be a 2-form F on M. The first pair of Maxwell equations says just that

$$dF = 0.$$

Sometimes — but not always — we can split spacetime up into space and time, that is, write M as $\mathbb{R} \times S$ for some manifold S we call 'space'. If so, we can write t for the usual coordinate on $\mathbb{R}$, and split F into an electric and magnetic field:

Exercise 50. *Show that any 2-form F on $\mathbb{R} \times S$ can be uniquely expressed as $B + E \wedge dt$ in such a way that for any local coordinates x^i on S we have $E = E_i dx^i$ and $B = \frac{1}{2} B_{ij} dx^i \wedge dx^j$*

We can also split the exterior derivative into spacelike and timelike parts as before:

Exercise 51. *Show that for any form ω on $\mathbb{R} \times S$ there is a unique way to write $d\omega = dt \wedge \partial_t \omega + d_S \omega$ such that for any local coordinates x^i on S, writing $t = x^0$, we have*

$$\begin{aligned} d_S \omega &= \partial_i \omega_I \, dx^i \wedge dx^I, \\ dt \wedge \partial_t \omega &= \partial_0 \omega_I \, dx^0 \wedge dx^I. \end{aligned}$$

When we split spacetime up into space and time, $dF = 0$ becomes equivalent to the pair of equations

$$d_S B = 0, \qquad \partial_t B + d_S E = 0.$$

In the static case, when $\partial_t E = \partial_t B = 0$, we can forget about the t coordinate entirely and treat E and B as forms on space satisfying the static equations
$$d_S B = 0, \qquad d_S E = 0.$$

Note that the electric and magnetic fields are only defined after we choose a way of splitting spacetime into space and time! If someone hands us a manifold M, it may be diffeomorphic to $\mathbb{R} \times S$ in many different ways, or in no way at all. In special relativity one learns that different inertial frames (corresponding to observers moving at constant velocity) will give different splittings of spacetime into $\mathbb{R} \times \mathbb{R}^3$, which are related by Lorentz transformations. This means that the electric and magnetic fields will get mixed up when we do a Lorentz transformation, as described in Chapter 1. More drastically, we could split spacetime into space and time in a wiggly way as in Figure 1 above. This may seem perverse, but there is usually no 'best' way to split spacetime into space and time, particularly in the context of general relativity.

The Metric

In the Space and Time marriage we have the greatest Boy meets Girl story of the age. To our great-grandchildren this will be as poetical as the ancient Greek marriage of Cupid and Psyche seems to us. — Lawrence Durrell, Balthazar

The first pair of Maxwell equations does not involve measuring distances in spacetime. That is why they are 'generally covariant', i.e., one can pull back a solution by any diffeomorphism, no matter how much it stretches or distorts spacetime, and get another solution. This is not the case for the second pair, which require for their formulation a way of measuring distances and times. The key idea of relativity is that distances and time intervals are two aspects of a single concept, the 'spacetime interval'. Mathematically, spacetime intervals are calculated using a 'metric' on spacetime.

In ordinary Euclidean $\mathbb{R}^3$ we measure distances and angles using

the dot product of vectors:
$$v \cdot w = v^1 w^1 + v^2 w^2 + v^3 w^3,$$
and the norm defined using the dot product:
$$\|v\|^2 = v \cdot v.$$
In Minkowski spacetime we measure 'intervals' using a generalization of the dot product. In units where the speed of light, c, is equal to 1, this is given by
$$v \cdot w = -v^0 w^0 + v^1 w^1 + v^2 w^2 + v^3 w^3.$$
If $x \in V$ has $x \cdot x > 0$, x is called **spacelike**, since it points more in the space directions than the time direction. If x is spacelike, the square root of $x \cdot x$ represents the length of a straight ruler that stretches from the origin to x. If x has $x \cdot x < 0$ we call it **timelike**, since it points more in the time direction than in the space directions. The velocity of a particle moving slower than the speed of light is timelike. If x is timelike, the square root of $x \cdot x$ measures the time a clock would tick off as it moved from the origin to x in a straight line. If $x \cdot x = 0$, x is called **null** or **lightlike**, since it points just as much in the time direction as in the space directions. We should add that sometimes people use the negative of our Minkowski metric and reverse the definitions of spacelike and timelike. This is just a matter of convention — but we will always follow the above convention!

The notion of a metric generalizes these concepts. A **semi-Riemannian metric** (or just 'metric') on a vector space V is a map
$$g \colon V \times V \to \mathbb{R},$$
that is **bilinear**, or linear in each slot:
$$\begin{aligned} g(cv + v', w) &= cg(v, w) + g(v', w) \\ g(v, cw + w') &= cg(v, w) + g(v, w'), \end{aligned}$$
symmetric:
$$g(v, w) = g(w, v),$$

The Metric

and **nondegenerate**: if $g(v,w) = 0$ for all $w \in V$, then $v = 0$. We say that $v \in V$ is spacelike, timelike or null depending on whether $g(v,v)$ is positive, negative or zero. If $g(v,w) = 0$, we say that v and w are orthogonal. Note that null vectors are orthogonal to themselves!

Given a metric on V, we can always find an **orthonormal basis** for V, that is, a basis $\{e_\mu\}$ such that $g(e_\mu, e_\nu)$ is 0 if $\mu \neq \nu$, and ± 1 if $\mu = \nu$. The number of $+1$'s and -1's is independent of the orthonormal basis, and if the number of $+1$'s is p and the number of -1's is q, we say the metric has **signature** (p,q). For example, Minkowski spacetime has signature $(3,1)$, with the **Minkowski metric** given by

$$\eta(v,w) = -v^0 w^0 + v^1 w^1 + v^2 w^2 + v^3 w^3.$$

So far we have been talking about spacetimes that are vector spaces. Now let M be a manifold and consider a situation where the metric depends on where one is. A **metric** g on M assigns to each point $p \in M$ a metric g_p on the tangent space $T_p M$, in a smoothly varying way. By 'smoothly varying' we mean that if v and w are smooth vector fields on M, the inner product $g_p(v_p, w_p)$ is a smooth function on M. By the way, we usually write this function simply as $g(v,w)$.

One can show that the smoothness condition implies that the signature of g_p is constant on any connected component of M. We are really only interested in cases where the signature is constant on all of M. If the signature of g is $(n,0)$, where $\dim M = n$, we say that g is a **Riemannian** metric, while if the signature is $(n-1, 1)$, we say that g is **Lorentzian**. By a semi-Riemannian manifold we mean a manifold equipped with a metric, and similarly for a **Riemannian manifold** and a **Lorentzian manifold**.

In relativity, spacetime is a Lorentzian manifold, which in the real world appears to be 4-dimensional, although other cases are certainly interesting. The easiest way to get ahold of a 4-dimensional Lorentzian manifold is to take a 3-dimensional manifold S, 'space', with a Riemannian metric ^3g, and let M, 'spacetime', be given by $\mathbb{R} \times S$. Then we can define a Lorentzian metric

$$g = -dt^2 + {}^3g$$

on M as follows. Let x^i ($i = 1,2,3$) be local coordinates on an open subset $U \subseteq S$, and let t or x^0 denote the coordinate on $\mathbb{R}$, that is, 'time'. Then x^μ ($\mu = 0, 1, 2, 3$) are local coordinates on $\mathbb{R} \times U \subseteq M$, and we can define the metric g to be that with components

$$g_{\mu\nu} = \begin{pmatrix} -1 & 0 & 0 & 0 \\ 0 & & & \\ 0 & & {}^3g_{ij} & \\ 0 & & & \end{pmatrix}$$

This represents a special sort of **static** spacetime, in which space has a metric that is independent of time.

The most basic use of a Lorentzian metric is to measure distances and times. For example, if a path $\gamma \colon [0, 1] \to M$ is spacelike, that is, if its tangent vector is everywhere spacelike, we define its **arclength** to be

$$\int_0^1 \sqrt{g(\gamma'(t), \gamma'(t))} \, dt.$$

If γ is timelike, we define the **proper time** along γ — that is, the time ticked off by a clock moving along γ — to be

$$\int_0^1 \sqrt{-g(\gamma'(t), \gamma'(t))} \, dt.$$

We will mainly be interested in some more sophisticated applications of the metric, however. The most fundamental of these is 'raising and lowering indices', that is, converting between tangent and cotangent vectors. If V is a vector space equipped with a metric g, there is a natural way to turn an element $v \in V$ into an element of V^*, namely the linear functional $g(v, \cdot)$ which eats another element of V and spits out a number.

Exercise 52. *Use the nondegeneracy of the metric to show that the map from V to V^* given by*

$$v \mapsto g(v, \cdot)$$

is an isomorphism, that is, one-to-one and onto.

It follows that if M is a semi-Riemannian manifold the metric defines an isomorphism between each tangent space T_pM and the corresponding cotangent space T_p^*M. We can picture this as follows: if the tangent vector v is a little arrow, the cotangent vector $\omega = g(v, \cdot)$ is

The Metric

a stack of hyperplanes perpendicular to v, as in Figure 2. The reason for this is that ω vanishes on vectors orthogonal to v. The key point is that one needs the metric to know what 'orthogonal' means!

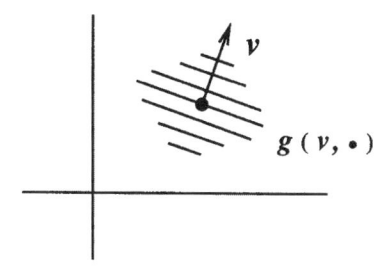

Fig. 2. Tangent vector v and cotangent vector $g(v, \cdot)$

Similarly, we can convert between vector fields and 1-forms on M. By using the metric on space, for example, we can think of the electric field as a vector field instead of a 1-form. We need to do this in order to think of the electric field as 'pointing' in some direction.

Suppose M is a semi-Riemannian manifold. Now that we can visualize 1-forms on M as fields of little arrows, there is a nice way for us to visualize p-forms for higher p as well. We can draw a wedge product $\omega \wedge \mu$ of two cotangent vectors at p as a little parallelogram, as in Figure 3. So we can visualize a 2-form on M as field of such 'area elements'. Similarly, we can draw a wedge product $\omega \wedge \mu \wedge \nu$ of three cotangent vectors at p as a little parallelepiped, as in Figure 4, and visualize a 3-form as a field of these 'volume elements' — and so on for higher p-forms.

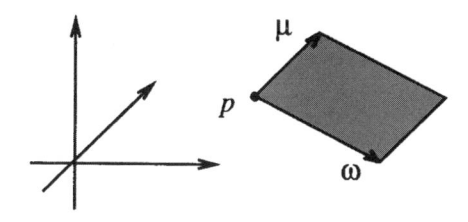

Fig. 3. Picture of $\omega \wedge \mu \in \Lambda^2 T_p^* M$

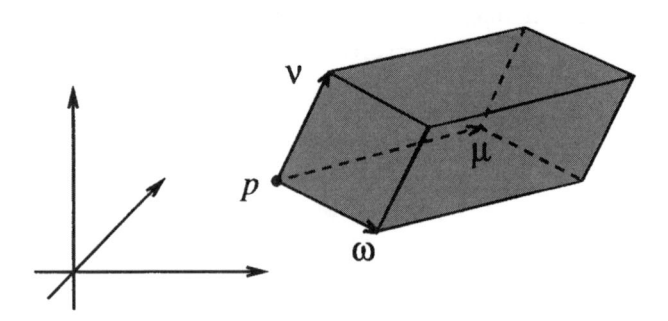

Fig. 4. Picture of $\omega \wedge \mu \wedge \nu \in \Lambda^3 T_p^* M$

We should not take these pictures *too* seriously. For example, if we drew

$$(dx + dy) \wedge (dy + dz)$$

and

$$(dy + dz) \wedge (dz - dx)$$

this way, we would get different-looking parallelograms, even though they are equal as elements of $\Lambda^2 T_p^* \mathbb{R}^3$. However, there is a lot about these parallelograms that is the same. First, they lie in the same plane $V \subseteq \mathbb{R}^3$. Second, they have the same area. And third, the pairs $(dx + dy, dy + dz)$, $(dy + dz, dz - dx)$ are bases for V that have the same orientation, in the usual sense of right-handed versus left-handed bases. This is true in general, which is why we think of $\omega \wedge \mu \in \Lambda^2 T_p^* M$ as an 'area element'. Similar statements are true for wedge products of more cotangent vectors. On the other hand, it is worth noting that there are usually plenty of elements of $\Lambda^i T_p^* M$ that are not wedge products of i cotangent vectors — we need to consider linear combinations, too. Still, with a little care one can get some good insights about differential forms using these pictures, as we will see.

Now let us say some things about how the metric looks in the language of indices. Let us work in a chart, and let e_μ be a basis of vector fields. Then we can define the components of the metric as follows:

$$g_{\mu\nu} = g(e_\mu, e_\nu).$$

The Metric

If M is n-dimensional, $g_{\mu\nu}$ is an $n \times n$ matrix. The nondegeneracy condition implies this matrix is invertible, so let $g^{\mu\nu}$ denote the inverse matrix. Then we have the following handy formulas, which explain why the process of converting between vector fields and 1-forms using the metric is called **raising and lowering indices**:

Exercise 53. *Let $v = v^\mu e_\mu$ be a vector field on a chart. Show that the corresponding 1-form $g(v,\cdot)$ is equal to $v_\nu f^\nu$, where f^ν is the dual basis of 1-forms and*
$$v_\nu = g_{\mu\nu} v^\mu.$$

Exercise 54. *Let $\omega = \omega_\mu f^\mu$ be a 1-form on a chart. Show that the corresponding vector field is equal to $\omega^\nu e_\nu$, where*
$$\omega^\nu = g^{\mu\nu} \omega_\mu.$$

Exercise 55. *Let η be the Minkowski metric on $\mathbb{R}^4$ as defined above. Show that its components in the standard basis are*
$$\eta_{\mu\nu} = \begin{pmatrix} -1 & 0 & 0 & 0 \\ 0 & 1 & 0 & 0 \\ 0 & 0 & 1 & 0 \\ 0 & 0 & 0 & 1 \end{pmatrix}.$$

In general, if we have any quantity with some indices, such as
$$A^{\alpha\beta\cdots\gamma}{}_{\delta\epsilon\cdots\zeta},$$
we can lower or raise any index with the metric and its inverse, using the Einstein summation convention. E.g., we can lower α and get
$$A_\alpha{}^{\beta\cdots\gamma}{}_{\delta\epsilon\cdots\zeta} = g_{\alpha\mu} A^{\mu\beta\cdots\gamma}{}_{\delta\epsilon\cdots\zeta},$$
or raise δ and get
$$A^{\alpha\beta\cdots\gamma\delta}{}_{\epsilon\cdots\zeta} = g^{\delta\mu} A^{\alpha\beta\cdots\gamma}{}_{\mu\epsilon\cdots\zeta}.$$

If we have a lot indices floating around it is important to keep track of their order when we raise and lower them; otherwise things get confusing. Note that we can even raise and lower indices on the metric itself:

Exercise 56. *Show that g_ν^μ is equal to the Kronecker delta δ_ν^μ, that is, 1 if $\mu = \nu$ and 0 otherwise. Note that here the order of indices does not matter, since $g_{\mu\nu} = g_{\nu\mu}$.*

We finish off this section by showing how to extend the idea of a metric to differential forms. Let M be a semi-Riemannian manifold. Recall that if v and w are vector fields on M, $g(v, w)$ is a function on M whose value at p is $g_p(v_p, w_p)$. This is bilinear,

$$g(fv + v', w) = fg(v, w) + g(v', w), \quad g(v, fw + w') = fg(v, w) + g(v, w'),$$

where now f is any function on M. It is also symmetric:

$$g(v, w) = g(w, v),$$

and nondegenerate :

$$\forall w \in V \ g(v, w) = 0 \implies v = 0.$$

We can define something with the same properties that works for 1-forms on M using the fact that the metric allows us to turn 1-forms into vector fields. Given two 1-forms ω and μ, we call the resulting function $\langle \omega, \mu \rangle$, the **inner product** of ω and μ. In terms of indices, if

$$g(v, w) = g_{\alpha\beta} v^\alpha w^\beta,$$

then for any 1-forms ω and μ we have

$$\langle \omega, \mu \rangle = g^{\alpha\beta} \omega_\alpha \mu_\beta.$$

Next, we define the inner product of p-forms. The **inner product** of two p-forms ω and μ on M will be a function $\langle \omega, \mu \rangle$ on M, and it is required to be bilinear, so it suffices to define it for p-forms that are wedge products of 1-forms. Say $e^1, \ldots, e^p$ and $f^1, \ldots, f^p$ are 1-forms on M. Then we define

$$\langle e^1 \wedge \cdots \wedge e^p, f^1 \wedge \cdots \wedge f^p \rangle = \det[\langle e^i, f^j \rangle]$$

where the right-hand side denotes the determinant of the $p \times p$ matrix of inner products $\langle e^i, f^j \rangle$.

The Metric

Exercise 57. *Show that the inner product of p-forms is nondegenerate by supposing that $(e^1, \ldots, e^n)$ is any orthonormal basis of 1-forms in some chart, with*
$$g(e^i, e^i) = \epsilon(i),$$
where $\epsilon(i) = \pm 1$. Show the p-fold wedge products
$$e^{i_1} \wedge \cdots \wedge e^{i_p}$$
form an orthonormal basis of p-forms with
$$\langle e^{i_1} \wedge \cdots \wedge e^{i_p}, e^{i_1} \wedge \cdots \wedge e^{i_p} \rangle = \epsilon(i_1) \cdots \epsilon(i_p).$$

Exercise 58. *Let $E = E_x dx + E_y dy + E_z dz$ be a 1-form on $\mathbb{R}^3$ with its Euclidean metric. Show that*
$$\langle E, E \rangle = E_x^2 + E_y^2 + E_z^2.$$
Similarly, let
$$B = B_x dy \wedge dz + B_y dz \wedge dx + B_z dx \wedge dy$$
be a 2-form. Show that
$$\langle B, B \rangle = B_x^2 + B_y^2 + B_z^2.$$
In physics, the quantity
$$\frac{1}{2}(\langle E, E \rangle + \langle B, B \rangle)$$
is called the **energy density** *of the electromagnetic field. The quantity*
$$\frac{1}{2}(\langle E, E \rangle - \langle B, B \rangle)$$
is called the **Lagrangian** *for the vacuum Maxwell's equations, which we discuss more in Chapter 4 of Part II, in greater generality.*

Exercise 59. *In $\mathbb{R}^4$ let F be the 2-form given by $F = B + E \wedge dt$, where E and B are given by the formulas above. Using the Minkowski metric on $\mathbb{R}^4$, calculate $-\frac{1}{2}\langle F, F \rangle$ and relate it to the Lagrangian above.*

The Volume Form

Since a metric allows us to measure distances on a manifold, it should allow us to measure volumes as well, and thus allow us to do integrals. This is in fact the case. We will postpone the study of integration on manifolds to Chapter 6, but we will define a basic ingredient of it here, the 'volume form'. This concept is needed to write down Maxwell's equations in differential form language. It turns out that a closely related concept is that of an 'orientation', that is, a globally well-defined way to tell the difference between left and right.

Given an n-dimensional vector space V with two bases $\{e_\mu\}$, $\{f_\mu\}$, there is always a unique linear transformation $T: V \to V$ taking one basis to the other:

$$Te_\mu = f_\mu.$$

This is necessarily invertible, so its determinant is nonzero. Let us say that $\{e_\mu\}$ and $\{f_\mu\}$ have the same orientation if $\det T > 0$, and the opposite orientation if $\det T < 0$. For example, any right-handed basis in $\mathbb{R}^3$ has the same orientation as the usual right-handed basis (e_1, e_2, e_3):

$$e_1 = (1,0,0), \quad e_2 = (0,1,0), \quad e_3 = (0,0,1),$$

while any left-handed basis, like $(-e_1, -e_2, -e_3)$, has the opposite orientation.

Exercise 60. *Show that any even permutation of a given basis has the same orientation, while any odd permutation has the opposite orientation.*

Let us define an **orientation** on V to be a choice of an equivalence class of bases of V, where two bases are deemed equivalent if they have the same orientation. E.g., on $\mathbb{R}^3$ there is the right-handed orientation, which contains the basis (e_1, e_2, e_3) and all other bases with the same orientation, and the left-handed orientation. There are always only two orientations on V.

There is another way to think about orientations. Suppose V is an n-dimensional vector space with basis $\{e_\mu\}$. Then

$$e_1 \wedge \cdots \wedge e_n$$

The Volume Form

is a nonzero element of $\Lambda^n V$ which we call the **volume element** associated to the basis $\{e_\mu\}$. We can picture it as a little parallepiped in n dimensions.

Let us see how the volume element depends on a change of basis. Note that any element $\omega \in \Lambda^n V$ can be written as

$$ce_1 \wedge \cdots \wedge e_n,$$

for some constant c, since a wedge product that contains any e_μ twice automatically vanishes. Suppose $\{f_\nu\}$ is another basis of V and let T^ν_μ be the matrix with

$$f_\nu = T^\mu_\nu e_\mu.$$

Then

$$\begin{aligned} f_1 \wedge \cdots \wedge f_n &= (T^1_1 e_1 + \cdots + T^n_1 e_n) \wedge \cdots \wedge (T^1_n e_1 + \cdots + T^n_n e_n) \\ &= (\det T)\, e_1 \wedge \cdots \wedge e_n \end{aligned}$$

since in the first line one is really summing over all expressions of the form

$$\text{sign}(\sigma) T_1^{\sigma(1)} \cdots T_n^{\sigma(n)} e_1 \wedge \cdots \wedge e_n$$

where σ is a permutation and $\text{sign}(\sigma)$ is its sign, which comes in from the anticommutativity of the wedge product. Thus two bases have the same orientation if the corresponding volume elements differ by a *positive* scalar multiple. Or, if we like, we can think of an orientation as being a choice of a volume form modulo positive scalar multiples.

Now let us turn from vector spaces to manifolds in general. As usual, let M be an n-dimensional manifold. We define a **volume form** ω on M to be a nowhere vanishing n-form. Thus for each point $p \in M$, ω_p is a volume element on $T^*_p M$. The standard volume form on $\mathbb{R}^n$ is

$$\omega = dx^1 \wedge \cdots \wedge dx^n.$$

As we will see, when we do a multiple integral like

$$\int_{\mathbb{R}^3} f\, dx\, dy\, dz$$

we are really integrating the 3-form $f dx \wedge dy \wedge dz$.

We say M is **orientable** if there exists a volume form on M. By an **orientation** on M we mean a choice of an equivalence class of volume forms on M, where two volume forms ω and ω' are equivalent if $\omega' = f\omega$ for some positive function f. Any volume form that is in the chosen equivalence class is said to be **positively oriented**; otherwise it is said to be **negatively oriented**. In particular, the **standard orientation** on $\mathbb{R}^n$ is the equivalence class containing the volume form $dx^1 \wedge \cdots \wedge dx^n$.

If we have an orientation on M, we can decide unambiguously whether any basis e^μ of a cotangent space T_p^*M is **right-handed** or **left-handed**, as follows. Just pick a volume form ω in the equivalence class, write $e^1 \wedge \cdots \wedge e^n$ as a constant times ω, and check to see whether the constant is positive or negative. This is the precise sense in which an orientation gives a global definition of right vs. left. Since a basis of the tangent space gives a dual basis of the cotangent space, we can also define right-handed and left-handed bases of the tangent space.

The classic example of a nonorientable manifold is the Möbius strip:

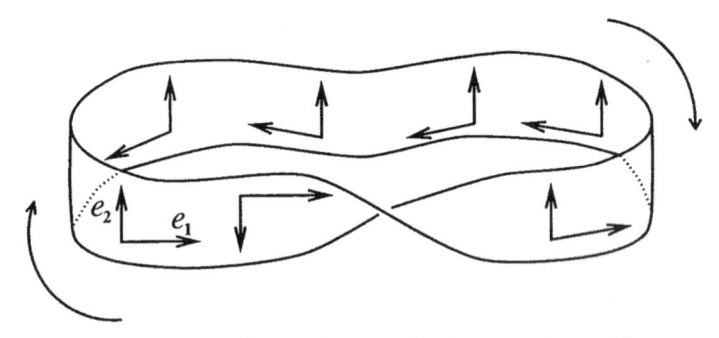

Fig. 5. The Möbius strip is nonorientable

As the figure indicates, there is no way to define the notion of a right-handed basis of T_p^*M for the Möbius strip in a smoothly varying way. Using a Riemannian metric we can identify T_p^*M with the tangent space T_pM. We have drawn a 'right-handed' basis of T_pM at one point, and show how if one drags it smoothly around a noncontractible loop it become 'left-handed'. If space was nonorientable, we might take a long journey in a spaceship around a noncontractible loop and come back home as a mirror-image version of ourselves. (However, we would not feel reflected; we would think everything *else* had been reflected.)

The Volume Form

A manifold equipped with an orientation is said to be **oriented**. One can also think of an oriented manifold as one having 'oriented charts' as follows:

Exercise 61. *Let M be an oriented manifold. Show that we can cover M with oriented charts $\varphi_\alpha\colon U_\alpha \to \mathbb{R}^n$, that is, charts such that the basis dx^μ of cotangent vectors on $\mathbb{R}^n$, pulled back to U_α by φ_α, is positively oriented.*

Exercise 62. *Given a diffeomorphism $\phi\colon M \to N$ from one oriented manifold to another, we say that ϕ is **orientation-preserving** if the pullback of any right-handed basis of a cotangent space in N is a right-handed basis of a cotangent space in M. Show that if we can cover M with charts such that the transition functions $\varphi_\alpha \circ \varphi_\beta^{-1}$ are orientation-preserving, we can make M into an oriented manifold by using the charts to transfer the standard orientation on $\mathbb{R}^n$ to an orientation on M.*

Now suppose that M is an oriented n-dimensional manifold with metric g. There is a canonical volume form on M which we can construct as follows. First, cover M with oriented charts $\varphi_\alpha\colon U_\alpha \to \mathbb{R}^n$. In any chart set

$$g_{\mu\nu} = g(\partial_\mu, \partial_\nu),$$

and define

$$\mathrm{vol} = \sqrt{|\det g_{\mu\nu}|}\, dx^1 \wedge \cdots \wedge dx^n.$$

Clearly this is a volume form on U. What we need to show is that given any overlapping chart $\varphi'\colon U' \to \mathbb{R}^n$, and defining

$$g'_{\mu\nu} = g(\partial'_\mu, \partial'_\nu),$$

then the volume form

$$\mathrm{vol}' = \sqrt{|\det g'_{\mu\nu}|}\, dx'^1 \wedge \cdots \wedge dx'^n,$$

agrees with vol on the overlap $U \cap U'$. This will imply the existence of a volume form on all of M, defined by this sort of formula, and independent of choice of chart.

On the overlap we have

$$dx'^\nu = T^\nu_\mu dx^\mu$$

where the matrix-valued function T is given by

$$T_\mu^\nu = \frac{\partial x'^\nu}{\partial x^\mu}.$$

Thus we have

$$dx'^1 \wedge \cdots \wedge dx'^n = (\det T) dx^1 \wedge \cdots \wedge dx^n,$$

so to show vol $=$ vol$'$ we need to show

$$\sqrt{|\det g'_{\mu\nu}|} = (\det T)^{-1} \sqrt{|\det g_{\mu\nu}|}.$$

To see this, note that

$$\begin{aligned} g'_{\mu\nu} &= g(\partial'_\mu, \partial'_\nu) \\ &= g(\frac{\partial x^\alpha}{\partial x'^\mu}\partial_\alpha, \frac{\partial x^\beta}{\partial x'^\nu}\partial_\beta) \\ &= (T^{-1})_\mu^\alpha (T^{-1})_\nu^\beta g_{\alpha\beta} \end{aligned}$$

or, taking determinants,

$$\det g'_{\mu\nu} = (\det T)^{-2} \det g_{\mu\nu}.$$

Since both charts are oriented, $\det T > 0$, so

$$\sqrt{|\det g'_{\mu\nu}|} = (\det T)^{-1} \sqrt{|\det g_{\mu\nu}|}$$

as desired.

We call vol the volume form on M associated to the metric g. People often write the volume form as $\sqrt{|\det g|} d^n x$. In the Lorentzian case, this is just

$$\text{vol} = \sqrt{-\det g}\, d^n x,$$

since the determinant of $g_{\mu\nu}$ is negative. In general relativity, people often write the volume form as simply $\sqrt{-g}\, d^n x$, using g to stand for the determinant of $g_{\mu\nu}$.

In Chapter 6 we will describe integration theory on an oriented manifold, and show how to integrate functions on an oriented semi-Riemannian manifold M. The basic idea is that when we integrate a function f over M, we are really doing the integral

$$\int_M f\, \text{vol},$$

The Hodge Star Operator

that is, integrating the n-form f vol. Right now our main goal is to describe the second pair of Maxwell equations in differential form language. For this, we need the volume form to define something called the Hodge star operator. The following fact will come in handy:

Exercise 63. *Let M be an oriented n-dimensional semi-Riemannian manifold and let $\{e_\mu\}$ be an oriented orthonormal basis of cotangent vectors at some point $p \in M$. Show that*

$$e_1 \wedge \cdots \wedge e_n = \text{vol}_p,$$

where vol *is the volume form associated to the metric on M, and* vol_p *is its value at p.*

The Hodge Star Operator

The Hodge star operator is the key to understanding the 'duality' symmetry of the vacuum Maxwell equations, as described in Chapter 1. This symmetry is the reason why the second pair of Maxwell equations look similar (but not quite the same) as the first pair. Think about these equations in ordinary Minkowski space. In old-fashioned notation, they are:

$$\begin{aligned} \nabla \cdot \vec{B} &= 0 \\ \nabla \times \vec{E} + \frac{\partial \vec{B}}{\partial t} &= 0 \\ \nabla \cdot \vec{E} &= \rho \\ \nabla \times \vec{B} - \frac{\partial \vec{E}}{\partial t} &= \vec{j}. \end{aligned}$$

In differential form notation, the first pair becomes:

$$\begin{aligned} d_S B &= 0 \\ \partial_t B + d_S E &= 0, \end{aligned}$$

where B is a 2-form on space and E is a 1-form on space (both functions of time). The funny thing is that the second pair seems to have the roles of E and B reversed (modulo the minus sign). This would amount to treating E as a 2-form and B as a 1-form! The Hodge star operator saves the day, since it converts 1-forms on 3-dimensional space into 2-forms, and vice versa. However, it does so at a price: it requires a choice of metric and also a choice of orientation.

How does the Hodge star operator do this? Here is where our way of drawing differential forms comes in handy. At any point p in a 3-dimensional Riemannian manifold M, the Hodge star operator maps a 1-form ν, which we draw as a little arrow, into a 2-form $\omega \wedge \mu$ that corresponds to an area element that is orthogonal to ν, as follows:

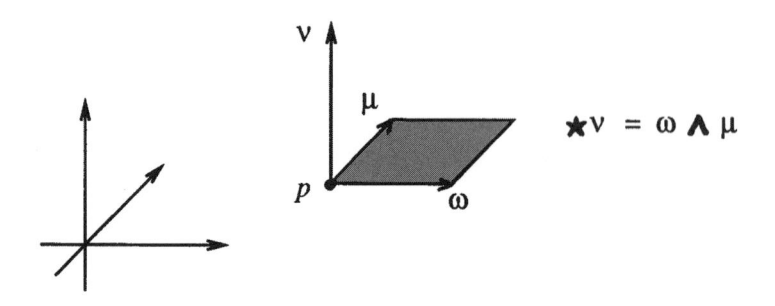

Fig. 6. The Hodge star of ν is $\omega \wedge \mu$

Conversely, it maps $\omega \wedge \mu$ to ν. In general, in n dimensions the Hodge star operator maps p-forms to $(n-p)$-forms in a very similar way, taking each little 'p-dimensional area element' to an orthogonal '$(n-p)$-dimensional area element'.

The precise definition of the Hodge star operator uses the inner product of differential forms. Let M be an n-dimensional oriented semi-Riemannian manifold. Then the inner product of two p forms ω and μ on M is a function $\langle \omega, \mu \rangle$ on M. We define the **Hodge star operator**

$$\star \colon \Omega^p(M) \to \Omega^{n-p}(M)$$

to be the unique linear map from p-forms to $(n-p)$-forms such that for all $\omega, \mu \in \Omega^p(M)$,

$$\omega \wedge \star \mu = \langle \omega, \mu \rangle \, \mathrm{vol}$$

Note that both sides of the equation are n-forms. We often call $\star \mu$ the **dual** of μ.

The Hodge Star Operator

It might not be obvious from this definition that the Hodge star operator really exists, or how to compute it! For this, it is nice to have a formula for it. Suppose that $e^1, \ldots, e^n$ are a positively oriented orthonormal basis of 1-forms on some chart. Thus

$$\langle e^\mu, e^\nu \rangle = 0$$

if $\mu \neq \nu$, and

$$\langle e^\mu, e^\mu \rangle = \epsilon(\mu)$$

where $\epsilon(\mu) = \pm 1$. Then we claim that for any distinct $1 \leq i_1, \ldots, i_p \leq n$,

$$\star(e^{i_1} \wedge \cdots \wedge e^{i_p}) = \pm e^{i_{p+1}} \wedge \cdots \wedge e^{i_n}$$

where $\{i_{p+1}, \ldots, i_n\}$ consists of the integers from 1 to n not included in $\{i_1, \ldots, i_p\}$:

$$\{i_{p+1}, \ldots, i_n\} = \{1, \ldots, n\} - \{i_1, \ldots, i_p\}.$$

The sign $\pm$ is given by

$$\text{sign}(i_1, \ldots, i_n)\epsilon(i_1) \cdots \epsilon(i_p),$$

where $\text{sign}(i_1, \ldots, i_n)$ denotes the sign of the permutation taking $(1, \ldots, n)$ to $(i_1, \ldots, i_n)$.

Exercise 64. *Show that if we define the Hodge star operator in a chart using this formula, it satisfies the property $\omega \wedge \star\mu = \langle \omega, \mu \rangle \text{vol}$. Use the result from Exercise 63.*

The formula for the Hodge star operator might seem complicated, so consider an example. Take dx, dy, dz as a basis of 1-forms on $\mathbb{R}^3$ with its usual Euclidean metric and orientation. Then we have

$$\star dx = dy \wedge dz, \quad \star dy = dz \wedge dx, \quad \star dz = dx \wedge dy,$$

and conversely

$$\star dx \wedge dy = dz, \quad \star dy \wedge dz = dx, \quad \star dz \wedge dx = dy.$$

If one interprets the definition correctly, one also can work out what the Hodge star operator does to the 0-form (or function) 1 and the volume form $dx \wedge dy \wedge dz$:

$$\star 1 = dx \wedge dy \wedge dz, \quad \star dx \wedge dy \wedge dz = 1.$$

Since Hodge star operator on $\mathbb{R}^3$ lets us turn 1-forms into 2-forms and vice versa, it sheds some new light on familiar operations like the cross product, curl and divergence. Given two 1-forms ω and μ on $\mathbb{R}^3$, their wedge product is a 2-form, and is perfectly well-defined without reference to a metric and orientation. But if we allow ourselves to use a metric and orientation, we can take the Hodge star of $\omega \wedge \nu$ and obtain a 1-form! If

$$\omega = \omega_i dx^i, \quad \nu = \nu_i dx^i,$$

then using the standard metric and orientation we get

$$\star(\omega \wedge \nu) = (\omega_y \nu_z - \omega_z \nu_y)dx + (\omega_z \nu_x - \omega_x \nu_z)dy + (\omega_x \nu_y - \omega_y \nu_x)dz.$$

This is basically just the cross product! The reader may wonder why we have done all this work to get back to concepts that everyone knows from basic vector calculus. Part of the point is that we now can work in spacetimes of arbitrary dimension, with arbitrary metrics and orientations. But it is also nice to see just where the metric and orientation are needed in the definition of the cross product in $\mathbb{R}^3$: only when we want to take a 2-form and convert it into a 1-form are they necessary.

Moreover, if ω is a 1-form on $\mathbb{R}^3$, $d\omega$ is a 2-form, but $\star d\omega$ is a 1-form again, and if we use the standard metric and orientation this is basically just the curl of ω:

Exercise 65. *Calculate $\star d\omega$ when ω is a 1-form on $\mathbb{R}^3$.*

Similarly, if ω is a 1-form on $\mathbb{R}^3$, $d\star\omega$ is a 3-form, but $\star d\star\omega$ is a 0-form, or function, and this basically amounts to taking the divergence of ω:

Exercise 66. *Calculate $\star d \star \omega$ when ω is a 1-form on $\mathbb{R}^3$.*

We encourage the reader to do the following exercises, too:

The Second Pair of Equations

Exercise 67. *Give $\mathbb{R}^4$ the Minkowski metric and the orientation in which (dt, dx, dy, dz) is positively oriented. Calculate the Hodge star operator on all wedge products of dx^μ's. Show that on p-forms*

$$\star^2 = (-1)^{p(4-p)+1}.$$

Exercise 68. *Let M be an oriented semi-Riemannian manifold of dimension n and signature $(s, n-s)$. Show that on p-forms*

$$\star^2 = (-1)^{p(n-p)+s}.$$

Exercise 69. *Let M be an oriented semi-Riemannian manifold of dimension n and signature $(s, n-s)$. Let e^μ be an orthonormal basis of 1-forms on some chart. Define the **Levi-Civita symbol** for $1 \le i_j \le n$ by*

$$\epsilon_{i_1 \ldots i_n} = \begin{cases} \text{sign}(i_1, \ldots, i_n) & \text{all } i_j \text{ distinct} \\ 0 & \text{otherwise} \end{cases}$$

Show that for any p-form

$$\omega = \frac{1}{p!} \omega_{i_1 \ldots i_p} e^{i_1} \wedge \cdots \wedge e^{i_p}$$

we have

$$(\star \omega)_{j_1 \ldots j_{n-p}} = \frac{1}{p!} \epsilon^{i_1 \ldots i_p}{}_{j_1 \ldots j_{n-p}} \omega_{i_1 \ldots i_p}.$$

The Second Pair of Equations

We now use the Hodge star operator to write the second pair of Maxwell equations in terms of differential forms. The key thing to understand is the effect of taking the dual $\star F$ of the electromagnetic field F.

First consider the case where M is Minkowski spacetime with its usual coordinates x^μ. We will sometimes write t for the time coordinate x^0. Then we can split F into electric and magnetic fields,

$$F = B + E \wedge dt,$$

where B is a time-dependent 2-form on space and E is a time-dependent 1-form on space. If one likes components, we have $F = \frac{1}{2} F_{\mu\nu} dx^\mu \wedge dx^\nu$

where

$$F_{\mu\nu} = \begin{pmatrix} 0 & -E_x & -E_y & -E_z \\ E_x & 0 & B_z & -B_y \\ E_y & -B_z & 0 & B_x \\ E_z & B_y & -B_x & 0 \end{pmatrix}$$

Now introduce the Minkowski metric on spacetime:

$$\eta(v,w) = -v^0 w^0 + v^1 w^1 + v^2 w^2 + v^3 w^3.$$

This allows us to define the Hodge star operator. A little calculation using Exercise 67 shows that

$$(\star F)_{\mu\nu} = \begin{pmatrix} 0 & B_x & B_y & B_z \\ -B_x & 0 & E_z & -E_y \\ -B_y & -E_z & 0 & E_x \\ -B_z & E_y & -E_x & 0 \end{pmatrix}.$$

In other words, taking the dual of F amounts to doing the replacements

$$E_i \mapsto -B_i, \qquad B_i \mapsto E_i.$$

This is the main difference between the first pair of Maxwell equations — which in old-fashioned form are

$$\nabla \cdot \vec{B} = 0 \qquad \nabla \times \vec{E} + \frac{\partial \vec{B}}{\partial t} = 0,$$

and the second pair:

$$\nabla \cdot \vec{E} = \rho \qquad \nabla \times \vec{B} - \frac{\partial \vec{E}}{\partial t} = \vec{j}.$$

The other difference between the first and second pairs is that the latter contain ρ and $\vec{j}$. To speak of these in the language of differential forms, we use the fact that the metric allows us to turn vector fields into 1-forms. Thus we can turn the good old current density

$$\vec{j} = j^1 \partial_1 + j^2 \partial_2 + j^3 \partial_3$$

into the 1-form

$$j = j_1 dx^1 + j_2 dx^2 + j_3 dx^3.$$

The Second Pair of Equations

Similarly, we can combine the current density and the electric charge density ρ in a single vector field on Minkowski spacetime:

$$\vec{J} = \rho \partial_0 + j^1 \partial_1 + j^2 \partial_2 + j^3 \partial_3,$$

and by using the Minkowski metric, we can turn this vector field into a 1-form

$$J = j - \rho dt$$

which we call the **current**.

Now we claim that just as the first pair of Maxwell equations are really

$$dF = 0,$$

the second pair are really

$$\star d \star F = J.$$

This is not so surprising, because at least on Minkowski space, the second pair of Maxwell equations

$$\nabla \cdot \vec{E} = \rho, \qquad \nabla \times \vec{B} - \frac{\partial \vec{E}}{\partial t} = \vec{j}$$

can be rewritten as

$$\star_S d_S \star_S E = \rho,$$
$$-\partial_t E + \star_S d_S \star_S B = j,$$

where $\star_S$ denotes the Hodge star operator on 'space', that is, $\mathbb{R}^3$ with its usual Euclidean metric.

Exercise 70. *Check this result.*

These look very similar to the version of the first pair of Maxwell equations in which we have split spacetime into space and time:

$$d_S B = 0,$$
$$\partial_t B + d_S E = 0.$$

The difference really amounts to using the Hodge star operator twice.

More generally, start by assuming that spacetime M is any manifold. Then the electromagnetic field F is a 2-form on M the current J is a 1-form on M, and the first Maxwell equation is $dF = 0$. We must assume M is semi-Riemannian and oriented to write down the second pair of Maxwell's equations, that is, $\star d \star F = J$. To introduce electric and magnetic fields we must assume $M = \mathbb{R} \times S$, where S is space, and write $F = B + E \wedge dt$. Similarly we write $J = j - \rho dt$. Then the first Maxwell equation splits into

$$d_S B = 0, \qquad \partial_t B + d_S E = 0.$$

Suppose also that space is 3-dimensional and that the metric on M is a static one of the form $g = -dt^2 + {}^3g$ where ${}^3g$ is a Riemannian metric on space, S. Let $\star_S$ denote the Hodge star operator on (time-dependent) differential forms on S. Then

$$\star F = \star_S E - \star_S B \wedge dt$$

so

$$d \star F = \star_S \partial_t E \wedge dt + d_S \star_S E - d_S \star_S B \wedge dt$$

and

$$\star d \star F = -\partial_t E - \star_S d_S \star_S E \wedge dt + \star_S d_S \star_S B.$$

Setting $\star d \star F = J$ and equating like terms, we obtain

$$\star_S d_S \star_S E = \rho, \qquad -\partial_t E + \star_S d_S \star_S B = j,$$

as desired.

Exercise 71. *Check the calculations above.*

It is interesting to note that in the **static Maxwell equations**, where E and B are independent of t, there is a pair involving only E:

$$dE = 0, \qquad \star_S d_S \star_S E = \rho,$$

and a pair involving only B:

$$dB = 0, \qquad \star_S d_S \star_S B = 0.$$

The Second Pair of Equations

This makes it clear that only when the electric and magnetic fields are time-dependent do they affect each other. Historically, it was Faraday who first discovered in 1831 that a changing magnetic field causes a nonzero curl in the electric field. He is responsible for the

$$\frac{\partial \vec{B}}{\partial t}$$

term in the equations of electromagnetism. Maxwell's brilliant contribution to the equations came when he hypothesized in 1861 that a changing electric field causes a nonzero curl in the magnetic field. In other words, he guessed there should be a

$$\frac{\partial \vec{E}}{\partial t}$$

term, too. It is only when both of these effects are taken into account that we get electromagnetic radiation, in which ripples in E cause ripples in B and vice versa, causing waves that move through space.

Interestingly enough, the reason Maxwell made his hypothesis was not an experiment, but a problem with the equations of electromagnetism as they stood at the time. This was the problem of charge conservation. Not only is the total electric charge of the world constant, the only way charge can get from one place to another is by moving through the intervening regions. This is called a 'local conservation law'. Mathematically, one can formulate it in Minkowski spacetime by saying that any increase or decrease in the charge density at any point is solely due to the divergence of the current density. In old-fashioned language one expresses this by the **continuity equation**

$$\frac{d\rho}{dt} = -\nabla \cdot \vec{j}.$$

Maxwell realized that the $\partial \vec{E}/\partial t$ term would make the continuity equation an automatic consequence of the laws of electromagnetism! This can be seen by starting with

$$\nabla \times \vec{B} - \frac{\partial \vec{E}}{\partial t} = \vec{j},$$

taking the divergence of both sides to obtain

$$-\nabla \cdot \frac{\partial \vec{E}}{\partial t} = \nabla \cdot \vec{j},$$

and then interchanging the order of the derivatives on the left hand side and using the fact that

$$\nabla \cdot \vec{E} = \rho.$$

In fact, the continuity equation can expressed more elegantly in differential form language as

$$d \star J = 0,$$

and this law is a simple consequence of Maxwell's equations in their most general modern form. Starting with $\star d \star F = J$ and taking the dual of both sides we obtain $d \star F = \pm \star J$, where the sign depends on the value of $\star^2$ on 1-forms (see Exercise 68). Taking the exterior derivative of both sides and using $d^2 = 0$, we get $d \star J = 0$. In terms of components, this equation is written $\partial^\mu J_\mu = 0$.

This is a good example of how the identity $d^2 = 0$ has powerful physical consequences. When we get to gauge theories we will see that Maxwell's equations are a special case of the Yang-Mills equations, which describe not only electromagnetism but also the strong and weak nuclear forces. A generalization of the identity $d^2 = 0$, the Bianchi identity, implies conservation of 'charge' in all of these theories — although these theories have different kinds of 'charge'. Similarly, we will see when we get to general relativity that due to the Bianchi identity, Einstein's equations for gravity automatically imply local conservation of energy and momentum! So what we are seeing here is only the tip of the iceberg.

It is also interesting to consider the **vacuum Maxwell equations,** that is, the case $J = 0$:

$$dF = 0, \qquad d \star F = 0.$$

These are preserved by duality:

$$F \mapsto \star F.$$

The Second Pair of Equations

Recall that when spacetime M is of the form $\mathbb{R} \times S$, so that $F = B + E \wedge dt$, we have $\star F = \star_S E - \star_S B \wedge dt$, so duality amounts to:

$$B \mapsto \star_S E, \qquad E \mapsto -\star_S B,$$

or when $S = \mathbb{R}^3$,

$$\vec{B} \mapsto \vec{E}, \qquad \vec{E} \mapsto -\vec{B}$$

in old-fashioned language.

In 4 dimensions something very interesting happens, since then the dual of a 2-form is a 2-form. Note from Exercise 67 that if M is a Lorentzian 4-dimensional manifold, the operator

$$\star: \Omega^2(M) \to \Omega^2(M)$$

has

$$\star^2 = -1,$$

while if M is Riemannian, we have

$$\star^2 = 1.$$

In the Riemannian case things are very nice: we say $F \in \Omega^2(M)$ is **self-dual** if $\star F = F$, and **anti-self-dual** if $\star F = -F$. Since $\star^2 = 1$, it is not surprising that the Hodge star operator has eigenvalues ± 1. That is, we can write any $F \in \Omega^2(M)$ as a sum of self-dual and anti-self-dual parts:

$$F = F_+ + F_-, \qquad \star F_\pm = \pm F_\pm.$$

Exercise 72. *Show this is true if we take*

$$F_\pm = \frac{1}{2}(F \pm \star F).$$

In the Lorentzian case things are not quite as nice, since $\star^2 = -1$ implies its eigenvalues are $\pm i$. This means that we should really consider complex-valued differential forms on M. If we do that, we can write any $F \in \Omega^2(M)$ as

$$F = F_+ + F_-$$

where

$$\star F_\pm = \pm i F_\pm.$$

Exercise 73. *Show that this result is true.*

Let us bend words a bit and say in this case too that F_+ is self-dual and F_- is anti-self-dual.

In either the Riemannian or Lorentzian case, if we have a self-dual (or anti-self-dual) 2-form F satisfying the first pair of vacuum Maxwell equations:
$$dF = 0,$$
it automatically satisfies the second pair:
$$d \star F = 0.$$

Of course, in the Lorentzian case F will need to be complex-valued, which is not very sensible physically. However, since Maxwell's equations are linear, we can always take the real part (or imaginary part) of a solution and get a real-valued solution.

The trick of turning two pairs of vacuum Maxwell equations into one turns out to be the tip of another iceberg. First, the Hodge star operator and the exterior derivative interact with each other in a very nice way that has a lot to do with topology. This leads to a subject called Hodge theory. Self-duality is also important in the Yang-Mills equations. These are a lot harder to solve than Maxwell's equations, because they are nonlinear, but using self-duality one can find some solutions in the Riemannian case. These self-dual (or anti-self-dual) solutions are called 'instantons', because they start out small near $t = -\infty$, get big for a little while, and then get small again near $t = +\infty$. Instantons are of importance both in the physics of the strong force and in studying the topology of 4-dimensional manifolds.

Self-duality also turns out to be important for the Einstein equations. This was emphasized by Penrose, who used a method called 'twistors' to find self-dual solutions to the Einstein equations. Self-duality of a somewhat different sort is also crucial in Ashtekar's reformulation of general relativity, which we discuss in Chapter 5 of Part III.

We can get a bit of the flavor of this business by using self-duality to find some solutions of the vacuum Maxwell's equations on Minkowski space. These solutions represent light moving around through empty

The Second Pair of Equations

space! If we write
$$F = B + E \wedge dt$$
we have
$$\star F = \star_S E - \star_S B \wedge dt,$$
so F will be self-dual if
$$\star_S E = iB, \qquad \star_S B = -iE.$$

Exercise 74. *Show that these equations are equivalent, and both hold if at every time t we have*
$$E = E_1 dx^1 + E_2 dx^2 + E_3 dx^3,$$
$$B = -i(E_1 dx^2 \wedge dx^3 + \text{cyclic permutations}).$$

Let us assume F is self-dual and that E is a **plane wave**, that is, of the form
$$E(x) = \mathbf{E} e^{i k_\mu x^\mu}$$
where $\mathbf{E} = \mathbf{E}_j dx^j$ is a constant complex-valued 1-form on $\mathbb{R}^3$ and $k \in (\mathbb{R}^4)^*$ is a fixed covector, called the **energy-momentum**. Recall that the covector k eats the vector $x \in \mathbb{R}^4$ corresponding to a point in Minkowski space and spits out a number $k(x)$ in a linear way: in coordinates this is just
$$k(x) = k_\mu x^\mu.$$

By self-duality, we have
$$B(x) = \mathbf{B} e^{i k_\mu x^\mu}$$
where $\mathbf{B} = -i \star_S \mathbf{E}$. Thus the first Maxwell equation, $d_S B = 0$, implies that
$$\mathbf{B} \wedge d_S e^{i k_\mu x^\mu} = 0$$
at all points x. Let us write $^3 k$ for $k_j dx^j$, the spatial part of the energy-momentum, called the **momentum** of the plane wave. Then
$$d_S e^{i k_\mu x^\mu} = e^{i k_\mu x^\mu} \, ^3 k,$$
so the first Maxwell equation holds precisely when
$$\mathbf{B} \wedge \, ^3 k = 0.$$

Expressing **B** in terms of **E**, this equation is equivalent to

$$\star_S \mathbf{E} \wedge {}^3k = 0,$$

or, by the definition of the Hodge star operator,

$$\langle \mathbf{E}, {}^3k \rangle = 0.$$

This says that the electric field must be orthogonal to the momentum of the plane wave.

Similarly, the second Maxwell equation, $\partial_t B + d_S E = 0$, says that

$$ {}^3k \wedge \mathbf{E} = k_0 \mathbf{B}.$$

Exercise 75. *Check the above result.*

This equation is really just a fancy way of saying that the cross product of the electric field and the momentum is proportional to the magnetic field. The number k_0 is called the **frequency** of the plane wave. Writing **B** in terms of **E**, we obtain an equation **E** must satisfy:

$$ {}^3k \wedge \mathbf{E} = -ik_0 \star_S \mathbf{E}.$$

Exercise 76. *Show this equation implies $k_\mu k^\mu = 0$. Thus the energy-momentum of light is light-like!*

If we solve the first pair of vacuum Maxwell's equations this way, duality automatically implies we have solved the second pair. A simple example of a solution is

$$k = dt - dx, \qquad \mathbf{E} = dy - i\,dz.$$

Note that ${}^3k$ and **E** are really orthogonal, and also

$$ {}^3k \wedge \mathbf{E} = -dx \wedge dy + i\,dx \wedge dz = -ik_0 \star_S \mathbf{E},$$

as required.

It is enlightening to express this solution in old-fashioned language. It gives:

$$\vec{E} = (0, e^{i(t-x)}, -ie^{i(t-x)}), \qquad \vec{B} = (0, -ie^{i(t-x)}, -e^{i(t-x)}).$$

The Second Pair of Equations

Exercise 77. *Check the above result.*

Of course, to get an honest, *real* solution of Maxwell's equations we can take the real part:

$$\vec{E} = (0, \cos(t-x), \sin(t-x)), \qquad \vec{B} = (0, \sin(t-x), -\cos(t-x)).$$

In other words, the plane wave moves in the x direction at the speed of light, with the electric and magnetic fields orthogonal to each other rotating counterclockwise in the yz plane. A plane wave in which $\vec{E}$ and $\vec{B}$ rotate counterclockwise when viewed as the wave moves towards one is said to be **left circularly polarized**. As it turns out, all the self-dual plane wave solutions of Maxwell's equations are left circularly polarized. To get right circularly polarized plane waves, we need the anti-self-dual plane wave solutions. General plane wave solutions will be linear combinations of self-dual and anti-self-dual ones.

One thing we see here is a close connection between the Hodge star operator and chirality, or handedness. In a more sophisticated quantum-field theoretic picture of light, we may think of it as made of photons that spin either clockwise or counterclockwise about their axis of motion. Light has no preferred chirality. However, a different sort of massless particle, the neutrino, *does* have a preferred chirality — one of the puzzles of nature.

Exercise 78. *Prove that all self-dual and anti-self-dual plane wave solutions are left and right circularly polarized, respectively.*

Exercise 79. *Let $P \colon \mathbb{R}^4 \to \mathbb{R}^4$ be parity transformation, that is,*

$$P(t, x, y, z) = (t, -x, -y, -z).$$

Show that if F is a self-dual solution of Maxwell's equations, the pullback P^*F is an anti-self-dual solution, and vice versa.

Chapter 6
DeRham Theory in Electromagnetism

I received your paper, and thank you very much for it. I do not say I venture to thank you for what you have said about "Lines of Force", because I know you have done it for the interests of philosophical truth; but you must suppose it is work grateful to me, and gives me much encouragement to think on. I was at first almost frightened when I saw such mathematical force made to bear upon the subject, and then wondered to see that the subject stood it so well. — Michael Faraday, to James Clerk Maxwell

Closed and Exact 1-forms

As we have seen, the first pair of Maxwell equations simply say that electromagnetic field F has $dF = 0$. In the static case, they say that the electric field has $dE = 0$ and the magnetic field B has $dB = 0$. Equations of this sort are especially charming because they are 'generally covariant', that is, independent of any fixed choice of metric or other geometrical structure on spacetime. This implies that they are preserved by any diffeomorphism. In other words, if ω is a form on a manifold M satisfying the equation $d\omega = 0$, the pullback of ω under any diffeomorphism of M again satisfies this equation. Since a diffeomorphism is a kind of change of coordinates, this means that the first pair of Maxwell equations is invariant, not just under Lorentz transformations, rotations, and translations, but under *all* coordinate transformations.

Now let us try to solve these equations. It is easy to come up with lots of solutions, because $d^2 = 0$. If F is d of something, it automatically satisfies $dF = 0$, and similarly for E and B in the static case. This simple observation is the basis of a surprisingly large amount of mathematics and physics. It leads to a very interesting question: can one get *all* the solutions of the first pair of Maxwell equations this way? The branch of mathematics that answers this sort of question is called deRham cohomology.

Let us first introduce some standard terminology. In general, if the exterior derivative of a differential form is zero, we say the differential form is **closed**. On the other hand, a differential form that is the exterior derivative of some other differential form is called **exact**. The equation $d^2 = 0$ may thus be expressed in words by saying 'all exact forms are closed'. For example, if the electric field E is d of some function on space, we will automatically have $dE = 0$. In physics one calls a function (or 0-form) ϕ with

$$E = -d\phi$$

a **scalar potential** for E; the minus sign is just a convention. Similarly, if the magnetic field B is d of some 1-form on space, we will automatically have $dB = 0$. One calls a 1-form A with

$$B = dA$$

a **vector potential** for B. Also, if the electromagnetic field F satisfies

$$F = dA$$

for some 1-form A on spacetime, we automatically have $dF = 0$, and we call A a vector potential for F.

Now let us study when a closed 1-form is exact. Say we have a manifold S, with a 1-form E on it satisfying $dE = 0$. Can we cook up a function ϕ on S with $E = -d\phi$? Let us try and see what, if anything, prevents us. We will attempt to find such a function ϕ by integrating the 1-form E along paths in S. Technically, a **path** γ in S is a piecewise smooth map from $\gamma\colon [0,T] \to S$, but in this section we will be lazy and only work with smooth paths. If γ is a path, $\gamma'(t)$ is a tangent vector

Closed and Exact 1-forms

at the point $\gamma(t)$, and applying the cotangent vector $E_{\gamma(t)}$ at the same point we get a number; then we integrate this from 0 to T. We write this as

$$\int_\gamma E = \int_0^T E_{\gamma(t)}(\gamma'(t))\, dt.$$

Our plan will be to define ϕ as follows: fix any point $p \in S$ and for any $q \in S$ let

$$\phi(q) = -\int_\gamma E$$

where γ is some path from p to q. The reader may be familiar with this strategy in the special case when $S = \mathbb{R}^3$; this is how one writes a curl-free vector field as the gradient of a function.

There are a number of potential problems with this plan. First, there might not be any path from p to q! It is rather odd to imagine in terms of physics, but mathematically there is nothing to stop S from being made of several pieces, or 'components', with no paths from one to another. For example, S might be the disjoint union of two copies of $\mathbb{R}^3$ — two separate universes, as it were — and there would be no path from one to the other. We will have to rule out this case. If there is a path between any two points in S, we say that S is **connected** (or more precisely, **arc-connected**). If not, a maximal connected subset of S is called a **connected component**. Henceforth in our quest to solve $dE = 0$ we will assume S is connected. (If not, it would be easy to apply our technique to each connected component separately.)

The next problem, which is more serious, is that the integral $\int_\gamma E$ will in general depend on the details of the path γ, not just its endpoints $\gamma(0) = p$ and $\gamma(T) = q$. We want to see what conditions are necessary to rule out this problem. First, let us see how the integral changes when we smoothly vary the path γ. In other words, suppose that we have a smoothly varying family of paths from p to q labelled by some parameter $s \in [0,T]$. We can describe all these by a function $\gamma(s,t)$. For each s, $\gamma(s,\cdot)$ is a smooth path with $\gamma(s,0) = p$ and $\gamma(s,T) = q$, and $\gamma(s,t)$ should depend smoothly on s as well as t.

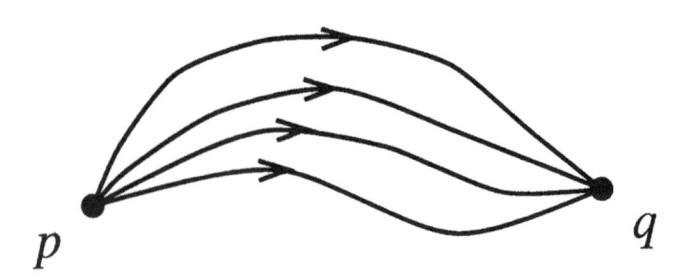

Fig. 1. A smoothly varying family of paths from p to q

To see how
$$I_s = \int_0^T E_{\gamma(s,t)}(\gamma'(s,t))\,dt$$
depends on s, let us differentiate it with respect to s. To do computations we can assume we are working in a coordinate chart on S – if not, break up the integral into pieces that each fit in a chart. Using coordinates to describe the pairing of the 1-form E and the tangent vector γ', we have
$$I_s = \int_0^T E_\mu(\gamma(s,t))\,\partial_t\gamma^\mu(s,t)\,dt,$$
Thus
$$\begin{aligned}\partial_s I_s &= \int \partial_s[E_\mu(\gamma(s,t))\,\partial_t\gamma^\mu(s,t)]\,dt \\ &= \int [\partial_s E_\mu(\gamma(s,t))\,\partial_t\gamma^\mu(s,t) + E_\mu(\gamma(s,t))\,\partial_s\partial_t\gamma^\mu(s,t)]\,dt \\ &= \int [\partial_s E_\mu(\gamma(s,t))\,\partial_t\gamma^\mu(s,t) - \partial_t E_\mu(\gamma(s,t))\,\partial_s\gamma^\mu(s,t)]\,dt \\ &= \int \partial_\nu E_\mu(\gamma(s,t))\,[\partial_s\gamma^\nu\,\partial_t\gamma^\mu - \partial_t\gamma^\nu\,\partial_s\gamma^\mu]\,dt\end{aligned}$$
using the product rule, then integration by parts, and then the chain rule. Recalling that
$$dE = (\partial_\mu E_\nu - \partial_\nu E_\mu)dx^\mu dx^\nu,$$
we obtain
$$\partial_s I_s = \int (dE)_{\mu\nu}\,\partial_s\gamma^\mu\,\partial_t\gamma^\nu\,dt.$$

Closed and Exact 1-forms

Thus I_s is independent of s when $dE = 0$. This shows that I_s will be the same for two different paths as long as we can find a smoothly varying family of paths interpolating between them.

In math jargon, we say two paths $\gamma_0, \gamma_1 \colon [0, T] \to S$ from p to q are **homotopic** if there exists a smooth function $\gamma \colon [0,1] \times [0, T] \to S$ such that $\gamma(s, \cdot)$ is a path from p to q for each s, and

$$\gamma(0, t) = \gamma_0(t), \qquad \gamma(1, t) = \gamma_1(t).$$

We call the function γ a **homotopy** between γ_0 and γ_1. In this terminology, what we have shown is that a closed 1-form has the same integral along any two homotopic paths.

There still may be a problem with defining

$$\phi(q) = -\int_\gamma E$$

where γ is any path from p to q. Perhaps not all paths from p to q are homotopic! A nice example is the plane with the origin removed: this is a manifold, and the two paths from $(-1, 0)$ to $(1, 0)$ shown below are not homotopic:

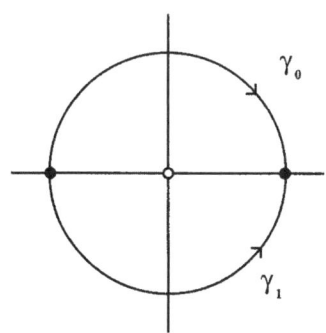

Fig. 2. Two paths that are not homotopic in $\mathbb{R}^2 - \{0\}$

It is pretty obvious that there is no way to smoothly deform the path γ_0 to the path γ_1 without getting snagged on the hole at the origin. Of course, being 'obvious' does not count as a proof! However, we can really *prove* this fact by finding a closed 1-form that has different integrals along the two paths. It is not hard: try

$$E = \frac{x\,dy - y\,dx}{x^2 + y^2}.$$

This 1-form 'wraps around the hole', so it has different integrals along γ_0 and γ_1:

Exercise 80. *Show that this 1-form E is closed. Show that $\int_{\gamma_0} E = -\pi$ and $\int_{\gamma_1} E = \pi$.*

This means that we cannot use $\phi(q) = -\int_\gamma E$ to define ϕ in a path-independent manner. We can visualize how E wraps around the hole if we draw E in the manner described in Chapter 4:

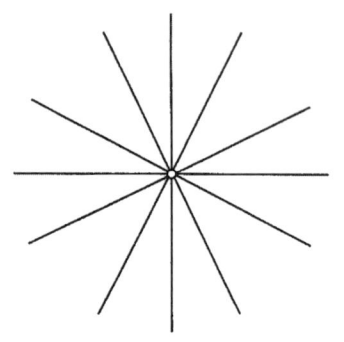

Fig. 3. Picture of $E = (x\,dy - y\,dx)/(x^2 + y^2)$

The fact that E is not exact simply means that there is no function whose level curves are the lines in the figure. If there were such a function, say $-\phi$, we would have $E = -d\phi$.

Given a connected manifold S, we say that S is **simply connected** if any two paths between two points p, q are homotopic. If S is simply connected, we can carry out our plan and define ϕ unambiguously when E is closed. In particular, things are fine on $\mathbb{R}^n$:

Exercise 81. *Show that $\mathbb{R}^n$ is simply connected by exhibiting an explicit formula for a homotopy between any two paths between arbitrary points $p, q \in \mathbb{R}^n$.*

Now let us show that when S is simply connected our plan really succeeds! Namely, suppose that S is simply connected and E is a closed 1-form on S. Pick any point $p \in S$ and define a function ϕ on S by

$$\phi(q) = -\int_\gamma E$$

Closed and Exact 1-forms

where γ is any path from p to $q \in S$. Let us show that

$$E = -d\phi.$$

To show that these 1-forms agree at some point q, it suffices to show that they agree when applied to any tangent vector $v \in T_q S$. By the definition of $d\phi$, this means we need to show

$$E(v) = -v(\phi).$$

To do this, pick a path $\gamma: [0, 2] \to S$ with $\gamma(0) = p$ and $\gamma(1) = q$, and such that $\gamma'(1) = v$, as shown below. Then we have

$$\begin{aligned} E(v) &= E(\gamma'(1)) \\ &= \frac{d}{ds} \int_0^s E(\gamma'(t)) \, dt \bigg|_{s=1} \\ &= -\frac{d}{ds} \phi(\gamma(s)) \bigg|_{s=1} \\ &= -v(\phi) \end{aligned}$$

using the fact that the derivative of $\phi(\gamma(s))$ with respect to s is the same as the derivative of ϕ in the direction $\gamma'(s) = v$.

Fig. 4. Proof that $E = -d\phi$

To summarize, we have shown that on a simply connected manifold, every closed 1-form is exact. In this case, we can always find a scalar potential for the electric field. Later, we will show how to generalize this result to p-forms for higher p. For 2-forms, this will let us understand when we can find a vector potential for the magnetic field or electromagnetic field.

Let us finish this section with a few words about loops! A path $\gamma\colon [0,T] \to S$ is a **loop** if it ends where it starts, that is, if $\gamma(0) = \gamma(T) = p$ for some point $p \in S$. We also say then that γ is a loop **based at** p, or that p is the **basepoint** of γ. Loops play a special role in electromagnetism, gauge theory and in the new approach to quantum gravity known as the 'loop representation', for which this book is intended as preparation. The basic idea is that we can understand fields in a very natural way by imagining a particle that goes around a loop and is altered somehow in the process. For example, we will explain later in this chapter how when we move a charged particle around a loop in space, its wavefunction is multiplied by a number $e^{i\theta}$, where θ is proportional to the integral of the vector potential around the loop! A similar fact holds for loops in spacetime, with the electromagnetic field F taking the place of the magnetic field. And a grand generalization of this fact holds for all the forces in the standard model — this is why we say they are all 'gauge fields'. Gravity is similar but in a sense even simpler: gravity is just a manifestation of the curvature of spacetime, where by 'curvature' we refer to the fact that if we take an object and move it around a loop, trying our best to 'parallel transport' it, nonetheless it comes back *rotated*.

We conclude this section by describing the role loops play in electrostatics. Let us suppose, as above, that space is some manifold S and the electric field on S is a 1-form E on S. Consider the integral of E around a loop γ, i.e. $\int_\gamma E$. If we wish to emphasize that γ is a loop we can write this as

$$\oint_\gamma E.$$

In certain important cases this will be zero! We say that a loop $\gamma\colon [0,T] \to S$ based at p is **contractible** if it is homotopic to a constant loop η that just stays at p:

$$\eta(t) = p$$

for all $t \in [0, T]$. Below we show a contractible loop γ and a noncontractible loop δ in $\mathbb{R}^2 - \{0\}$.

Closed and Exact 1-forms

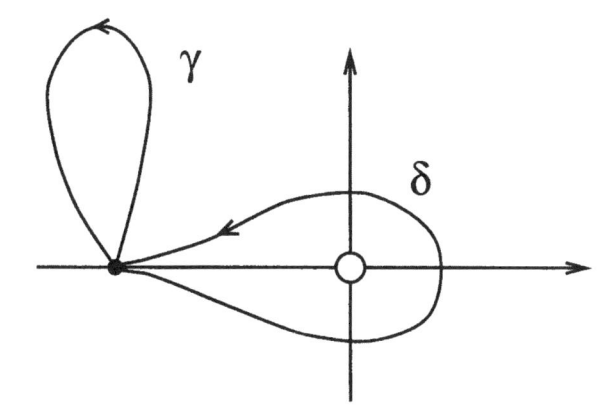

Fig. 5. A contractible loop γ and a noncontractible loop δ

By the result established earlier, if $dE = 0$ then we must have $\int_\gamma E = 0$ if γ is contractible, since the integral of E around a constant loop is zero. In particular, if S is simply connected, $\int_\gamma E = 0$ for *all* loops if $dE = 0$. This is definitely *not* true when S is not simply connected; for example, our friend the 1-form

$$E = \frac{x\,dy - y\,dx}{x^2 + y^2}$$

on $\mathbb{R}^2 - \{0\}$ gives an integral of 2π around the loop δ shown above. More generally, it gives 2π times the **winding number** of the loop, that is, the number of times the loop goes around the origin, counted with a plus sign when it goes around counterclockwise, and with a minus sign when it goes around clockwise.

There is a converse, too, that allows us to rephrase the electrostatic equation $dE = 0$ purely in terms of integrals around loops. This converse is a consequence of Stokes' theorem relating the curl of a vector field to its integral around a loop bounding a surface. Let us pick a chart giving coordinates x^μ about some point $p \in S$, and consider the integral of E around a square loop γ in the x^μ-x^ν plane:

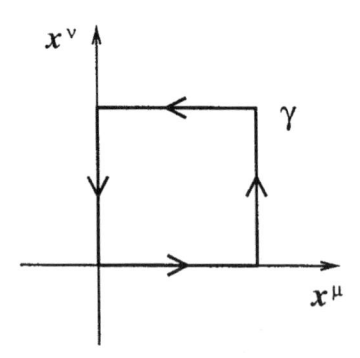

Fig. 6. The integral of E around a small square

Suppose this square is given by

$$\{0 \leq x^\mu \leq \epsilon,\ 0 \leq x^\nu \leq \epsilon\}.$$

Then by Green's theorem,

$$\int_\gamma E = \int_0^\epsilon \int_0^\epsilon (\partial_\mu E_\nu - \partial_\nu E_\mu)\, dx^\mu dx^\nu$$

and in the limit as $\epsilon \to 0$ this is equal to

$$\epsilon^2 (\partial_\mu E_\nu - \partial_\nu E_\mu) = \epsilon^2 (dE)_{\mu\nu}$$

evaluated at p, plus terms of order ϵ^3. So if $\int_\gamma E$ vanishes for all contractible loops in S, then $dE = 0$.

In short, a 1-form E is closed if and only if $\int_\gamma E = 0$ for all contractible loops γ. Similarly, it follows from things we have already shown that E is exact if and only if $\int_\gamma E = 0$ for *all* loops. In the next sections we will generalize this result to p-forms. For this, we will need to generalize Stokes' theorem.

Exercise 82. *Show that a 1-form E is exact if and only if $\int_\gamma E = 0$ for all loops γ. (Hint: if ω is not exact, show that there are two smooth paths γ, γ' from some point $x \in M$ to some point $y \in M$ such that $\int_\gamma \omega \neq \int_{\gamma'} \omega$. Use these paths to form a loop, perhaps only piecewise smooth.)*

Exercise 83. *For any manifold M, show the manifold $S^1 \times M$ is not simply connected by finding a 1-form on it that is closed but not exact.*

Stokes' Theorem

The objects which we shall study are called exterior differential forms. *These are the things which occur under integral signs.* — Harley Flanders

We have been so busy showing what differential forms have to do with Maxwell's equations that we have neglected to properly emphasize that differential forms are just things that one integrates! This is a terrible omission, which we now correct. We will see that n-forms can be integrated over n-manifolds, or more generally n-manifolds with a 'boundary', and that the concepts of exterior derivative and boundary are tied together by the modern version of Stokes' theorem.

The modern version of Stokes' theorem is beautiful because it shows that a number of important theorems of calculus are really all aspects of the same thing. Let us give rough statements of these to point out how similar they are. First, there is the fundamental theorem of calculus. This says that if one has a function $f: [a, b] \to \mathbb{R}$, then

$$\int_a^b f'(x)dx = f(b) - f(a).$$

It relates the integral of the derivative of f over the closed interval $[a, b]$ to the values of f on the 'boundary', that is, the endpoints. Second, there is the good old version of Stokes' theorem. This says that if one has a 2-dimensional surface S in $\mathbb{R}^3$ whose boundary ∂S is traced out by a loop $\gamma: [0, T] \to \mathbb{R}^3$, and $\vec{A}$ is a vector field on $\mathbb{R}^3$, then

$$\int_S (\nabla \times \vec{A}) \cdot \vec{n} = \int_\gamma \vec{A},$$

where $\vec{n}$ is the unit normal to S. Again, this relates the integral of the derivative of $\vec{A}$ over S to the integral of $\vec{A}$ over the boundary ∂S. Third, there is Gauss' theorem. This says that if one has a 3-dimensional region $R \subset \mathbb{R}^3$ with smooth boundary ∂R, and $\vec{A}$ is a vector field defined on R, then

$$\int_R \nabla \cdot \vec{A} = \int_{\partial R} \vec{A} \cdot \vec{n}$$

where $\vec{n}$ is the outwards-pointing unit normal to ∂R. This too relates the integral of the derivative of $\vec{A}$ over R to the integral of $\vec{A}$ over the boundary ∂R. In physics, we call $\int_{\partial R} \vec{A} \cdot \vec{n}$ the **flux** of $\vec{A}$ through the surface ∂R.

Now that we know about differential forms, it is clear that in the fundamental theorem of calculus we are starting with a function, or 0-form, f, forming the 1-form $df = f'(x)dx$, and integrating it over a closed interval. A closed interval is not quite a manifold, since the two endpoints do not have neighborhood that looks like $\mathbb{R}$, but we will see that it is a 1-dimensional 'manifold with boundary'. We have also seen that the curl really amounts to d of a 1-form. Thus in Stokes' theorem we are really taking d of a 1-form, obtaining a 2-form, and integrating it over a 2-dimensional manifold with boundary, S. We have also seen that the divergence in $\mathbb{R}^3$ is really d of a 2-form. So in Gauss' theorem we are really taking d of a 2-form, obtaining a 3-form, and integrating it over a 3-dimensional manifold with boundary, R.

Roughly speaking, the general Stokes' theorem says that under certain conditions, if M is a $n+1$-dimensional manifold with boundary and ω is an n-form on M, then

$$\int_M d\omega = \int_{\partial M} \omega.$$

We will not prove this theorem, but we will make sense of all the pieces involved. To do this, we will define a manifold with boundary, and then explain how to integrate differential forms over manifolds with boundary. We refer the reader to the notes at the end of Part I for books that prove the theorem — it is not really all that hard!

The concept of a manifold with boundary is a simple generalization of that of an ordinary manifold. A simple example would be the annulus

$$\{(x,y) \in \mathbb{R}^2 \colon 1 \leq x^2 + y^2 \leq 2\}.$$

Stokes' Theorem 115

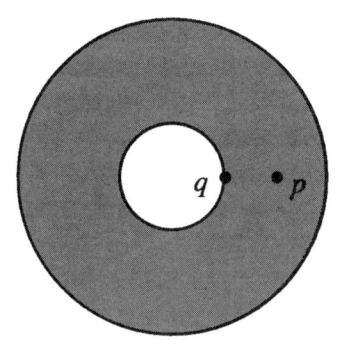

Fig. 7. A manifold with boundary: the annulus

The point $p = (3/2, 0)$ has a neighborhood that looks just like $\mathbb{R}^2$, but the point $q = (1, 0)$, which is on the boundary, does not. It does, however, have a neighborhood that looks like the closed half-plane

$$\mathbf{H}^2 = \{(x, y): y \geq 0\}.$$

Thus in a manifold with boundary we want to allow charts that look like the closed half-space

$$\mathbf{H}^n = \{(x^1, \ldots, x^n): x^n \geq 0\}.$$

We have to worry a bit about the fact that we have not yet defined what it means for a function on $\mathbf{H}^n$ to be smooth! We want such functions to be smooth 'up to and including the boundary'. Perhaps the simplest way to say this is that a function on $\mathbf{H}^n$ is **smooth** if it extends to a smooth function on the manifold

$$\{(x^1, \ldots, x^n): x^n > -\epsilon\}$$

for some $\epsilon > 0$.

So: we define a n-dimensional manifold **with boundary** to be a topological space M equipped with charts of the form $\varphi_\alpha: U_\alpha \to \mathbb{R}^n$ or $\varphi_\alpha: U_\alpha \to \mathbf{H}^n$, where U_α are open sets covering M, such that the transition function $\varphi_\alpha \circ \varphi_\beta^{-1}$ is smooth where it is defined. (We also assume some technical conditions, namely that M is Hausdorff and paracompact. We will have a bit more to say about these in a bit.) Note that a plain old manifold is automatically a manifold with boundary, but not vice versa. If M is a manifold with boundary, we define the **boundary** of M to be the set of $p \in M$ such that some chart $\varphi_\alpha : U_\alpha \to \mathbf{H}^n$ maps p to a point in

$$\partial \mathbf{H}^n = \{(x^1, \ldots, x^n) : x^n = 0\}.$$

We write ∂M for the boundary of M.

Exercise 84. *Let the n-disk D^n be defined as*

$$D^n = \{(x_1, \cdots, x_n) : x_1^2 + \cdots + x_n^2 \le 1\}.$$

Show that D^n is an n-manifold with boundary in an obvious sort of way.

We say that a function $f : M \to \mathbb{R}$ is **smooth** if for any chart φ_α, $f \circ \varphi_\alpha$ is smooth as a function on $\mathbb{R}^n$ or $\mathbf{H}^n$. Similarly, smooth maps, vector fields, differential forms, and so on are defined just as in the 'without boundary' case. In particular, the tangent space at a point in the boundary of a manifold works out being a vector space as usual. One should imagine something like this:

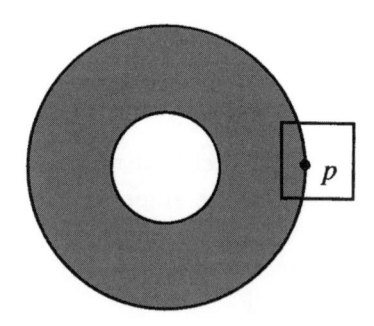

Fig. 8. Tangent space of a point on the boundary

Exercise 85. *Check that the definition of tangent vectors in Chapter 3 really does imply that the tangent space at point on the boundary of an n-dimensional manifold with boundary is an n-dimensional vector space.*

Now let us explain how to integrate differential forms. The main idea is that when we do an integral on $\mathbb{R}^n$ like

$$\int_{\mathbb{R}^n} f(x^1, \ldots, x^n) \, dx^1 \cdots dx^n,$$

Stokes' Theorem

we should think of it as integrating the n-form

$$f dx^1 \wedge \cdots \wedge dx^n,$$

not the function f. The reason is that when we change coordinates, an n-form picks up exactly the right factor of the determinant of the Jacobian of the coordinate transformation that we need in the change of variables formula for multiple integrals.

More precisely, suppose ω is any n-form on $\mathbb{R}^n$. We can write

$$\omega = f dx^1 \wedge \cdots \wedge dx^n,$$

so let us define

$$\int_{\mathbb{R}^n} \omega = \int_{\mathbb{R}^n} f \, dx^1 \cdots dx^n,$$

assuming the integral on the right side converges. Now let us see whether this definition is coordinate-independent. Suppose that x'^μ are another set of coordinate functions on $\mathbb{R}^n$, and write

$$\omega = f' dx'^1 \cdots dx'^n$$

for some other function f'. We saw in the section on the volume form in Chapter 5 that

$$dx'^1 \wedge \cdots \wedge dx'^n = (\det T) \, dx^1 \wedge \cdots \wedge dx^n$$

where

$$T^\mu_\nu = \frac{\partial x'^\mu}{\partial x^\nu}$$

is the Jacobian of the coordinate transformation from the unprimed to the primed coordinates. Thus

$$f = (\det T) f'.$$

This implies

$$\int_{\mathbb{R}^n} f \, dx^1 \cdots dx^n = \int_{\mathbb{R}^n} f' \, (\det T) \, dx^1 \cdots dx^n,$$

but by the change of variables formula for multiple integrals we have

$$\int_{\mathbb{R}^n} f' \, |(\det T)| \, dx^1 \cdots dx^n = \int_{\mathbb{R}^n} f' dx'^1 \cdots dx'^n.$$

Thus we do indeed have coordinate-independence,

$$\int_{\mathbb{R}^n} f\, dx^1 \cdots dx^n = \int_{\mathbb{R}^n} f' dx'^1 \cdots dx'^n,$$

if $\det T > 0$. Recall from Chapter 5 that when $\det T > 0$, the volume forms $dx^1 \wedge \cdots \wedge dx^n$ and $dx'^1 \wedge \cdots \wedge dx'^n$ define the same orientation.

Now let M be an oriented manifold with boundary, and let ω be an n-form on M. The obvious way to define

$$\int_M \omega$$

is to break the integral up into a sum of integrals over charts. So let $\{\varphi_\alpha\}$ be an atlas of charts for M, with either $\varphi_\alpha\colon U_\alpha \to \mathbb{R}^n$ or $\varphi_\alpha\colon U_\alpha \to \mathbf{H}^n$. We can assume that all these charts are oriented. (See the section on the volume form in Chapter 5.) As it turns out, we can always find a collection of smooth functions $\{f_\alpha\}$ on M such that:

1. f_α is zero outside U_α.

2. Any point $p \in M$ has an open set containing it on which only finitely many of the functions f_α are nonzero.

3. For any $p \in M$,
$$\sum_\alpha f_\alpha = 1.$$

These functions $\{f_\alpha\}$ are called a **partition of unity**. This technical result uses the fact that M is 'paracompact and Hausdorff' — see the references in the notes for details — but this is the case in all reasonable examples, and we have been implicitly assuming this hypothesis all along.

Using this device we have

$$\omega = \sum_\alpha f_\alpha \omega$$

where $f_\alpha \omega$ vanishes outside U_α. We may thus write

$$f_\alpha \omega = g_\alpha(x^1, \ldots, x^n) dx^1 \wedge \cdots \wedge dx^n$$

Stokes' Theorem

where x^μ are the local coordinates on U_α associated to the chart φ_α, and the function g_α vanishes outside U_α. We then define

$$\int_M \omega = \sum_\alpha \int g_\alpha(x^1, \ldots, x^n) dx^1 \wedge \cdots \wedge dx^n$$

whenever the integrals and sum on the right hand side converge absolutely. Using the fact that all the charts are oriented, one can show that this definition of $\int_M \omega$ is independent of the choices we have made.

Exercise 86. *For the mathematically inclined reader: prove that $\int_M \omega$ is independent of the choice of charts and partition of unity.*

The other thing to note is that if M is an oriented manifold with boundary, the boundary ∂M is an oriented manifold in a natural way. Take an atlas of charts for M and only consider those charts $\varphi_\alpha: U_\alpha \to \mathbf{H}^n$ that map to the half-space $\mathbf{H}^n$. Let $V_\alpha = U_\alpha \cap \partial M$, so that V_α is an open subset of ∂M, and let ψ_α denote the restriction of φ_α to V_α. Then

$$\psi_\alpha: V_\alpha \to \mathbb{R}^{n-1}$$

is continuous with a continuous inverse, and the transition functions $\psi_\alpha \circ \psi_\beta^{-1}$ are smooth and orientation-preserving. Thus $\{\psi_\alpha\}$ form a collection of charts for ∂M, making it into an $(n-1)$-manifold, and it becomes an oriented manifold by Exercise 62.

Exercise 87. *Show that $\partial D^n = S^{n-1}$, where the n-disk D^n is defined as in Exercise 84.*

Now we can state **Stokes' theorem** again, and everything in it should make sense. Namely, let M be a compact oriented n-manifold with boundary and let ω be an $(n-1)$-form on M. Then

$$\int_M d\omega = \int_{\partial M} \omega.$$

Alternatively, we can drop the hypothesis that M be compact if we assume that ω vanishes outside of some compact set.

The simplest example is the following:

Exercise 88. *Let $M = [0,1]$. Show that Stokes' theorem in this case is equivalent to the fundamental theorem of calculus:*

$$\int_0^1 f'(x)\,dx = f(1) - f(0).$$

Exercise 89. *Let $M = [0,\infty)$, which is not compact. Show that without the assumption that f vanishes outside a compact set, Stokes' theorem may not apply. (Hint: in this case Stokes' theorem says $\int_0^1 f'(x)\,dx = -f(0)$.)*

For fancier examples, it is nice to consider 'submanifolds'. Given a subset S of a n-manifold M, we say that S is a k-dimensional **submanifold** of M if for each point $p \in S$ there is an open set U of M containing p and a chart $\varphi\colon U \to \mathbb{R}^n$ such that

$$S \cap U = \varphi^{-1}\mathbb{R}^k.$$

In other words, just as M looks locally like $\mathbb{R}^n$, S locally looks like a k-dimensional hyperplane in $\mathbb{R}^n$, as below:

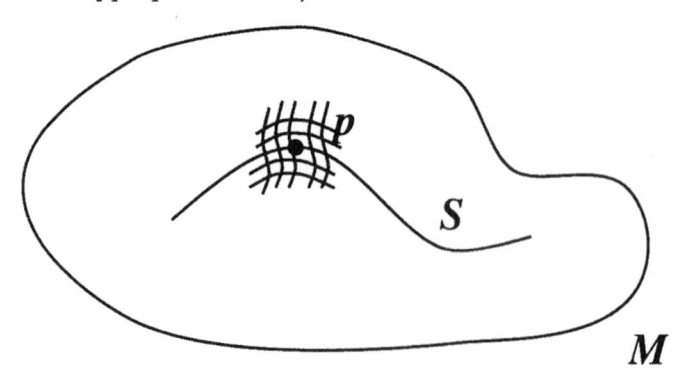

Fig. 9. A submanifold $S \subseteq M$

Exercise 90. *Show that any submanifold is a manifold in its own right in a natural way.*

Exercise 91. *Show that S^{n-1} is a compact submanifold of $\mathbb{R}^n$.*

Exercise 92. *Show that any open subset of a manifold is a submanifold.*

Stokes' Theorem

There is a similar definition for S to be a **submanifold with boundary** of an n-manifold M; here for some points $p \in S$,

$$S \cap U = \varphi^{-1} \mathbf{H}^k.$$

If N is a manifold (possibly with boundary) and $\phi: N \to M$ is a smooth map such that $\phi(N)$ is a submanifold of M (possibly with boundary), we say ϕ is an **embedding** of N in M, and we say N is **embedded** in M. Applying our generalized Stokes' theorem to such submanifolds of $\mathbb{R}^n$ for n equal to 2 or 3, we get several classic theorems of vector calculus: the original version of Stokes' theorem, as well as Green's and Gauss' theorems.

Exercise 93. *Show that if S is a k-dimensional submanifold with boundary of M, then S is a manifold with boundary in a natural way. Moreover, show that ∂S is a $(k-1)$-dimensional submanifold of M.*

Exercise 94. *Show that D^n is a submanifold of $\mathbb{R}^n$ in this sense.*

Exercise 95. *Suppose that $S \subset \mathbb{R}^2$ is a 2-dimensional compact orientable submanifold with boundary. Work out what Stokes' theorem says when applied to a 1-form on S. This is sometimes called Green's theorem.*

Exercise 96. *Suppose that $S \subset \mathbb{R}^3$ is a 2-dimensional compact orientable submanifold with boundary. Show Stokes' theorem applied to S boils down to the classic Stokes' theorem.*

Exercise 97. *Suppose that $S \subset \mathbb{R}^3$ is a 3-dimensional compact orientable submanifold with boundary. Show Stokes' theorem applied to S is equivalent to Gauss' theorem, also known as the divergence theorem.*

Next, we will apply Stokes' theorem to the problem of closed versus exact forms.

DeRham Cohomology

The boundary of a boundary is zero. — *John Archibald Wheeler*

All exact forms are closed but not vice versa. The study of this 'vice versa' is called deRham cohomology, after the inventor of differential forms. In the previous section we saw that the closed 1-forms on a manifold are automatically exact if a certain *topological* condition held, namely that the manifold was simply connected. If a manifold is not simply connected, it has some sort of 'holes' in it: think of the example $\mathbb{R}^2 - \{0\}$, or more generally the plane with some finite set of points removed. One might call these '1-holes' (this is not standard terminology!) because they prevent closed 1-forms that 'wrap around them' from being exact. They also deserve that name because they prevent certain 1-dimensional objects, namely paths, from being homotopic.

There are, however, various sorts of holes besides 1-holes. For example, the space $\mathbb{R}^3 - \{0\}$ clearly has some sort of 'hole' in it. However, this space is simply connected; it is easy to visualize how any two paths in the space are homotopic by a homotopy that 'dodges the hole'. So this hole is not a 1-hole. In fact, this hole deserves to be called a '2-hole', because it prevents certain 2-dimensional surfaces from being deformed into one another, namely the upper and lower hemispheres of the unit sphere:

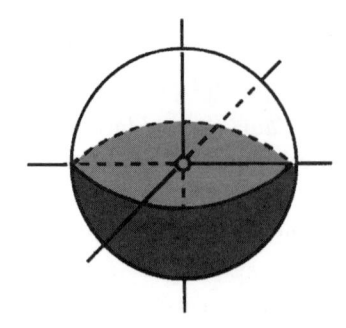

Fig. 10. Two surfaces that are not homotopic in $\mathbb{R}^3 - \{0\}$

Cohomology is basically the study of holes by algebraic methods. Having holes is a topological property of a space, that is, a property pre-

DeRham Cohomology

served by all continuous mappings with continuous inverses, so cohomology theory is a branch of algebraic topology. We will barely scratch the surfaces of this subject, which is becoming ever more important in physics, but we provide a list of basic references in the Notes at the end of Part I.

The 'pth deRham cohomology' of a manifold M is a vector space, written $H^p(M)$, whose dimension is the number of 'p-holes' in M. To define this vector space, first write $Z^p(M)$ for the set of closed p-forms on M. This is a vector space, since the sum of closed forms, or any number times a closed form, is again closed. Similarly, let us write $B^p(M)$ for the vector space of exact p-forms. The exact p-forms are a subspace of the closed p-forms:

$$B^p(M) \subseteq Z^p(M)$$

so the most natural way to see how many closed forms there are that are not exact is to take the quotient space

$$H^p(M) = Z^p(M)/B^p(M),$$

called the **pth deRham cohomology group** of M. This is really a vector space, not just a group (every vector space is a group under addition), but the term group is used because other sorts of cohomology theories only give groups.

It might not hurt to remind the reader what this quotient space business really means. An element of $H^p(M)$ is an equivalence class of closed p-forms, where two closed forms ω, ω' are equivalent if they differ by an exact p-form, or in other words, if there is a $(p-1)$-form μ such that

$$\omega - \omega' = d\mu.$$

As part of the jargon of cohomology theory, when ω and ω' are equivalent in this way we say they are **cohomologous**, and we call the equivalence class of ω its **cohomology class**:

$$[\omega] = \{\omega': \exists \mu \ \omega - \omega' = d\mu\}.$$

The simplest case of these definitions is H^0. Let M be a manifold. When is a 0-form on M closed? Recall that the 0-forms on M are just functions. In local coordinates,

$$df = \partial_\mu f \, dx^\mu$$

for any function f on M. Thus a 0-form f is closed if and only if all its first partial derivatives vanish, that is, if it is **locally constant**. A function can be locally constant but not constant if M is not connected. For example, suppose M has connected components $M_1, \ldots, M_k$. Then the most general locally constant function on M is one that takes the constant value c_i on the ith component. When is a 0-form on M exact? When it is d of a (-1)-form, presumably, but there are no (-1)-forms! Thus by convention we say that the space of exact 0-forms is the trivial vector space $\{0\}$ consisting only of the zero function. We thus have

$$H^0(M) = Z^0(M)/B^0(M) = Z^0(M)/\{0\} \cong Z^0(M),$$

or in words, the 0th deRham cohomology of M is isomorphic to the space of locally constant functions on M. This is a vector space whose dimension is the number of connected components of M. Thus $H^0(M)$ conveys some very basic information about the topology of M. In particular, $H^0(M) = \mathbb{R}$ if and only if M is connected.

Similarly, as we have said before in other ways, $H^1(M) = 0$ if M is simply connected, since then every closed 1-form is exact. But how do we calculate $H^1(M)$ when it is not zero? There are lots of ways, but going into these would require a long digression on algebraic topology. Here we will only describe how to show a given closed 1-form is not exact. This is a step in the right direction, since if we can find a set of closed 1-forms $\omega^1, \ldots, \omega^d$ on M such that no nontrivial linear combination of them is exact:

$$\left[\sum_i c_i \omega^i\right] = 0 \implies c_1, \ldots, c_d = 0,$$

then we know that $H^1(M)$ is at least d-dimensional.

The trick is to use Stokes' theorem. Suppose $S \subseteq M$ is a circle embedded in M. If $\omega \in \Omega^1(M)$ equals df for some function f, then Stokes' theorem implies

$$\int_S \omega = \int_S df = \int_{\partial S} f = 0$$

DeRham Cohomology

because ∂S is the empty set! So if we can find a circle $S \subseteq M$ with

$$\int_S \omega \neq 0$$

we automatically know that ω is not exact. In fact, we saw this a different way in the first section of this chapter: there we showed that if the integral of ω around any loop is nonzero, ω is not exact.

In fact, this trick can be substantially generalized. Suppose $\omega = d\mu$ is an exact p-form on M. Then for every compact p-dimensional manifold S and map $\phi \colon S \to M$, we have

$$\int_S \phi^* \omega = \int_S \phi^*(d\mu) = \int_S d(\phi^*\mu) = \int_{\partial S} \phi^*\mu = 0$$

since S has no boundary. In particular, if $S \subseteq M$ is any compact orientable submanifold, we have

$$\int_S \omega = 0.$$

There is, in fact, a remarkable converse: if $\int_S \phi^*\omega = 0$ for every map $\phi \colon S \to M$ of a p-dimensional manifold S to M, then ω is exact. For $p = 1$ this fact is only a slight refinement of Exercise 82, but it is considerably trickier when $p > 1$. We will not prove this fact, referring the reader instead to the notes.

This fact gives us a new outlook on exact differential forms: they are the ones whose integrals are zero! In the following sections we apply this idea to electromagnetism. We leave the reader with a few handy facts presented as exercises, and the following pretty formula that explains the quote by Wheeler at the beginning of this section:

$$0 = \int_M d^2\omega = \int_{\partial M} d\omega = \int_{\partial\partial M} \omega = 0$$

since ∂M has no boundary.

Exercise 98. *Show that the pullback of a closed form is closed and the pullback of an exact form is exact.*

Exercise 99. *Show that given any map $\phi\colon M \to M'$ there is a linear map from $H^p(M')$ to $H^p(M)$ given by*

$$[\omega] \mapsto [\phi^*\omega]$$

where ω is any closed p-form on M'. Call this linear map

$$\phi^*\colon H^p(M') \to H^p(M).$$

Show that if $\psi\colon M' \to M''$ is another map, then

$$(\psi\phi)^* = \phi^*\psi^*.$$

Gauge Freedom

Just as one can get solutions for one of the two equations of electrostatics more cheaply if the electric field comes from a scalar potential, one can simplify work in magnetostatics if the magnetic field comes from a vector potential. Remember that the equations of magnetostatics are

$$dB = 0, \qquad \star d \star B = j$$

where the magnetic field B is a 2-form on space and the current density j is a 1-form. If B is exact:

$$B = dA,$$

the first equation is automatically true, and the second one reduces to the following equation for the 1-form A:

$$\star d \star dA = j.$$

Given the magnetic field B, a 1-form A is called a vector potential for B. We say 'a' vector potential rather than 'the' vector potential because A is not uniquely determined, since we can add any closed 1-form to A without changing dA. In particular, we can change A to $A + df$ for any function f without changing B. This way of changing A is called a **gauge transformation**. Our freedom in choosing A is called **gauge freedom**.

Gauge Freedom

As we have noted, the same remarks hold in the spacetime context for the electromagnetic field F. Maxwell's equations say that this 2-form satisfies
$$dF = 0, \quad \star d \star F = J.$$
We say A is a vector potential for F if $dA = F$; if this is the case, Maxwell's equations reduce to
$$\star d \star dA = J.$$

It can be handy to use the gauge freedom to make the vector potential satisfy various extra conditions. Choosing such a condition is called **choosing a gauge**. Many physicists have gauges named after them, the most well-known being Coulomb gauge, Lorentz gauge, Feynman gauge and Landau gauge. We do not want to get very deep into this issue, but, particularly for mathematicians (who tend to be scared of this for some reason), we want to give an example.

The simplest gauge is temporal gauge. Suppose we are working on a spacetime of the form $\mathbb{R} \times S$, where S is 'space', and $\mathbb{R} \times S$ is given the Lorentzian metric $dt^2 - {}^3g$, where ${}^3g$ is a Riemannian metric on S and t, 'time', is the coordinate on $\mathbb{R}$. Differentiation with respect to t can be thought of as a vector field ∂_t on $\mathbb{R} \times S$. If the 1-form A on $\mathbb{R} \times S$ satisfies
$$A(\partial_t) = 0,$$
we say A is in **temporal gauge**. For example, in Minkowski spacetime, $\mathbb{R}^4$, any 1-form A can be written as
$$A = A_0 dt + A_1 dx + A_2 dy + A_3 dz,$$
and temporal gauge is simply the condition that $A_0 = 0$. To keep our notation simple, let us define
$$A_0 = A(\partial_t)$$
for *any* spacetime of form $\mathbb{R} \times S$, so that A is in temporal gauge if $A_0 = 0$.

Given any exact 2-form F on $\mathbb{R} \times S$, we can find some A in temporal gauge such that $dA = F$. To see this, start with A, not necessarily in temporal gauge, such that $dA = F$. Let f be the function on $\mathbb{R} \times S$ such that for any point $(t,p) \in \mathbb{R} \times S$,

$$f(t,p) = \int_0^t A_0(s,p)\,ds.$$

Let

$$A' = A - df.$$

We claim that $dA' = F$ and that A' is in temporal gauge. For the former, simply note that

$$dA' = d(A - df) = dA = F.$$

For the latter, note that

$$\begin{aligned} A'_0(t,p) &= A_0(t,p) - (df(\partial_t))(t,p) \\ &= A_0(t,p) - (\partial_t f)(t,p) \\ &= A_0(t,p) - \partial_t \int_0^t A_0(s,p)\,ds \\ &= 0. \end{aligned}$$

Let us see what Maxwell's equations on $\mathbb{R} \times S$ look like when the vector potential A is in temporal gauge. Since $A_0 = 0$, we can think of A as just a 1-form on S that is a function of time. Moreover, since $F = B + E \wedge dt$ and

$$F = dA = dt \wedge \partial_t A + d_S A,$$

we have

$$E = -\partial_t A, \qquad B = d_S A.$$

We will rewrite Maxwell's equations in terms of the **Cauchy data** (A, E) on a spacelike surface $\{t\} \times S$. The first pair of Maxwell equations, namely

$$d_S B = 0, \qquad \partial_t B + d_S E = 0,$$

become tautologies in terms of A:

$$d_S^2 A = 0, \qquad \partial_t d_S A - d_S \partial_t A = 0,$$

Gauge Freedom

while the second pair, namely

$$\star_S d_S \star_S E = \rho, \qquad -\partial_t E + \star_S d_S \star_S B = j,$$

become two equations about the Cauchy data. The first equation, the **Gauss law**, is a **constraint** that the Cauchy data (A, E) must satisfy at any given time:

$$\star_S d_S \star_S E = \rho.$$

The second equation, together with the fact $\partial_t A = -E$, can be summarized as an **evolutionary equation** that says how the Cauchy data change with time:

$$\partial_t (A, E) = (-E, \star_S d_S \star_S d_S A - j).$$

If we are good at differential equations, we can use the evolutionary equation to determine (A, E) at any later (or earlier) time provided we know it at time t. Typically one starts with $t = 0$.

It is worth noting that as long as the continuity equation

$$\partial_t \rho + \star_S d_S \star_S j = 0$$

holds (recall that this expresses local conservation of electric charge), the Gauss law at $t = 0$ together with the evolutionary equation imply the Gauss law at later times. We say that the Gauss law is 'preserved by time evolution'. The basic idea of the proof is that the evolutionary equation and continuity equation at any time t imply that

$$\partial_t (\star_S d_S \star_S E - \rho) = 0,$$

so if the Gauss law holds at time t it will continue to hold later. To see this, just compute:

$$\begin{aligned}
\partial_t(\star_S d_S \star_S E - \rho) &= \star_S d_S \star_S \partial_t E - \partial_t \rho \\
&= \star_S d_S \star_S (\star_S d_S \star_S d_S A - j) - \partial_t \rho \\
&= -\star_S d_S \star_S j - \partial_t \rho \\
&= 0
\end{aligned}$$

using the facts that $\star_S^2 = \pm 1$ and $d_S^2 = 0$.

One other thing to note is that if we have an exact 2-form F on $\mathbb{R} \times S$, the 1-form A in temporal gauge such that $F = dA$ is not *unique*. In other words, there is still some gauge freedom. The reason is that if ω is any fixed closed 1-form on space, $A' = A + \omega$ will again be a 1-form on $\mathbb{R} \times S$ that is in temporal gauge and has $dA' = F$. In particular, we can take $\omega = df$ for some function f on space. Getting rid of this remaining gauge freedom, if for some reason we want to, is more work.

The Bohm-Aharonov Effect

The Bohm-Aharonov effect is important because it dramatizes the importance of the vector potential in electromagnetism, especially in the context of quantum mechanics, and also shows how funny things can happen in regions of space that are not simply connected. It also has technological applications, as we shall see.

First let us do a little problem in magnetostatics that we will need the answer to later: determining the magnetic field produced by a current running through an infinitely long cylindrical wire. Suppose the wire runs along the z axis. We will use cylindrical coordinates (r, θ, z) on $\mathbb{R}^3$. Here we should note that z is a smooth function on $\mathbb{R}^3$, so dz is a 1-form defined on all $\mathbb{R}^3$, but r is smooth only away from the z axis, that is, $r = 0$, so dr is defined only away from this line. Moreover, the 'coordinate' θ, in addition to being ill-defined on the z axis, is really only defined modulo 2π. Nonetheless, it is customary to define a 1-form '$d\theta$', which we can do in rectangular coordinates by

$$d\theta = \frac{x\, dy - y\, dx}{x^2 + y^2}.$$

The calculation in Exercise 80 shows that '$d\theta$' is closed, but we have seen that it is not exact, so the name '$d\theta$' is very misleading. We will bow to tradition and call it $d\theta$, however.

Now, suppose the current is cylindrically symmetric and flows in the z direction, so that

$$j = f(r)\, dz.$$

The Bohm-Aharonov Effect

Then one can calculate that (away from the z axis)

$$\star j = f(r) r\, dr \wedge d\theta.$$

Exercise 100. *Do this. (Hint: show that $\star dz = r\, dr \wedge d\theta$.)*

Since the curl of the magnetic field must be the current, the magnetic field should look something like this:

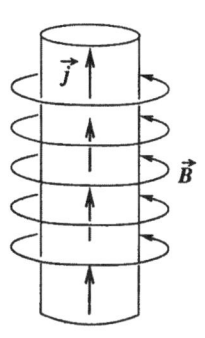

Fig. 11. Magnetic field produced by a current running through a wire

Thus we will assume B has the form

$$\star B = g(r) d\theta$$

or

$$B = \frac{g(r)}{r} dz \wedge dr.$$

Exercise 101. *Show that $\star d\theta = \frac{1}{r} dz \wedge dr$.*

Then the first equation of magnetostatics, $dB = 0$, is automatic, while the second, $d \star B = \star j$, is equivalent to $g'(r) = rf(r)$.

Exercise 102. *Check that $d \star B = \star j$ holds if and only if $g'(r) = rf(r)$.*

Integrating, we obtain:

$$g(r) = g(0) + \int_0^r s f(s)\, ds.$$

We will assume $g(0) = 0$ so that B does not blow up as $r \to 0$. Suppose that the wire is of radius R and f is zero outside the wire. Then for $r > R$ we have

$$\star B = \frac{I}{2\pi} d\theta, \qquad B = \frac{I}{2\pi r} dz \wedge dr,$$

where the total current I flowing through the wire is given by

$$I = 2\pi \int_0^R f(r) r \, dr.$$

Actually there is a certain amount of ambiguity to the field B. In other words, B is not completely determined by the equations $d \star B = \star j$ and $dB = 0$, since we can add to it any 1-form C such that

$$dC = d \star C = 0.$$

Note that such 1-forms actually exist. For example, a 1-form $C_i dx^i$ with constant coefficients C_i has this property. Why then did we feel entitled to speak of 'the' magnetic field produced by the wire? The reason is that no 1-form C on $\mathbb{R}^3$ with $dC = d \star C = 0$ goes to zero at infinity. Our solution for this problem is the unique one for which B goes to zero as $r \to \infty$.

Next, let us do a different problem that turns out to be mathematically very similar. One can build a **solenoid** by winding a wire around a cylinder in a tight spiral. Say the cylinder is centered on the z axis. If one flows a current through the wire, one obtains a constant magnetic field inside the solenoid, and a zero magnetic field outside (in the idealized situation where the solenoid is infinitely long and the wire is infinitely thin). That is,

$$\star B = f(r) dz, \qquad B = f(r) r \, dr \wedge d\theta$$

where f is a constant for $r < R$ and zero for $r > R$. Now, what is a vector potential A for this magnetic field?

The Bohm-Aharonov Effect

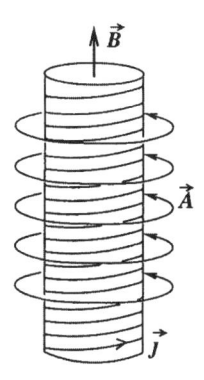

Fig. 12. Vector potential produced by magnetic field in a solenoid

Note that in our previous problem we had a current j running along a wire and sought a magnetic field with $dB = 0$, $d \star B = \star j$. Now we have a magnetic field B with the same form as the previous $\star j$, and seek a vector potential with $dA = B$. It follows that we can borrow the answer to our previous problem, and take

$$A = g(r)d\theta$$

where

$$g(r) = \int_0^r sf(s)\,ds.$$

In particular, outside the solenoid we have

$$A = \frac{\Phi}{2\pi}d\theta$$

where Φ is the **magnetic flux** through the solenoid, that is, the integral of B over the disc $r \le R$ in any plane of constant z:

$$\Phi = 2\pi \int_0^R f(r)r\,dr.$$

Of course, in this problem there is even more ambiguity in our answer for A, but all we want is *some* vector potential for B.

The Bohm-Aharonov effect occurs when a charged particle passes around a solenoid. It is a purely quantum-mechanical effect, so we need to explain a small amount of quantum mechanics. Our treatment will be *very* brief, so we refer the reader to some books on quantum mechanics in the notes for more details. In quantum theory the states of a physical system are typically described as unit vectors in some Hilbert space $\mathcal{H}$, called **state vectors**. The inner product of $\mathcal{H}$ is closely related to the probabilistic nature of quantum theory. More precisely, if one prepares the system in a state represented by a unit vector $\psi \in \mathcal{H}$, and immediately does an experiment to see if it is in the state represented by the unit vector $\phi \in \mathcal{H}$, the probability that one receives the answer 'yes' is

$$|\langle \phi, \psi \rangle|^2.$$

Such a quantity is called a **transition probability**, while the inner product

$$\langle \phi, \psi \rangle$$

itself is called the **transition amplitude**. Moreover, **observables** (that is, measurable quantities) are represented by self-adjoint operators on $\mathcal{H}$, and the **expected value** (average measured value) of an observable A in the state represented by the vector ψ is given by

$$\langle \psi, A\psi \rangle.$$

While we represent states as unit vectors, it is important to note that if two states ψ and ψ' **differ by a phase**, that is, if

$$\psi' = e^{i\theta} \psi$$

for some real number θ, then they describe the same state. The reason is that no transition probabilities are affected by using ψ' instead of ψ:

$$|\langle \phi, \psi' \rangle|^2 = |\langle \phi, \psi \rangle|^2$$

for all $\phi \in \mathcal{H}$. Similarly, expected values are unaffected by the phase.

Magnetism has a remarkable relationship to the phase in quantum mechanics, which is roughly as follows. First let us be *very* sloppy, just to get the idea across quickly! Suppose we have a particle in $\mathbb{R}^3$ with electric charge q in the state described by the vector ψ. Suppose there is a magnetic field B with vector potential A. If we drag the particle around a loop γ, ψ is multiplied by the phase

$$e^{-\frac{i}{\hbar} q \int_\gamma A}$$

The Bohm-Aharonov Effect

where $\hbar$ is **Planck's constant**, equal to about $1.055 \cdot 10^{-34}$ joule-seconds. In particular, if there is some oriented 2-disk D embedded in $\mathbb{R}^3$ such that γ runs counterclockwise around the boundary of D, by Stokes' theorem we have

$$\int_\gamma A = \int_D B,$$

and the latter quantity is called the **magnetic flux** through D. Then as we drag the particle around γ, ψ is multiplied by the phase

$$e^{-\frac{i}{\hbar} q \int_D B}.$$

In fact, the same formulas hold for a loop in *spacetime* rather than space, except that B must be replaced by the electromagnetic field F.

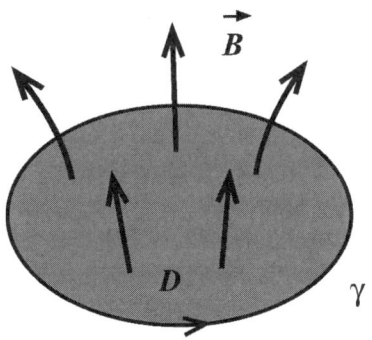

Fig. 13. Moving a particle around a loop

The alert reader will note that we are being *too* sloppy! First we said that two vectors that differ by a phase describe the same physical state, and then we said that if one drags a particle in a magnetic field around a loop, its vector ψ is only multiplied by a phase! What possible *physical* significance could this have? Also, it is an oversimplification to speak of dragging a particle along a path γ, since in quantum mechanics a particle does not really follow a well-defined trajectory.

To answer this we need to briefly mention path integrals, which are an approach to quantum mechanics developed by Richard Feynman. In classical mechanics, a particle moves along some path γ in $\mathbb{R}^3$. There is an important quantity called the **Lagrangian**, which is the kinetic energy minus the potential energy, and is a function of time that can be calculated from the particle's position and velocity:

$$L = L(\gamma(t), \gamma'(t)).$$

The exact formula for the Lagrangian depends on the forces acting upon the particle. If we consider the particle's path only from time 0 to time T and integrate the Lagrangian over this interval of time, we get a quantity called the **action**,

$$S(\gamma) = \int_0^T L.$$

The amazing thing about the action is that in classical mechanics, a particle going from some point p at time 0 to some point q at time T will always follow a path γ that is a critical point for the action. That is, if we change the path a little bit to a new path from p to q, the action will be unaffected to first order. Often the path simply *minimizes* the action, as if nature were lazy, but this is not always the case. We will derive the basic equation of classical mechanics, $F = ma$, from this 'action principle' in Chapter 4 of Part II.

In quantum mechanics the action also plays an important role. Here the state of the particle is described by a vector in a Hilbert which in the simplest case is just $L^2(\mathbb{R}^3)$, the space of all complex functions on $\mathbb{R}^3$ such that

$$\int |\psi(x)|^2 \, d^3x < \infty$$

This space has the inner product

$$\langle \phi, \psi \rangle = \int \overline{\phi}(x) \psi(x) \, d^3x.$$

We call a state vector $\psi \in L^2(\mathbb{R}^3)$ a **wavefunction**. Given the wavefunction $\psi \in L^2(\mathbb{R}^3)$, we can think of $\psi(x)$ as being the amplitude density for the particle to be at the point $x \in \mathbb{R}^3$, or $|\psi(x)|^2$ as the probability density, meaning that the probability for it being in some set $U \subseteq \mathbb{R}^3$ is

$$\int_U |\psi(x)|^2 \, d^3x.$$

The Bohm-Aharonov Effect

Now suppose that we have a quantum-mechanical particle that starts out in the state ψ at $t = 0$ and we wish to compute its state ϕ at some other time T. Suppose first that there is no magnetic field. Let
$$\mathcal{P} = \{\gamma\colon [0,T] \to \mathbb{R}^3\colon \gamma(0) = a,\ \gamma(T) = b\}$$
denote the space of all paths that start at the point a at time 0 and end at the point b at time T. Then
$$\phi(b) = \int_{\mathcal{P}} e^{\frac{i}{\hbar}S(\gamma)}\psi(a)\,\mathcal{D}\gamma$$
where $\mathcal{D}\gamma$ is some sort of mysterious 'Lebesgue measure' on the space $\mathcal{P}$. In other words, we can think of the particle as taking *all* paths from a to b, weighted by the phase factor
$$e^{\frac{i}{\hbar}S(\gamma)}.$$

One can show that as $\hbar \to 0$, this phase factor oscillates very rapidly except near the paths that are critical points of the action, cancelling out in such a way that only the classical path contributes.

We emphasize, however, that doing these integrals over $\mathcal{P}$, or **path integrals**, is highly nontrivial, primarily because the 'Lebesgue measure' $\mathcal{D}\gamma$ is not really a measure according to the standard mathematical definition. Figuring out what $\mathcal{D}\gamma$ really means and how to compute with it is a serious challenge! There are much easier ways to make quantum mechanics rigorous than via path integrals — for rigor, it is easier to use the 'Hamiltonian' approach. Path integrals are especially useful, however, for qualitative *insight* into quantum theory and for practical perturbative calculations. While many mathematicians have torn out their hair trying to provide a rigorous foundation for path integrals, with only partial success, physicists sail right along using them very effectively.

Next let us suppose that, in addition to whatever forces were already acting on our particle, there is also a magnetic field B on $\mathbb{R}^3$ with vector potential A. For simplicity let us consider the case when A and B are independent of time. Then the path-integral formula for the state ϕ at time T should be modified as follows:
$$\phi(b) = \int_{\mathcal{P}} e^{\frac{i}{\hbar}(S(\gamma) - q\int_\gamma A)}\psi(a)\,\mathcal{D}\gamma.$$

In other words, the phase factor is multiplied by the additional phase

$$e^{-\frac{i}{\hbar}q\int_\gamma A}.$$

Alternatively we can say that the Lagrangian L is replaced by the Lagrangian

$$L - qA(\gamma'(t)).$$

In particular, if we are interested in the case when $a = b$, so that the path γ is a loop based at a, the extra phase factor is just

$$e^{-\frac{i}{\hbar}q\oint_\gamma A}$$

or if γ bounds the disk D,

$$e^{-\frac{i}{\hbar}q\int_D B}.$$

This phase factor *does* have physical effects, since it can differ for different loops γ, producing constructive or destructive interference in the path integral.

All this about path integrals and the magnetic field applies equally well to any manifold S we wish to use as 'space', not just $\mathbb{R}^3$. The Bohm-Aharonov effect is an interesting phenomenon that occurs when S is not simply connected. For example, suppose we have a cylindrical solenoid of radius $1/2$ centered on the z-axis, as in Figure 12. If the solenoid completely excludes the electron, we might as well take space to be the manifold

$$S = \mathbb{R}^3 - \{r < 1/2\},$$

which is not simply connected. The magnetic field vanishes in S, but the vector potential does not. Now suppose that we send an electron from the point $a = (-1, 0, 0)$ to the point $b = (1, 0, 0)$ in S. Since this is quantum mechanics, the electron can take any path in S from a to b:

The Bohm-Aharonov Effect

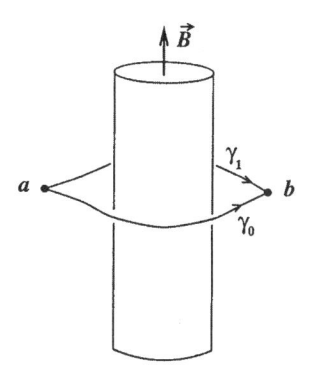

Fig. 14. Bohm-Aharonov effect: two paths from a to b

However, due to the vector potential, the electron can pick up a different phase depending on which path it takes from a to b. This gives rise to interference, which is the Bohm-Aharonov effect. In short, in quantum mechanics the vector potential can affect the wavefunction in significant ways even in regions where the magnetic field is zero!

To see this more precisely, first note that when the magnetic flux flowing through the solenoid is Φ, the vector potential is (up to gauge freedom)

$$A = \frac{\Phi}{2\pi} d\theta$$

so the phase factor

$$e^{-\frac{i}{\hbar} q \int_\gamma A}$$

equals $\exp(-iq\Phi/2\hbar)$ for the path γ_0 shown in Figure 14, while it equals $\exp(iq\Phi/2\hbar)$ for the path γ_1. By adjusting Φ to the appropriate value we can arrange for this phase factor to be i for the path γ_0 and $-i$ for γ_1. Similarly, by symmetry, every path from a to b has a reflected version for which the phase factor has the opposite sign! On the other hand, by symmetry, the standard action S will be the same for these two reflected paths. Thus

$$\int_\mathcal{P} e^{\frac{i}{\hbar}(S(\gamma) - q \int_\gamma A)}$$

vanishes, where now

$$\mathcal{P} = \{\gamma\colon [0,T] \to S\colon \gamma(0) = a,\ \gamma(T) = b\}.$$

In other words, for the right value of Φ there is *complete destructive interference*: an electron starting at a will never go to b! This effect has been observed and is in fact the basis for the technology of SQUIDs — superconducting quantum interference devices — which are used to accurately measure magnetic flux.

It is crucial here that the space S is not simply connected: this is what allows integrals of a closed form along different paths from a to b to give different answers. While one may object that space is still really $\mathbb{R}^3$, which is simply connected — and this is true — the point is that the Bohm-Aharonov effect is most easily understood using a model in which space is not simply connected.

Wormholes

A more detailed scrutiny of a surface might disclose that what we had considered an elementary piece in reality has tiny handles attached to it which change the connectivity character of the piece, and that a microscope of ever greater magnification would reveal ever new topological complications of this type, ad infinitum. *The Riemann point of view allows, also for real space, topological conditions entirely different from those realized by Euclidean space.* — Hermann Weyl.

Wormholes and monopoles live at the speculative end of theoretical physics, uneasily close to science fiction. They have never been observed, nor even firmly predicted from some well-established physical theory. They are, however, quite fun to think about, and very nice illustrations of deRham theory. We urge the reader to take this section and the next in that spirit.

A 'wormhole' is a kind of 'handle' in space that makes it non-simply-connected. This is easiest to visualize in 2 dimensions. In 2 dimensions, we can get a wormhole by taking $\mathbb{R}^2$, cutting out two disks, and gluing on a handle, that is, a cylinder $[0,1] \times S^1$:

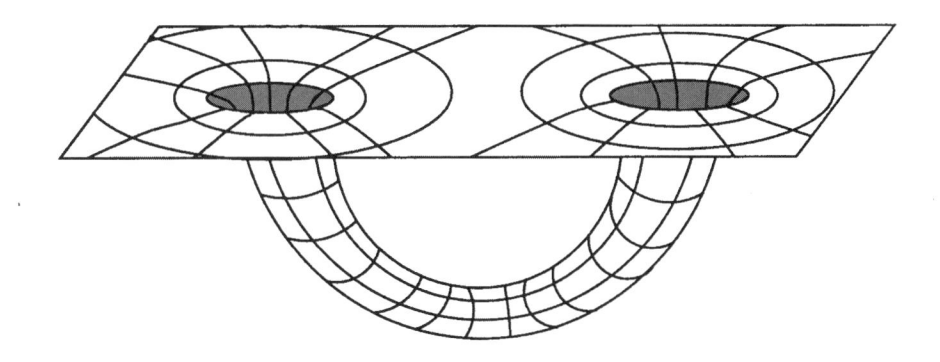

Fig. 15. A wormhole

Alternatively, we can start with the **torus** T^2, which is the 2-manifold that looks like the surface of a doughnut, i.e., $S^1 \times S^1$, and remove a point. Suitable stretching reveals this to be the same as the plane with a handle attached:

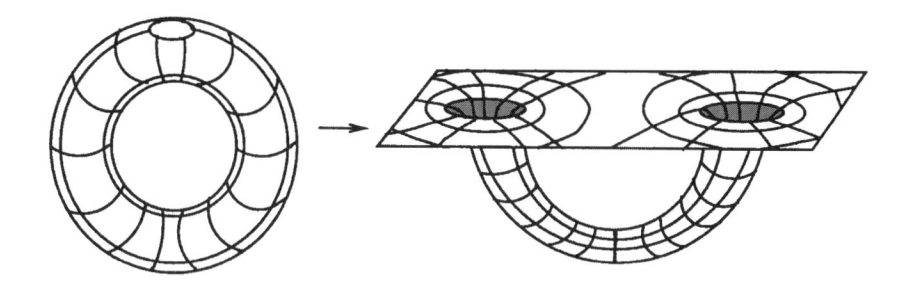

Fig. 16. Making a wormhole from a torus

This viewpoint makes it easy to prove that our wormhole is indeed a manifold; first one proves that the circle S^1 is a manifold, then that $S^1 \times S^1$ is a manifold, and then that removing a point from a manifold leaves a manifold. All of these were exercises in Chapter 1. One can also develop a very useful rigorous theory of cutting and pasting manifolds, called 'surgery theory', but we will not do this here.

Wormholes also make mathematical sense when space has 3 or more dimensions. One can either start with $\mathbb{R}^n$, cut out two disks D^n, and glue on a handle $[0,1] \times S^{n-1}$, or start with $S^1 \times S^{n-1}$, that is, the product of a circle and an $(n-1)$-sphere, and remove a point. Note that we can give this manifold a metric for which the handle forms a very good shortcut between two otherwise distant points! Whether such metrics are physically possible is another matter, of course.

An interesting idea, advocated by the relativist John A. Wheeler, is that the 'mouths' of wormholes can act somewhat like charged particles. Electric field lines can flow in one mouth and out the other, so that one mouth looks like a negatively charged particle and the other looks like a positively charged one, with equal and opposite charge. If we had a theory that could describe the interaction of the wormhole metric and the electric field (or other gauge fields) flowing through it, we might be able to see that such wormholes would have various stable states, corresponding to the different generations of particles. We could even imagine calculating their masses. Unfortunately, all this is just a dream at present, because to treat phenomena accurately at very small distance scales requires quantum theory, and we have no quantum theory of gravity of the sort required to treat the dynamics of the wormhole metric. In fact, in standard general relativity, which ignores quantum effects, wormholes tend to 'pinch off' very rapidly. In what follows we will completely ignore this problem and simply treat the wormhole metric as a given. We will also completely ignore quantum theory and consider only the classical Maxwell equations that we have been discussing so far.

First, recall that if we think of S^1 as the unit circle in the $\mathbb{R}^2$, there is a closed but not exact 1-form on it that goes by the misleading name $d\theta$. This gives rise to a closed but not exact 1-form on the space $S^1 \times S^{n-1}$, which we call ω.

Exercise 103. *Work out the details. (Hint – define a map $p: S^1 \times S^{n-1} \to S^1$ corresponding to projection onto the first factor, and let the 1-form ω on $S^1 \times S^{n-1}$ be the pullback of $d\theta$ by p.)*

Wormholes

Now consider the 1-form $E = \omega$. If we draw this using the method described in Chapter 5, it looks like Figure 17, at least in the 2-dimensional case.

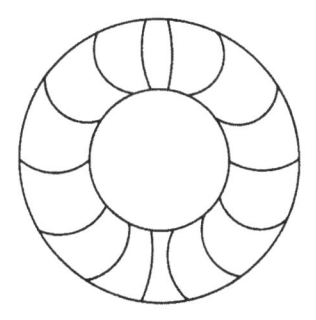

Fig. 17. Electric field on a torus — 1-form picture

Alternatively, we can use a metric to convert E into a vector field. Then it looks as follows:

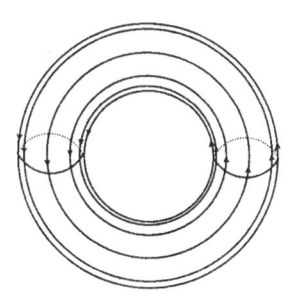

Fig. 18. Electric field on a torus — vector field picture

The arrows show how the 'electric field lines' wrap around the torus. Finally, if we remove one point from $S^1 \times S^{n-1}$ and do some stretching, we obtain our wormhole and an electric field E on it that is closed but still not exact, since its integral around a loop threading the wormhole is nonzero. This is shown in Figure 19. Each mouth of the wormhole will appear somewhat like a charged particle, and the two ends will appear to have equal and opposite charge, since electric field lines are flowing in one end and out the other.

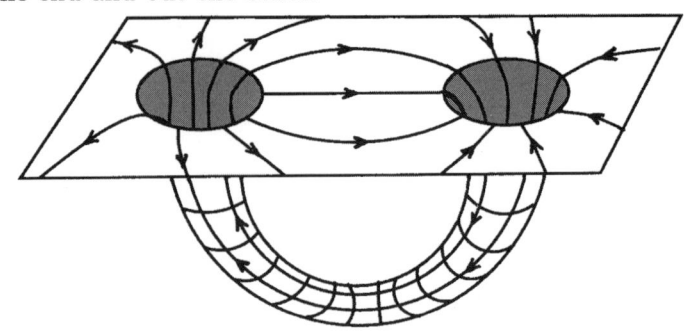

Fig. 19. Electric field on a wormhole — vector field picture

Now let us concentrate on the 3-dimensional case. We have been rather sketchy so far, since we have not specified metrics on our spaces, and the full equations of vacuum electrostatics

$$dE = d \star E = 0$$

require a metric. The equations actually work out more easily if we work with a slightly different kind of wormhole, namely one connecting two different 'universes', as in Figure 20. Here we have drawn the 2-dimensional case, which is the manifold $\mathbb{R} \times S^1$ — that is, just a cylinder, but with a funny metric on it. In 3 dimensions the corresponding space is $\mathbb{R} \times S^2$, with a metric of the form

$$g = dr^2 + f(r)^2(d\phi^2 + \sin^2 \phi \, d\theta^2)$$

By this, we simply mean that with respect to the coordinate basis of vector fields $\partial_r, \partial_\phi, \partial_\theta$, we have

$$g_{\mu\nu} = \begin{pmatrix} 1 & 0 & 0 \\ 0 & f(r)^2 & 0 \\ 0 & 0 & f(r)^2 \sin^2 \phi \end{pmatrix}.$$

The reason for this sort of notation will become clearer in Part III.

Here we are working in a modified version of spherical coordinates in which the coordinate $r \in \mathbb{R}$ ranges from 0 to $+\infty$ in this 'universe' and from 0 down to $-\infty$ in the other 'universe'. The function f should be

Wormholes

positive for all r, and it should equal r^2 for $|r|$ sufficiently large, so that each universe looks like flat Euclidean space when $|r|$ is large enough. One should attempt to visualize this space. Perhaps the easiest way is to imagine a bunch of concentric copies of S^2, starting out large for r large, narrowing down to a 'neck' of radius $f(0)$ at $r = 0$, and then becoming large again at $r \to -\infty$.

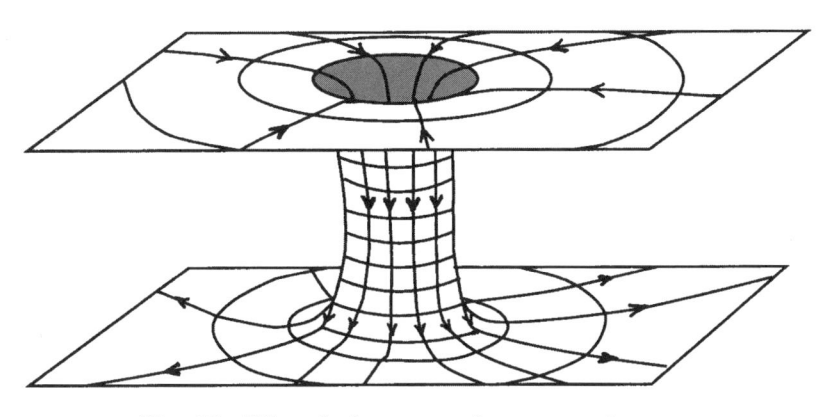

Fig. 20. Wormhole connecting two universes

An electric field flowing in one mouth of this wormhole and out the other would look like a positively charged particle in one universe and a negatively charged one in the other! We let the reader work out the details:

Exercise 104. *In the space $\mathbb{R} \times S^2$ with the metric g given above, let E be the 1-form*
$$E = e(r)dr.$$
Show that $dE = 0$ holds no matter what the function $e(r)$ is, and show that $d \star E = 0$ holds when
$$e(r) = \frac{q}{4\pi f(r)^2}.$$

Exercise 105. *Find a function ϕ with $E = -d\phi$.*

Note that now E is exact, unlike in the previous case. Heuristically, the reason is that there are no loops threading the wormhole in this case. In fact, one can show that $\mathbb{R} \times S^2$ is simply connected, so every closed 1-form is automatically exact. Second, we have

$$E = \frac{q\,dr}{4\pi r^2}$$

for $|r|$ large, so the usual inverse square law holds with the constant q playing the role of the electric charge of the wormhole. Third, the integral of $\star E$ over any 2-sphere centered about the mouth of the wormhole equals q. To do this integral, we need to pick an orientation for S^2. The two choices of orientation correspond to the two volume forms $\pm r^2 \sin\theta\, d\theta \wedge d\phi$, and we pick the standard choice, the one with the plus sign.

Exercise 106. *Let S^2 denote any of the 2-spheres of the form $\{r\} \times S^2 \subset \mathbb{R} \times S^2$, equipped with the above volume form. Show that*

$$\int_{S^2} \star E = q.$$

It should actually be no surprise that the integral of $\star E$ over a surface should measure the flow of the electric field through that surface. For recall that if we use the metric to forget the distinction between vectors and covectors, the $\star$ operator in 3 dimensions can be thought of as turning a vector into a little area element orthogonal to it, as in Figure 6 of Chapter 5. Thus the integral of $\star E$ over a surface is just a slick way of talking about the integral of the normal component of the electric field over that surface.

It is indeed quite natural to call

$$\int_{S^2} \star E$$

the charge of the wormhole. For suppose that instead of a wormhole we simply had an electric charge density ρ in a region R of Euclidean $\mathbb{R}^3$ with boundary $\partial R = S^2$. Then by Maxwell's equations and Stokes' theorem,

$$\int_{S^2} \star E = \int_R d\star E = \int_R \star \rho = \int_R \rho\, dx\, dy\, dz$$

where the final integral of the charge density over R is what we normally call the total electric charge of the region.

Wormholes 147

Now for a riddle: since electric field lines are flowing in one mouth of the wormhole and out the other, we would expect that one mouth would look positively charged and the other negatively charged. But in Exercise 106 we saw that the integral of $\star E$ over *any* 2-sphere of constant r gave the answer q. What integral gives the answer $-q$?

We will give the reader a clue: we must give S^2 an orientation in order to integrate over it. For the integral at hand, the standard orientation, given by the volume form $r^2 \sin\theta\, d\theta \wedge d\phi$ on the unit 2-sphere, gives the answer q. The opposite orientation, given by $-r^2 \sin\theta d\theta \wedge d\phi$, gives the answer $-q$. The amount of electric field flowing *in* one mouth is the same as that flowing *out* the other mouth, and it is the orientation of the 2-sphere that keeps track of this distinction between 'in' and 'out'.

Exercise 107. *With this clue, work out a careful answer to the riddle.*

It is also worth thinking about these integrals in terms of cohomology. The 2-form $\star E$ is closed, but since its integral over certain 2-spheres is nonzero it must not be exact. The fact that $\mathbb{R} \times S^2$ has closed 2-forms on it that are not exact implies that $H^2(\mathbb{R} \times S^2)$ is nonzero. Actually, using some algebraic topology one can show that while the space $\mathbb{R} \times S^2$ has $H^1 = 0$, it has $H^2 = \mathbb{R}$. In 3 dimensions, a space must have nonzero H^2 in order for there to be a surface S with $\int_S \star E \neq 0$ when $\rho = 0$. This phenomenon is what Wheeler called 'charge without charge'.

Exercise 108. *Describe how this result generalizes to spaces of other dimensions.*

Let us return to our wormhole and restrict our attention to one of the two 'universes', namely, the region $r > 0$. It is easy to see that this region, which is the manifold $(0, \infty) \times S^2$, is diffeomorphic to $\mathbb{R}^3 - \{0\}$. That is because in this region we can pretend that r, θ, and ϕ are the usual spherical coordinates on $\mathbb{R}^3 - \{0\}$. It follows that we can work with Cartesian coordinates related to these spherical coordinates by the usual formulas. Just for fun, let us calculate $\star E$ in terms of these Cartesian coordinates on the part of the $r > 0$ region where $f(r) = r$. As usual,

$$r = \sqrt{x^2 + y^2 + z^2},$$

so

$$dr = \frac{\partial r}{\partial x}dx + \frac{\partial r}{\partial y}dy + \frac{\partial r}{\partial z}dz$$
$$= \frac{xdx + ydy + zdz}{\sqrt{x^2 + y^2 + z^2}}.$$

Since $f(r) = r$, the metric looks like the usual Euclidean metric on $\mathbb{R}^3$, so the usual formulas for the Hodge star operator apply, and

$$\star E = \star \frac{qdr}{4\pi r^2}$$
$$= \frac{q \star (xdx + ydy + zdz)}{4\pi (x^2 + y^2 + z^2)^{3/2}}$$
$$= \frac{q(xdy \wedge dz + ydz \wedge dx + zdx \wedge dy)}{4\pi (x^2 + y^2 + z^2)^{3/2}}.$$

In fact, we can define a 2-form

$$\omega = \frac{xdy \wedge dz + ydz \wedge dx + zdx \wedge dy}{(x^2 + y^2 + z^2)^{3/2}}$$

on $\mathbb{R}^3 - \{0\}$, which blows up as $r \to 0$. One can, if one likes, check directly that ω is closed:

Exercise 109. *Show using Cartesian coordinates that ω is closed on $\mathbb{R}^3 - \{0\}$.*

Since the integral of ω over the unit sphere is nonzero, it is not exact. This implies that $H^2(\mathbb{R}^3 - \{0\})$ is nonzero. Using some algebraic topology, one can show that it is 1-dimensional; in other words, ω forms a basis for the closed 2-forms modulo exact 2-forms.

Clearly this 2-form ω is the natural analog in one higher dimension of the 1-form on $\mathbb{R}^2 - \{0\}$ that we discussed in the first section of this chapter, namely

$$\omega = \frac{xdy - ydx}{x^2 + y^2}.$$

This generalizes to all dimensions:

Monopoles

Exercise 110. *Generalize these examples and find an $(n-1)$-form on $\mathbb{R}^n - \{0\}$ that is closed but not exact. Conclude that $H^{n-1}(\mathbb{R}^n - \{0\})$ is nonzero.*

In fact, H^{n-1} of $\mathbb{R}^n - \{0\}$ is 1-dimensional. This makes precise the idea that $\mathbb{R}^n - \{0\}$ has a single hole in it, which is an '$(n-1)$-hole', since we can wrap an $(n-1)$-sphere around it as in Figure 10.

Monopoles

I think it's a peculiarity of myself that I like to play about with equations, just looking for beautiful mathematical relations which maybe don't have any physical meaning at all. Sometimes they do. — Paul Dirac

Just as we saw in the previous section a situation where the electric field is closed but not exact, so there is no scalar potential, there are situations, at least mathematically speaking, where the magnetic field is closed but not exact, so there is no vector potential. In physics, this sort of situation goes by the name of a magnetic monopole. Let us first give an example of this situation, and then talk a bit about the physics of it.

In fact, it is very easy for us to get a solution of the equations of magnetostatics in a vacuum:

$$dB = 0, \qquad d \star B = 0$$

where B is not exact, because in the previous section we found a solution of the equations of electrostatics in a vacuum:

$$dE = 0, \qquad d \star E = 0$$

where $\star E$ was not exact. We simply need to take $B = \star E$ — in other words, duality comes to the rescue!

Namely, suppose space is given by $\mathbb{R} \times S^2$, as in Figure 20, with the metric

$$g = dr^2 + f(r)^2(d\phi^2 + \sin^2 \phi \, d\theta^2)$$

as in the previous section. Then taking

$$B = \star \frac{m \, dr}{4\pi f(r)^2},$$

we know from the previous section that B satisfies the equations of vacuum magnetostatics but is not d of any vector potential. The same thing applies if we consider only the region $r > 0$, which gives the space $\mathbb{R}^3 - \{0\}$. We call this kind of field configuration a **magnetic monopole**, since if we do the integral

$$\int_{S^2} B = m$$

over a sphere of any radius about S^2, we get the **magnetic charge** m, unlike in ordinary $\mathbb{R}^3$, where the equation $dB = 0$ implies that B is exact, hence

$$\int_{S^2} B = 0$$

for any embedded 2-sphere.

Exercise 111. *Check this. (Hint: show that $B = (m/4\pi)\sin\phi\, d\theta \wedge d\phi$.)*

Monopoles were first seriously studied by Paul Dirac. They have not been repeatably detected, but certain 'grand unified theories' predict their existence. Perhaps the most interesting thing about them is Dirac's original argument that if *one* monopole existed, it would imply that the electric charge of all particles must be an integral multiple of a certain fundamental unit. All *free* particles have charges that are an integer multiple of the electron charge. However, quarks, which according to the standard model are 'confined' components of the proton, neutron and other hadrons (particles interacting via the strong force), have charges that are multiples of 1/3 the electron charge. Nonetheless, it appears that all particles do have charges that are integer multiples of *some* basic charge. It would be nice to find some reason for this fact.

How does Dirac's argument go? Recall the relationship between magnetic fields and the phase. Namely (and we are being sloppy again), if we drag a particle with electric charge q around a loop γ that bounds a 2-disk D embedded in space, its wavefunction is multiplied by a phase

$$e^{-\frac{i}{\hbar}q\int_D B}.$$

If there is a vector potential A with $dA = B$, this phase is clearly independent of the disk D we pick, since

$$\int_D B = \int_\gamma A.$$

But in the present situation there may be an ambiguity, since there is no vector potential. For example, say γ is a loop that goes around the equator of the unit sphere:

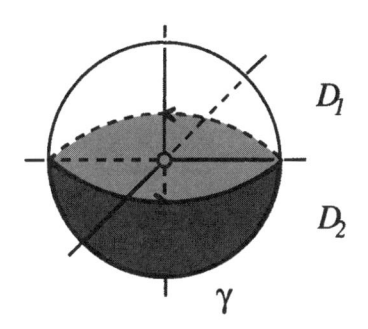

Fig. 21. Transporting a charged particle around a monopole

Then we can calculate the phase in two different ways! We can use the disk D_1, the northern hemisphere, or D_2, the southern hemisphere. By insisting that these two ways give the same answer, we will derive Dirac's result.

We have to be a bit careful about orientations. The standard orientation of the northern hemisphere is the right one to use when computing $\int_{D_1} B$, since that is the one compatible with the orientation on γ. (See the section on Stokes' theorem; the simple way to think about it is that we need the orientation such that γ runs counterclockwise around the disk.) Then

$$\int_{D_1} B = \frac{m}{4\pi} \int_0^{\pi/2} \int_0^{2\pi} \sin\phi \, d\theta \wedge d\phi = \frac{m}{2}$$

giving a phase of $\exp(-iqm/2\hbar)$. On the other hand, we need to use the opposite orientation on the southern hemisphere, so we get

$$\int_{D_2} B = -\frac{m}{4\pi} \int_{\pi/2}^{\pi} \int_0^{2\pi} \sin\phi \, d\theta \wedge d\phi = -\frac{m}{2}$$

giving a phase of $\exp(iqm/2\hbar)$. For these to be equal we need

$$e^{iqm/\hbar} = 1$$

or in other words, q must be an integer multiple of $2\pi\hbar/m$. Now Planck's original constant h was defined so that $\hbar = h/2\pi$, so we have the result that
$$q = Nh/m$$
for some integer N.

Thus, a fixed monopole charge forces quantization upon the electric charge; but we can equally well think of it the other way around. A more symmetrical way of putting it is that for any particle of electric charge q and any monopole of magnetic charge m we must have the relation
$$qm = Nh.$$
Indeed, if you compare the previous section on wormholes to the present section, it is clear that if particles were mouths of wormholes, there would be a complete symmetry between electrically charged particles and magnetic monopoles! This is, of course, nothing but duality.

However, whether or not we ever find magnetic monopoles, it is pretty clear that for some reason there is *not* much symmetry between electric and magnetic charge; the former is common while the latter, if it exists at all, is very rare. The version of Maxwell's equations in terms of F emphasizes the symmetry between magnetic and electric fields, and in 4 dimensions we can easily introduce a magnetic current J_m by analogy with the usual electric current J_e:
$$dF = J_m, \qquad \star d \star F = J_e.$$
On the other hand, the standard version in terms of the vector potential A with $dA = F$ makes $dF = 0$ a tautology and rules out the possibility of a magnetic current. Indeed, it is a generalization of the latter version, the so-called Yang-Mills equation, that describes the weak and strong forces as well as electromagnetism. However, as we shall see, there is room for monopoles even in the Yang-Mills equations if we work with 'nontrivial vector bundles'. But that is the subject of the next part!

Notes to Part I

1. Maxwell's equations

Maxwell's prescient remark appears in his book *Matter and Motion*, the first edition of which appeared in 1876, but which has been reprinted by Dover, New York, 1952. We found this quote in *Genesis of Relativity* by Loyd S. Swenson, Jr., Burt Franklin, New York, 1979, which is a short, readable history of how special relativity was born out of electromagnetism. Another interesting history of electromagnetism is *A History of the Theories of Aether and Electricity* by E. T. Whittaker, Tomash, New York, 1987. A detailed consideration of the prehistory of relativity theory can be found in *Absolute or Relative Motion?* by Julian B. Barbour, Cambridge U. Press, Cambridge, 1989. This will be of special interest to anyone interested in the philosophical aspects of relativity.

One of the best ways to get a gut feeling for Maxwell's equations is to read the second volume of *The Feynman Lectures on Physics*, by Richard P. Feynman, Robert B. Leighton and Matthew Sands, Addison-Wesley, Redwood City, 1989. The canonical text on electromagnetism, where everything is worked out in detail, is John David Jackson's *Classical Electrodynamics*, Wiley, New York, 1975.

2. Manifolds

The quote by Einstein is from 'Die Grundlage der allgemeinen Relativitätstheorie', which appeared in *Annalen der Physik* in 1916. It is reprinted in *The Principle of Relativity*, translated by W. Perrett and G. B. Jeffery, Dover, New York, 1923.

Topology is the study of topological spaces and continuous maps between them. The canonical text is *General Topology* by John L. Kelley, Van Nostrand, New York, 1955.

Differential topology is the study of smooth manifolds and smooth maps between them. A good friendly introduction is *Differential Topology* by Victor Guillemin and Alan Pollack, Prentice-Hall, Englewood Cliffs, 1974. As a reference work, *Differential Topology* by Morris W. Hirsch, Springer-Verlag, New York, 1976, is good to have on hand.

Differential geometry is closely related, but tends to emphasize various structures one can build up on smooth manifolds. Some we will be discussing are vector bundles, connections, and metrics. There are many good books on this subject. Two of our favorites are Frank W. Warner's *Foundations of Differentiable Manifolds and Lie Groups*, Springer-Verlag, New York, 1983, and *Lectures on Differential Geometry* by Shlomo Sternberg, Chelsea, New York, 1983. The canonical reference work is *Foundations of Differential Geometry*, two volumes by Shoshichi Kobayashi and Katsumi Nomizu, Interscience, New York, 1963-69.

For people interested in physics, an excellent overview of vast amounts of differential geometry and other mathematics, with applications to physics described, is *Analysis, Manifolds, and Physics*, by Yvonne Choquet-Bruhat, Cecile DeWitt-Morette, and Margaret Dillard-Bleick, North Holland, New York, 1982. A second part, by Choquet-Bruhat and Cecile DeWitt-Morette, appeared in 1989, with the same publisher. This covers many examples and somewhat more advanced topics. It is good to keep these books by ones bedside until one learns everything in them.

If the texts above are too intimidating, it might be a good idea to try *A Course in Mathematics for Students of Physics*, two volumes by Paul Bamberg and Shlomo Sternberg, Cambridge University, Cambridge, 1988-1990. This is an excellent gentle introduction to the mathematics modern physicists need.

3. Vector Fields

The quote is from Oliver Heaviside's *Electromagnetic Theory*, published in 1893, but we found it in Michael J. Crowe's *A History of Vector Analysis*, University of Notre Dame, Notre Dame, 1967. Heaviside was a fiery polemicist in favor of notation for vectors similar to that used in current undergraduate mathematics and physics courses. This notation was very close to that developed by Gibbs, and around the turn of the twentieth century there was a battle between this notation and the quaternionic notation developed by Hamilton and advocated by Tait. Vector fields are a basic aspect of differential geometry and are treated in all the texts listed in the notes

Notes to Part I 155

for Chapter 2.

4. Differential Forms

The quote from Goethe's *Faust* was translated by George Madison Priest. The quote from Bishop Berkeley is from 'The Analyst: A Discourse Addressed to an Infidel Mathematician', written in 1734, part of which is reprinted in Volume 1 of James R. Newman's *The World of Mathematics*, Simon and Schuster, New York, 1956. The quote from Grassmann is from his *Theorie der Ebbe und Flut*, in which he applied his ideas on linear algebra to physics; we found it in *A History of Vector Analysis*, as cited above. Grassmann's works were regarded as very difficult in his day, and his development of exterior algebra went almost unread, but Gibbs later cited it as influencing his ideas on vectors. The quote from Weinberg is from his book *Gravitation and Cosmology*, Wiley, New York, 1972. His book approaches general relativity in a pragmatic fashion that downplays differential geometry, indeed, he states that "the passage of time has taught us not to expect that the strong, weak, and electromagnetic interactions can be understood in geometrical terms, and too great an emphasis on geometry can only obscure the deep connections between gravitation and the rest of physics." Our own attitude is in direct contradiction to this, and is represented by the quote from Weyl's *Philosophy of Mathematics and Natural Science*, which appeared in 1949, and has been reprinted by Atheneum, New York, 1963.

There are many ways to become acquainted with differential forms. A highly readable introduction is *Differential Forms with Applications to the Physical Sciences* by Harley Flanders, Dover, New York, 1989. They are also treated in most of the texts on differential topology and differential geometry listed in the notes for Chapter 2.

5. Rewriting Maxwell's Equations

The quote from Minkowski is from his 1908 address to the 80th Assembly of German Natural Scientists and Physicians, titled 'Space and Time'; it is reprinted in English translation in *The Principle of Relativity*, Dover, 1923.

6. DeRham Theory in Electromagnetism

The quote by Faraday appears in C. N. Yang's *Selected Papers*, where Yang warns experimentalists not to be intimidated by theorists; see the notes to

Chapter 1 of Part II. The quote by Wheeler appears in *Gravitation*, as cited in the notes to Chapter 2 of Part III. The quote by Dirac is from Abraham Pais' essay 'Playing with equations, the Dirac way', in *Paul Adrien Maurice Dirac*, eds. Behram N. Kursunoglu and Eugene P. Wigner, Cambridge U. Press, Cambridge, 1987. The quote by Weyl is from his *Philosophy of Mathematics and the Natural Sciences*, trans. Olaf Helmer, Princeton U. Press, New Jersey, 1949.

DeRham cohomology can only be fully appreciated if one knows some other homology and cohomology theories. A homology theory associates to a space X a **chain complex** C_p, that is, a sequence of vector spaces (or more generally groups) together with linear maps (or homomorphisms) $d_p \colon C_p \to C_{p-1}$ satisfying $d_{p-1} d_p = 0$. The pth **homology** of the chain complex is then

$$H_p = \frac{\ker d_p}{\operatorname{im} d_{p+1}}.$$

Typically this contains topological information about the space X. The first homology theory one should learn about is probably singular homology, which is based on maps from simplices or cubes into the space X. This is discussed in many books on algebraic topology; for example, William Massey's *Singular Homology Theory*, Springer-Verlag, New York, 1980. The canonical text on algebraic topology is Edwin Spanier's *Algebraic Topology*, Springer-Verlag, New York, 1981. This may be slightly terrifying to the uninitiated.

Given any homology theory one can get a cohomology theory, that is, something that associates to X a **cochain complex** C^p, which consists of vector spaces with linear maps $d_p \colon C^p \to C^{p+1}$ satisfying $d_{p+1} d_p = 0$. The pth **cohomology** of the cochain complex is then

$$H^p = \frac{\ker d_p}{\operatorname{im} d_{p-1}}.$$

DeRham's theorem says that under certain conditions the singular cohomology of a manifold agrees with its deRham cohomology, and it is the basis of many applications of algebraic topology to geometry and physics. This is proved in the books by Flanders and Warner cited above. For a deeper look at how differential forms are used in topology see *Differential Forms in Algebraic Topology* by Raoul Bott and Loring W. Tu, Springer-Verlag, New York, 1982. For a grand tour of geometry and algebraic topology, try the 3-volume text, *Modern Geometry – Methods and Applications*, by B. A.

Notes to Part I

Dubrovin, A. T. Fomenko, and S. P. Novikov, Springer-Verlag, New York, 1990.

Stokes' theorem is proved in any good book on differential forms, such as the books by Flanders, Choquet-Bruhat *et al*, Guillemin and Pollack, or Warner, mentioned above. A nice book that treats Stokes' theorem in an elementary fashion and in great detail is *Calculus on Manifolds* by Micheal Spivak, Benjamin, New York, 1965.

The fact that a form on M is exact only if its pullback by all maps $\phi\colon S \to M$ integrate to zero is really a statement about cobordism theory with real coefficients. Cobordism theory is a cohomology theory which is based on maps from manifolds into a topological space. This is treated nicely in Volume III of the book by Dubrovin, Fomenko, and Novikov cited above.

Our favorite introduction to quantum mechanics is the third volume *The Feynman Lectures on Physics*, by Richard P. Feynman, Robert B. Leighton and Matthew Sands, Addison-Wesley, Redwood City, 1989. This needs to be supplemented by texts that work out lots of problems in detail; a very thorough introduction to quantum mechanics is A. Galindo and P. Pascual's *Quantum Mechanics*, in two volumes, Springer-Verlag, New York, 1990-1991. For more on the path-integral approach to quantum mechanics, R. P. Feynman and A. R. Hibbs' *Quantum Mechanics and Path Integrals*, McGraw-Hill, New York, 1965, is an excellent place to start. A detailed *rigorous* treatment of path integrals in quantum mechanics can be found in Barry Simon's *Functional Integration and Quantum Physics*. This requires a fair amount of competence in analysis; a good place to acquire such competence and see its relevance to physics is the 4-volume series *Methods of Modern Mathematical Physics* by Michael Reed and Barry Simon, Academic Press, New York, 1980.

It is also good to spend some time pondering the mathematical and conceptual foundations of quantum theory; for this, Josef M. Jauch's *Foundations of Quantum Mechanics*, Addison-Wesley, Reading, 1968, is a nice place to start. However, we urge the reader not to get too entangled in the endless debate about the philosophy of quantum mechanics until he or she is rather competent at using it to solve physics problems!

A brief introduction to SQUIDs appears in Chapter 21 of Feynman's first volume, but more can be found in the books on superconductivity cited in the notes for Chapter 5 of Part II. A nice place to start learning about recent work on wormholes is 'Wormholes in spacetime and their use for interstellar travel: a tool for teaching general relativity', by Michael S. Morris and Kip

S. Thorne, *Am. J. Phys.* **56** (1988), 395-412.

A good introduction to the Dirac theory of monopoles can be found in 'Magnetic monopoles, fiber bundles and gauge fields', by C. N. Yang, reprinted in his *Selected Papers*, as cited in the notes to Chapter 2 of Part II. To understand the role monopoles play in grand unified theories one must become well acquainted with gauge theory. Monopoles also play a significant role as part of pure mathematics; see for example *The Geometry and Dynamics of Magnetic Monopoles* by Michael Atiyah and Nigel Hitchin, Princeton U. Press, Princeton, 1988, as well as the book by Jaffe and Taubes cited in the notes for Chapter 5 of Part II.

Part II

Gauge Fields

Chapter 1

Symmetry

Einstein in his lifetime had toiled incessantly to construct a "complete system of theoretical physics." He searched for "the concepts and fundamental principles" that would allow for a grand synthesis of the structure of the physical world. Central to this synthesis are the forces, or interactions, that hold matter together, that produce the multitude of reactions that constitute natural phenomena.

I believe we are today still very far from this grand synthesis that Einstein dreamed about. But we do have one of its key elements: the principle that symmetry dictates interactions, first used by Einstein himself. — C. N. Yang

Lie Groups

Group theory is the study of symmetry. For mathematicians, symmetry is worth studying simply for the sake of its beauty, but symmetry is also very important in physics, because it allows us to at least partially understand situations that would otherwise be too complicated. Gauge theories are among the most beautiful, symmetrical laws of physics we know, and our current theories of electromagnetism, the strong and weak forces, and gravity are all gauge theories. The first three forces are described by a kind of gauge theory called Yang-Mills theory, a generalization of Maxwell's equations, which we describe in this part. Gravity, the odd man out, is described by a rather different sort of gauge theory, general relativity, which is the topic of Part III of this book. In this section we give a brief introduction to group theory, emphasizing the aspects used in gauge theory.

What is a group? A good example is the group of all rotations in 3-dimensional space. Given any two rotations, we can compose them — do one after another — to obtain a third. Composition is associative, but not commutative. Why? Take a book and lay it on the table facing towards one. Rotate it 90 degrees clockwise about the x axis, then 90 degrees clockwise around the y axis. Alternatively, rotate it 90 degrees clockwise about the y axis and then 90 degrees clockwise about the x axis! The results are not the same. Note also that there is a particularly boring rotation, the 'identity', which consists of rotating *not at all*. If we compose the identity with any rotation we get that rotation back again. Moreover, any given rotation has an 'inverse' rotation such that if we compose the rotation with its inverse we get the identity. For example, the inverse of rotating 90 degrees clockwise about some axis consists of rotating 90 degrees counterclockwise about that axis.

Abstracting these properties, we define a **group** G to be a set equipped with a binary operation $\cdot : G \times G \to G$, often called the **product**, an operation $^{-1}: G \to G$, called the **inverse**, and a special element $1 \in G$, called the **identity**, such that for all $g, h, k \in G$ we have

1) $(g \cdot h) \cdot k = g \cdot (h \cdot k)$.
2) $g \cdot 1 = 1 \cdot g = g$.
3) $g \cdot g^{-1} = g^{-1} \cdot g = 1$.

We usually leave out the $\cdot$ and write the product of g and h simply by gh. However, there are some groups where the product is called 'addition'. In these groups we write the product as $g + h$, the identity as 0, and the inverse of g as $-g$. For example, the real or complex numbers form a group with addition as the product.

Many of the groups useful in physics are **matrix groups,** that is, sets of matrices closed under matrix multiplication and inverse, and containing the identity matrix. For example, the group of all invertible $n \times n$ matrices with real entries is called the **general linear group** $\mathrm{GL}(n, \mathbb{R})$. Similarly, the group of all invertible $n \times n$ matrices with complex entries is denoted $\mathrm{GL}(n, \mathbb{C})$. A **subgroup** of a group is a subset closed under multiplication and inverse, and containing the identity, so

Lie Groups

we may say that a matrix group is a subgroup of $GL(n, \mathbb{R})$ or $GL(n, \mathbb{C})$. Of course, $GL(n, \mathbb{R})$ is a subgroup of $GL(n, \mathbb{C})$.

Some other important matrix groups are as follows. The **special linear group** is the set of matrices with determinant 1; we write $SL(n, \mathbb{R})$ for the $n \times n$ real matrices with determinant 1, and $SL(n, \mathbb{C})$ for the $n \times n$ complex matrices with determinant 1. We can think of $SL(n, \mathbb{R})$ as the group of all volume-preserving linear transformations of $\mathbb{R}^n$, since the Jacobian of a linear transformation is simply its determinant.

Just as it is interesting to consider groups of linear transformations that preserve volume, it is interesting to consider groups that preserve distances and angles — that is, the metric. If p and q are nonnegative integers with $p + q = n$, let g be a metric on $\mathbb{R}^n$ of signature (p, q), for example,

$$g(v, w) = v^1 w^1 + \cdots + v^p w^p - v^{p+1} w^{p+1} - \cdots - v^{p+q} w^{p+q}.$$

We define the **orthogonal group** $O(p, q)$ to be the set of $n \times n$ real matrices T that preserve g, that is, such that

$$g(Tv, Tw) = g(v, w)$$

for all $v, w \in \mathbb{R}^n$. The **special orthogonal group**, $SO(p, q)$, is set of matrices in $O(p, q)$ that also have determinant 1. If $p = n$, so that g is the usual Euclidean metric, we simply call these groups $O(n)$ and $SO(n)$. Thus $SO(3)$ is the official name for the group of all rotations in 3-dimensional Euclidean space. Note that parity, the linear transformation

$$P: (x, y, z) \mapsto (-x, -y, -z)$$

of $\mathbb{R}^3$, lies in $O(3)$ but not $SO(3)$, because it has determinant -1.

The group $SO(3, 1)$ does for Minkowski spacetime more or less what $SO(3)$ does for Euclidean space. It is called the **Lorentz group**. (More generally, we can think of any group $SO(n, 1)$ as a Lorentz group.) We prefer to think of it as the group of 4×4 matrices preserving the standard Minkowski metric

$$\eta(v, w) = -v^0 w^0 + v^1 w^1 + v^2 w^2 + v^3 w^3.$$

It contains the spatial rotations in an obvious way, but also contains the Lorentz transformations that mix up space and time coordinates:

Exercise 1. *Show that* $SO(3,1)$ *contains the Lorentz transform mixing up the t and x coordinates:*

$$\begin{pmatrix} \cosh\phi & -\sinh\phi & 0 & 0 \\ -\sinh\phi & \cosh\phi & 0 & 0 \\ 0 & 0 & 1 & 0 \\ 0 & 0 & 0 & 1 \end{pmatrix}$$

as well as Lorentz transformations mixing up t and y, or t and z coordinates.

Exercise 2. *Show that* $SO(3,1)$ *contains neither parity,*

$$P\colon (t,x,y,z) \mapsto (t,-x,-y,-z),$$

nor **time-reversal**,

$$T\colon (t,x,y,z) \mapsto (-t,x,y,z),$$

but that these lie in $O(3,1)$. *Show that the product PT lies in* $SO(3,1)$.

A very important Lie group in particle physics is the **Poincaré group**. This is the group of symmetries of Minkowski space, that is, the group of all diffeomorphisms of Minkowski space that preserve spacetime intervals. It turns out that any such diffeomorphism is a product of a translation, a Lorentz transformation, and possibly parity and/or time reversal.

The orthogonal groups have complex analogs. The most important is $U(n)$, the **unitary group**, consisting of all **unitary** $n \times n$ complex matrices, that is, those that preserve the usual inner product on $\mathbb{C}^n$ given by

$$\langle v, w \rangle = \sum_{i=1}^{n} \bar{v}^i w^i.$$

Finally, $SU(n)$, the **special unitary group**, denotes the subgroup of $U(n)$ consisting of matrices that have determinant 1. Of course, one should check the following:

Exercise 3. *Show that* $SL(n,\mathbb{R})$, $SL(n,\mathbb{C})$, $O(p,q)$, $SO(p,q)$, $U(n)$ *and* $SU(n)$ *are really matrix groups, that is, that they are closed under matrix multiplication, inverses, and contain the identity matrix.*

Lie Groups

The matrix groups we have listed above happen to be submanifolds of the the vector space of $n \times n$ matrices. Moreover, the product and inverse operations can be shown to be smooth maps, using the explicit formulas for them. Groups of this type turn out to be the most important ones in physics. We say a group G is a **Lie group** if it is a manifold, and the product and inverse operations $\cdot: G \times G \to G$ and $^{-1}: G \to G$ are smooth maps. These are named after Sophus Lie, who began their study in the 1880s; however, the modern definition came much later.

Exercise 4. *Show that the groups* $\mathrm{GL}(n, \mathbb{R})$, $\mathrm{GL}(n, \mathbb{C})$, $\mathrm{SL}(n, \mathbb{R})$, $\mathrm{SL}(n, \mathbb{C})$, $\mathrm{O}(p, q)$, $\mathrm{SO}(p, q)$, $\mathrm{U}(n)$ *and* $\mathrm{SU}(n)$ *are Lie groups. (Hint: the hardest part is to show that they are submanifolds of the space of matrices.)*

Exercise 5. *Given a Lie group G, define its* **identity component** *G_0 to be the connected component containing the identity element. Show that the identity component of any Lie group is a subgroup, and a Lie group in its own right.*

Exercise 6. *Show that every element of $\mathrm{O}(3)$ is either a rotation about some axis or a rotation about some axis followed by a reflection through some plane. Show that the former class of elements are all in the identity component of $\mathrm{O}(3)$, while the latter are not. Conclude that the identity component of $\mathrm{O}(3)$ is $\mathrm{SO}(3)$.*

Exercise 7. *Show that there is no path from the identity to the element PT in $\mathrm{SO}(3, 1)$. Show that $\mathrm{SO}(3, 1)$ has two connected components. The identity component is written $\mathrm{SO}_0(3, 1)$; we warn the reader that sometimes this group is called the Lorentz group. We prefer to call it the* **connected Lorentz group.**

Just as the notion of a vector space would be of little use without the notion of a linear map, and the notion of a manifold would be worthless without the notion of a smooth map, playing around with groups requires the idea of a 'homomorphism'. Given two groups G and H, we say a function $\rho: G \to H$ is a **homomorphism** if

$$\rho(gh) = \rho(g)\rho(h).$$

As it turns out, this automatically implies some other good things.

Exercise 8. *Show that if $\rho: G \to H$ is a homomorphism of groups, then*
$$\rho(1) = 1$$
and
$$\rho(g^{-1}) = \rho(g)^{-1}.$$
(Hint: first prove that a group only has one element with the properties of the identity element, and for each group element g there is only one element with the properties of g^{-1}.)

A homomorphism that is one-to-one and onto is called an **isomorphism**. Generally speaking one can get away with being sloppy and regarding isomorphic groups as 'the same'. For example:

Exercise 9. *A 1×1 matrix is just a number, so show that*
$$U(1) = \{e^{i\theta}: \theta \in \mathbb{R}\}.$$
In physics, an element of $U(1)$ is called a **phase**. *Show that $U(1)$ is isomorphic to $SO(2)$, with an isomorphism being given by*
$$\rho(e^{i\theta}) = \begin{pmatrix} \cos\theta & \sin\theta \\ -\sin\theta & \cos\theta \end{pmatrix}$$
(Hint: rotations of the 2-dimensional real vector space $\mathbb{R}^2$ are the same as rotations of the complex plane, $\mathbb{C}$.)

We have said that groups describe symmetries, and given one example: $SO(3)$ describes the rotational symmetries of 3-dimensional Euclidean space. In other words, $SO(3)$ 'acts' on $\mathbb{R}^3$, meaning that any element of $SO(3)$ defines a linear transformation of $\mathbb{R}^3$. More generally, we say a group G **acts** on a vector space V if there is a map ρ from G to linear transformations of V such that
$$\rho(gh)v = \rho(g)\rho(h)v$$
for all $v \in V$. We also say that ρ is a **representation** of G on V. A representation is really just a special kind of homomorphism. If we define the **general linear group** $GL(V)$ to be the group of all

Lie Groups

invertible linear transformations of V, a representation of G on V is nothing but a homomorphism

$$\rho: G \to \mathrm{GL}(V).$$

Henceforth, when G is a Lie group, we will restrict attention to representations $\rho: G \to \mathrm{GL}(V)$ where V is finite-dimensional and ρ is smooth as a map between manifolds, so that we can apply the tools of differential geometry.

The beautiful fact, which will take a while to fully explain, is that different Lie groups give different equations, called Yang-Mills equations, which describe various forces in the standard model. The group is called the 'symmetry group' or 'gauge group' of the force in question. As we will see, electromagnetism has U(1) as its gauge group. In other words, the Yang-Mills equations with gauge group U(1) are simply Maxwell's equations. What makes this case so special is that U(1) is commutative, or 'abelian' — where we say that a group G is **abelian** if

$$gh = hg$$

for all $g, h \in G$. The Yang-Mills equations are linear precisely when the gauge group is abelian! In the standard model the strong nuclear force has as its gauge group the group SU(3), which is nonabelian. This makes the strong force behave in a nonlinear manner that is far more subtle than electromagnetism. Furthermore, for any two groups G and H there is a way to cook up a group $G \times H$ called the 'direct product' of G and H, and in the standard model the electromagnetic and weak forces are treated in a unified manner, with the so-called **electroweak** force having gauge group SU(2) × U(1). This group is also nonabelian.

Exercise 10. *Given groups G and H, let $G \times H$ denote the set of ordered pairs (g, h) with $g \in G$, $h \in H$. Show that $G \times H$ becomes a group with product*

$$(g, h)(g', h') = (gg', hh'),$$

identity element

$$1 = (1, 1),$$

and inverse

$$(g, h)^{-1} = (g^{-1}, h^{-1}).$$

The group $G \times H$ is called the **direct product** *or* **direct sum** *of G and H, depending on who you talk to. (When called the direct sum, it is written $G \oplus H$.) Show that if G and H are Lie groups so is $G \times H$. Show that $G \times H$ is abelian if and only if G and H are abelian.*

The gauge group of the entire standard model is $SU(3) \times SU(2) \times U(1)$. This is a rather funny group, so there has been much work on **grand unified theories** or 'GUTs', in which this group is treated as a subgroup of a nicer group. The simplest choice is $SU(5)$. Gravity is not described by quite the same sort of 'gauge theory' as the other forces, but as we will see, there is a sense in which it can be construed as a theory with gauge group $SO_0(3,1)$ or $SL(2, \mathbb{C})$.

In the standard model every particle has a charge, and by this we mean not only the usual electric charge but also charges that determine how the particle interacts with the weak and strong nuclear forces. (The strong force charge is usually called 'color'.) The wonderful connection between group theory and charge is that the charge of a particle really just amounts to a choice of a representation for the gauge group in question. This is one of many reasons why group representations have been extensively studied and, for some groups, completely classified. The Notes contain some basic references on Lie groups and their representations; here we will only scratch the surface of this beautiful subject, concentrating on the simplest examples, namely $U(1)$ and $SU(2)$.

In order to classify group representations we need to know two representations are essentially 'the same'. In other words, we need a notion of 'equivalence' of representations. Say we have two representations

$$\rho: G \to \mathrm{GL}(V), \qquad \rho': G \to \mathrm{GL}(V').$$

Then we say they are **equivalent** if there is a one-to-one and onto linear map $T: V \to V'$ with

$$\rho'(g)T = T\rho(g)$$

for all $g \in G$.

The next thing to do is find some representations! Of course, since every matrix group is already sitting inside $GL(V)$, where V is $\mathbb{R}^n$ or $\mathbb{C}^n$, matrix groups come pre-equipped with a special representation

Lie Groups

called the **fundamental representation**. However, a group typically has lots of different representations. There are, for example, various ways to get new representations from old ones. The simplest is called taking a 'direct sum'. Let G be a group and let ρ be a representation of G on V and ρ' be a representation of G on V'. Let $\rho \oplus \rho'$, the **direct sum** of the representations ρ and ρ', be the representation of G on the direct sum $V \oplus V'$ given by

$$(\rho \oplus \rho')(g)(v, v') = (\rho(g)v, \rho'(g)v')$$

for all $v \in V$, $v' \in V$. (Recall that the direct sum $V \oplus V'$ is the space of all pairs (v, v') with $v \in V$, $v' \in V'$.)

Exercise 11. *Show that direct sum of representations is really a representation.*

A subtler way to form new representations is by taking the 'tensor product' of old ones. First let us recall the notion of the tensor product of vector spaces. A quick and dirty way to define it is as follows. Let V and V' be vector spaces. Pick a basis $\{e_i\}$ for V and a basis $\{e'_j\}$ for V'. Then the **tensor product** $V \otimes V'$ is the vector space whose basis is given by all expressions of the form $e_i \otimes e'_j$. Thus, the dimension of $V \otimes V'$ is the dimension of V times the dimension of V'. Given $v = v^i e_i \in V$ and $v' = v'^j e'_j \in V'$, we define the tensor product of v and v', written $v \otimes v'$, by

$$v \otimes v' = v^i v'^j e_i \otimes e'_j.$$

The problem with this definition is that it depends on an arbitrary choice of basis for V and V'. The remedy is to realize that the tensor product has a certain basis-independent 'universal property'. Namely, given any bilinear function

$$f: V \times V' \to W$$

to some other vector space W — that is, a function $f(\cdot, \cdot)$ that is linear in each slot — there is a unique linear function

$$F: V \otimes V' \to W$$

such that

$$f(v, v') = F(v \otimes v').$$

Exercise 12. *Prove that the above is true.*

The slick definition of tensor product of V and V' is that it is any vector space having this universal property.

Now suppose that ρ is a representation of G on V and ρ' is a representation of G on V'. Then the **tensor product** $\rho \otimes \rho'$ of the representations ρ and ρ' is the representation of G on $V \otimes V'$ given by

$$(\rho \otimes \rho')(g)(v \otimes v') = \rho(g)v \otimes \rho'(g)v'.$$

Exercise 13. *Show that this is well-defined and indeed a representation.*

The direct sum and tensor product are both recipes for making big representations out of smaller ones. Alternatively, one can look for a small representation in a big one. Namely, suppose that ρ is a representation of a group G on the vector space V. Suppose that V' is an **invariant** subspace of V, that is, if $v \in V'$ then $\rho(g)v \in V'$ for all $g \in G$. Then we can define a representation ρ' of G on V' by setting

$$\rho'(g)v = \rho(g)v$$

for all $v \in V'$. We call ρ' a **subrepresentation** of ρ.

Exercise 14. *Given two representations ρ and ρ' of G, show that ρ and ρ' are both subrepresentations of $\rho \oplus \rho'$.*

A representation ρ of a group G on a vector space V always has the subspaces $\{0\}$ and V itself as invariant subspaces. If ρ has no other invariant subspaces we say it is **irreducible**. Irreducible representations are like the elementary building blocks from which one can build up other representations. More precisely, if G is compact, every representation of G is equivalent to a direct sum of irreducible ones. In fact, the gauge groups appearing in physics are usually compact — with general relativity being an exception! — and elementary particles typically do correspond to *irreducible* representations of these groups.

To get a little feeling for this sort of thing, let us take a good look at the groups U(1) and SU(2). We begin with U(1), which is just the unit circle in the complex plane with multiplication as the group operation. Note that for any integer n, U(1) has a representation ρ_n on $\mathbb{C}$ given by

$$\rho_n(e^{i\theta})v = e^{in\theta}v.$$

Lie Groups

Exercise 15. *Check that this is indeed a representation.*

It is clear that the representations ρ_n are irreducible, since the vector space $\mathbb{C}$ has no subspaces other than $\{0\}$ and the whole space. In fact, any irreducible representation of U(1) (on a *complex* vector space) is equivalent to one of these representations ρ_n. To see this, one first notes the following:

Exercise 16. *Show that any complex 1-dimensional representation of* U(1) *is equivalent to one of the representations ρ_n.*

Then one can use a basic result called **Schur's lemma**. This says that if we have an irreducible representation $\rho\colon G \to \mathrm{GL}(V)$, any linear operator $T\colon V \to V$ that commutes with all the operators $\rho(g)$ must be a scalar multiple of the identity operator. Now, if G is abelian, any $\rho(g)$ commutes with all the rest, so in an irreducible representation ρ of G all the operators $\rho(g)$ must be multiples of the identity. This implies that every subspace of V is invariant, so the only way ρ can be irreducible is by being 1-dimensional. In short, every irreducible representation of an abelian group is 1-dimensional. The previous exercise thus means that any irreducible representation of U(1) is equivalent to one of the representations ρ_n. Since U(1) is compact, this means that all finite-dimensional representations of U(1) can be built up as direct sums of the representations ρ_n.

We have mentioned that the group U(1) is the gauge group for electromagnetism. In fact, this is the correct attitude only if we assume in advance that the electric charge of any particle is a multiple of a certain unit charge q. Then a particle with charge equal to nq transforms according to the representation ρ_n of U(1)! In fact, we have already seen a bit of this in Chapter 6 of Part I. There we saw that there is a deep relationship between electromagnetism and the *phase* of a quantum particle. Namely, if we move a quantum particle of charge nq around a loop γ in spacetime, its wavefunction is multiplied by a certain phase, or element of U(1), namely

$$e^{-\frac{i}{\hbar}nq \oint_\gamma A}$$

where A is the vector potential. We can work in units where $q = \hbar = 1$, and then, thinking of

$$e^{-i \oint_\gamma A}$$

as an element of U(1), the above phase is just

$$\rho_n(e^{-i\oint_\gamma A}).$$

The point of gauge theory is to generalize these ideas to groups G that are more interesting than U(1). There will be a thing generalizing the vector potential, called a 'connection', and if we integrate the connection around a loop in a certain way we will get an element $g \in G$ called the 'holonomy'. If we have a particle corresponding to some representation ρ of G, moving it around the loop corresponds to applying the linear transformation $\rho(g)$ to its wavefunction.

To complete our picture of U(1) and its representations, we should mention how tensor products of representations of U(1) behave:

Exercise 17. *Show that the tensor product of the representations ρ_n and ρ_m is equivalent to the representation ρ_{n+m}.*

Physically, if we have two particles corresponding to two different representations of a group, a 'bound state' in which the two particles are held together by attractive forces corresponds to the tensor product of the two representations. The previous exercise simply means that the electric charge of such a bound state is the sum of the charges of the constituents!

Now let us get a good picture of SU(2) and its representations. Recall that SU(2) consists of all 2×2 unitary complex matrices with determinant 1. We will show that just as U(1) is really the circle in disguise, the group SU(2) is really the 3-sphere! First, define the **Pauli matrices** as follows:

$$\sigma_x = \begin{pmatrix} 0 & 1 \\ 1 & 0 \end{pmatrix}, \quad \sigma_y = \begin{pmatrix} 0 & -i \\ i & 0 \end{pmatrix}, \quad \sigma_z = \begin{pmatrix} 1 & 0 \\ 0 & -1 \end{pmatrix}.$$

The matrices $\sigma_x, \sigma_y,$ and σ_z are also called $\sigma_1, \sigma_2,$ and σ_3. Together with the identity matrix, sometimes called σ_0 in this context, these form a basis for the 2×2 hermitian matrices.

Exercise 18. *Show that any 2×2 matrix may be uniquely expressed as a linear combination of Pauli matrices $\sigma_0, \ldots, \sigma_3$ with complex coefficients, and that the matrix is hermitian if and only if these coefficients are real. Show that the matrix is **traceless**, that is, its trace (sum of diagonal entries) is zero, if and only if the coefficient of σ_0 vanishes.*

Lie Groups

They also satisfy some very nice algebraic relations:

Exercise 19. *For $i = 1, 2, 3$ show that*

$$\sigma_i^2 = 1,$$

and show that if (i, j, k) is a cyclic permutation of $(1, 2, 3)$ then

$$\sigma_i \sigma_j = -\sigma_j \sigma_i = \sqrt{-1}\, \sigma_k.$$

Note that if we let

$$I = -\sqrt{-1}\,\sigma_1, \ J = -\sqrt{-1}\,\sigma_2, \ K = -\sqrt{-1}\,\sigma_3$$

the exercise above implies:

$$I^2 = J^2 = K^2 = -1,$$

$$IJ = -JI = K, \quad JK = -KJ = I, \quad KI = -IK = J.$$

The algebra

$$\mathbb{H} = \{a + bI + cJ + dK : a, b, c, d \in \mathbb{R}\},$$

with the multiplication rules given above is called the **quaternions**, and it was invented by Hamilton when he was trying to generalize the complex numbers to create algebras of higher dimension. For a while they were popular as an approach to what we now handle with vectors. (In particular, the relation to the vector cross product should be clear!) They fell out of favor around 1900, when Gibbs' vector notation took over, but in the age of quantum mechanics they were reinvented by Pauli to explain 'spin', that is, the intrinsic angular momentum of the electron and other particles. The word 'spin' comes from the relationship with rotations in 3-dimensional space, which we will clarify in a bit.

The relation between the quaternions (or Pauli matrices) and SU(2) is as follows:

$$\mathrm{SU}(2) = \{a + bI + cJ + dK : a, b, c, d \in \mathbb{R},\ a^2 + b^2 + c^2 + d^2 = 1\}$$

In other words, SU(2) is just S^3, the unit sphere in the quaternions!

Exercise 20. *Show that the determinant of the 2×2 matrix $a+bI+cJ+dK$ is $a^2+b^2+c^2+d^2$. Show that if a,b,c,d are real and $a^2+b^2+c^2+d^2=1$, this matrix is unitary. Conclude that $\mathrm{SU}(2)$ is the unit sphere in $\mathbb{H}$.*

Now let us turn to the representation theory of $\mathrm{SU}(2)$. This group has one irreducible representation (up to equivalence) of each dimension. We will construct them, which is easy to do, but we will not prove that they are irreducible, nor that they are *all* of the irreducible representations. (Any good book on representation theory will do this; see the Notes.) In physics, these representations are called the spin-0 representation, the spin-1/2 representation, the spin-1 representation, and so on, with the spin-j representation having dimension $2j+1$.

Let us denote the spin-j representation by U_j, and define it as follows. Let $\mathcal{H}_j$ be the space of polynomial functions on $\mathbb{C}^2$ that are homogeneous of degree $2j$. In other words, if we write a vector in $\mathbb{C}^2$ as a pair of complex numbers (x,y), an element of $\mathcal{H}_j$ is just a polynomial in x and y that is a linear combination of polynomials

$$f(x,y) = x^p y^q$$

where the total degree $p+q$ is $2j$. Note that $\mathcal{H}_j$ has dimension $2j+1$, since it has a basis given by

$$x^{2j},\ x^{2j-1}y,\ x^{2j-2}y^2,\ \ldots,\ y^{2j}.$$

Now, for any $g \in \mathrm{SU}(2)$, let $U_j(g)$ be the linear transformation of $\mathcal{H}_j$ given by

$$(U_j(g)f)(v) = f(g^{-1}v)$$

for all $f \in \mathcal{H}_j$ and $v \in \mathbb{C}^2$. It is easy to check that this is really a representation: $U_j(1)$ is the identity, and for any $g, h \in \mathrm{SU}(2)$ we have

$$\begin{aligned}(U_j(g)U_j(h)f)(v) &= (U_j(h)f)(g^{-1}v) = f(h^{-1}g^{-1}v) = f((gh)^{-1}v) \\ &= (U_j(gh)f)(v)\end{aligned}$$

for all $f \in \mathcal{H}_j$, $v \in \mathbb{C}^2$. (Note that the annoying inverses are really needed to make things work!)

This definition is clever, but it takes a bit of work to see its significance. To start, the reader should do the following exercises:

Lie Groups

Exercise 21. *Show that the spin-0 representation of* SU(2) *is equivalent to the* **trivial** *representation in which every element of the group acts on* $\mathbb{C}$ *as the identity.*

Exercise 22. *Show that the spin-1/2 representation of* SU(2) *is equivalent to the* **fundamental** *representation, in which every element* $g \in$ SU(2) *acts on* $\mathbb{C}^2$ *by matrix multiplication.*

Exercise 23. *Show that for any representation* ρ *of a group* G *on a vector space* V *there is a* **dual** *or* **contragredient** *representation* ρ^* *of* G *on* V^*, *given by*

$$(\rho^*(g)f)(v) = f(\rho(g^{-1})v)$$

for all $v \in V$, $f \in V^*$. *Show that that all the representations* U_j *of* SU(2) *are equivalent to their duals.*

However, it is the spin-1 representation, which is 3-dimensional, that is the most closely related to the familiar geometry of rotations in 3-dimensional space. This relationship is based on the existence of a beautiful two-to-one homomorphism

$$\rho: \text{SU}(2) \to \text{SO}(3)$$

which we will describe shortly. If we take its existence on faith for a moment, and note that SO(3) is a subgroup of GL(3, $\mathbb{C}$), we can think of ρ as a homomorphism from SU(2) to GL(3, $\mathbb{C}$), that is, a 3-dimensional complex representation of SU(2). This turns out to be equivalent to the spin-1 representation.

In fact, the homomorphism

$$\rho: \text{SU}(2) \to \text{SO}(3)$$

is the reason why physicists became so interested in SU(2) in the first place! They started out being interested in rotations in $\mathbb{R}^3$; the study of SU(2) was eventually forced on them. We can construct this homomorphism as follows. Let V be the space of traceless hermitian matrices. We can identify this space with $\mathbb{R}^3$, since by Exercise 18 any such matrix is of the form

$$T = T^1\sigma_1 + T^2\sigma_2 + T^3\sigma_3$$

with real coefficients T^i. Now if $T \in V$ and $g \in \mathrm{SU}(2)$, we have

$$\mathrm{tr}(gTg^{-1}) = \mathrm{tr}(T) = 0$$

and

$$(gTg^{-1})^* = (g^{-1})^* T^* g^* = gTg^{-1},$$

so gTg^{-1} is again in V. Let us write

$$\rho(g)T = gTg^{-1}.$$

We claim ρ is a representation of $\mathrm{SU}(2)$ on V; to check this, simply note that

$$\rho(g)\rho(h)T = ghTh^{-1}g^{-1} = \rho(gh)T$$

and

$$\rho(1)T = 1\,T\,1^{-1} = T.$$

In other words, we have a homomorphism $\rho\colon \mathrm{SU}(2) \to \mathrm{GL}(V)$. If we identify V with $\mathbb{R}^3$ using the basis of Pauli matrices, we can think of this as a homomorphism

$$\rho\colon \mathrm{SU}(2) \to \mathrm{GL}(3, \mathbb{R}).$$

In fact, we claim ρ maps $\mathrm{SU}(2)$ into $\mathrm{O}(3)$. To show this, we just need to show that the transformation $\rho(g)$ preserves the lengths of vectors in V. Note that Exercise 20 implies that $\det T$ is just minus the Euclidean length of the vector $(T^1, T^2, T^3) \in \mathbb{R}^3$. Since

$$\det(\rho(g)T) = \det(gTg^{-1}) = \det(g)\det(T)\det(g^{-1}) = \det(T),$$

it follows that $\rho(g)$ is length-preserving.

The only thing left to show is that $\rho(g)$ actually lies, not just in $\mathrm{O}(3)$, but in $\mathrm{SO}(3)$. One approach is to calculate the determinant of $\rho(g)$ by brute force and show that it equals 1. There is, however, a more enlightening way to see this using some topology, as follows. We will not make this argument rigorous, but it is not too hard. Begin by pondering how $\mathrm{SO}(3)$ sits in $\mathrm{O}(3)$. The determinant of any element of $\mathrm{O}(3)$ is either 1 or -1. By Exercise 6, every element of $\mathrm{O}(3)$ is either a rotation or a rotation followed by a reflection. The former have

Lie Groups

determinant 1 and the latter have determinant -1; each kind forms a single connected component of O(3). Thus SO(3) is the connected component of O(3) that contains the identity element — the so-called **identity component** of O(3).

On the other hand, since SU(2) is just S^3 as a manifold, SU(2) is connected. The map $\rho\colon \mathrm{SU}(2) \to \mathrm{SO}(3)$ is continuous, and it maps the identity to the identity, so it must map all of SU(2) to the identity component of O(3). Thus

$$\rho\colon \mathrm{SU}(2) \to \mathrm{SO}(3),$$

as we wanted to show.

The interesting thing is that while ρ is not an isomorphism, it is close — it is two-to-one and onto, or what one calls a **double cover**. Certainly ρ is at least two-to-one, for we always have $\rho(g) = \rho(-g)$:

$$\rho(-g)T = (-g)T(-g)^{-1} = gTg^{-1} = \rho(g)T.$$

In fact, ρ is exactly two-to-one. Suppose $\rho(g) = \rho(h)$. Then

$$\rho(gh^{-1}) = \rho(g)\rho(h)^{-1} = 1.$$

Now the only way we can have $\rho(gh^{-1}) = 1$ is if gh^{-1} commutes with all matrices $T \in V$, but this can only happen if gh^{-1} is a scalar multiple of the identity:

Exercise 24. *Show that if S is a 2×2 matrix commuting with all 2×2 traceless hermitian matrices, S is a scalar multiple of the identity matrix. (One approach is to suppose S commutes with the Pauli matrices $\sigma_1, \sigma_2, \sigma_2$ and derive equations its matrix entries must satisfy.)*

However, the only scalar multiples of the identity that lie in SU(2) are ± 1, so we must have $h = \pm g$. Thus ρ is two-to-one. We will see that ρ is onto in the next section.

Exercise 25. *Using the fact that $\mathrm{GL}(3, \mathbb{R})$ is a subgroup of $\mathrm{GL}(3, \mathbb{C})$, we can think of ρ as a homomorphism from $\mathrm{SU}(2)$ to $\mathrm{GL}(3, \mathbb{C})$, or in other words, a representation of $\mathrm{SU}(2)$ on $\mathbb{C}^3$. Show that this is equivalent to the spin-1 representation of $\mathrm{SU}(2)$.*

The physical implications of the double cover ρ are profound. In quantum theory the states of a physical system are typically described as unit vectors in some Hilbert space $\mathcal{H}$. As we mentioned in Chapter 6 of Part I, the inner product of $\mathcal{H}$ is used to calculate probabilities in quantum theory; if one prepares the system in the state represented by the unit vector ψ and does an experiment to see if it is in the state represented by ϕ, the probability one gets the answer 'yes' is

$$|\langle \phi, \psi \rangle|^2.$$

In the simplest situations in which group G acts as symmetries of the system, there is a **unitary representation** ρ of G on $\mathcal{H}$, that is, a representation for which $\rho(g)$ is unitary for all $g \in G$. The idea is that each group element $g \in G$ corresponds to some sort of operation on the states of the system, given by

$$\psi \mapsto \rho(g)\psi.$$

For example, if $G = \mathrm{SO}(3)$ and the system is a particle of some sort, the operators $\rho(g)$ describe what happens when one rotates the particle in space. The point of requiring $\rho(g)$ to be unitary is that then it preserves transition amplitudes:

$$\langle \rho(g)\phi, \rho(g)\psi \rangle = \langle \phi, \psi \rangle.$$

The point of requiring ρ to be a representation is that, for example, first rotating the particle by some amount $h \in \mathrm{SO}(3)$ and then rotating it by some amount $g \in \mathrm{SO}(3)$ should have the same effect as rotating it by the amount gh:

$$\rho(g)\rho(h) = \rho(gh).$$

However, there is a further subtlety. For many purposes, the vector $\psi \in \mathcal{H}$ is indistinguishable from any other vector $e^{i\theta}\psi$ differing from ψ by a phase $e^{i\theta} \in \mathrm{U}(1)$. This is because the transition probabilities are insensitive to the phase:

$$|\langle \phi, \psi \rangle|^2 = |\langle \phi, e^{i\theta}\psi \rangle|^2.$$

Indeed, the probability that if the system is prepared in the state ψ then it will be detected in the state $e^{i\theta}\psi$ is equal to 1:

$$|\langle \phi, e^{i\theta}\psi \rangle|^2 = 1.$$

Lie Groups

What this means is that states are not really the same as unit vectors in $\mathcal{H}$; instead, they are *equivalence classes* of unit vectors

$$[\psi] = \{e^{i\theta}\psi \colon \theta \in \mathbb{R}\}.$$

Note, however, that even if ψ and ψ' differ by a phase, it will not usually be true that $\psi + \phi$ and $\psi' + \phi$ differ by a phase, so in path-integral calculations phases are very significant — see the section on the Bohm-Aharonov effect in Chapter 6 of Part I.

Because state vectors differing only by a phase are the same for certain purposes, it turns out that symmetries need not correspond to unitary representations; they can correspond to **projective** unitary representations, in which the rules of a representation hold only 'up to phases':

$$\rho(1) = e^{i\theta},$$

$$\rho(g)\rho(h) = e^{i\theta(g,h)}\rho(gh).$$

Here θ is a fixed real number, while the **cocycle** $e^{i\theta(g,h)}$ is any function of g and h.

Exercise 26. *Show that the cocycle automatically satisfies the* **cocycle condition**

$$e^{i\theta(g,h)}e^{i\theta(gh,k)} = e^{i\theta(g,hk)}e^{i\theta(h,k)}.$$

Now, for the same reasons, one can change a projective representation ρ to another one ρ' by throwing in an extra phase without changing the physics:

$$\rho'(g) = e^{i\varphi(g)}\rho(g).$$

One then has

$$\rho'(1) = e^{i\theta'},$$

$$\rho'(g)\rho'(h) = e^{i\theta'(g,h)}\rho'(gh),$$

where

$$\theta' = \theta + \varphi(1),$$

$$\theta'(g,h) = \theta(g,h) + \varphi(g) + \varphi(h) - \varphi(gh).$$

Clearly by a suitable choice of the function φ one can make $\rho'(1) = 1$, but one cannot always obtain $\theta'(g,h) = 0$ for all $g, h \in G$. If no choice of φ makes $\theta'(g,h) = 0$ for all g, h, we say the cocycle $e^{i\theta(g,h)}$ is **essential**. What this means is that it is impossible to 'straighten out' the projective representation ρ into an actual representation.

Essential cocycles turn up quite often in quantum physics, and perhaps the most important example arises when one is studying how particles transform under rotations in 3-dimensional space. As we will see, the spin-j representation of SU(2) gives rise to a unitary representation of SO(3) when j is an integer $0, 1, 2, \ldots$ — the so-called **bosonic** case — but only a projective unitary representation when j is a 'half-integer' $1/2, 3/2, 5/2, \ldots$ — the **fermionic** case. In fact, this gives a complete list of the irreducible projective unitary representations of SO(3) (where irreducibility is defined for projective representations just as for honest ones). Every particle in nature transforms under rotations according to one of these projective representations, and the 'bosons' (named after Bose) are utterly unlike the 'fermions' (named after Fermi). In particular, all the quarks and leptons in the standard model are fermions of spin 1/2, while the force-carrying particles, or gauge fields (the photons, W, Z, and gluons) are bosons of spin 1. For this reason, the force-carrying particles are usually called 'gauge bosons'.

The difference between bosons and fermions is a spinoff of the double cover

$$\rho \colon \mathrm{SU}(2) \to \mathrm{SO}(3).$$

Recall that this maps both 1 and -1 to the identity of SO(3). Now, an element of the spin-j representation space $\mathcal{H}_j$ is a homogeneous polynomial f of degree $2j$ on $\mathbb{C}^2$. This implies that

$$(U_j(-1)f)(v) = f(-v) = (-1)^{2j} f(v)$$

so

$$U_j(-1) = (-1)^{2j}.$$

Thus when j is an integer, U_j also maps both 1 and -1 to the identity, while when j is a half-integer it does not. This gives rise to a vast difference between the bosonic and fermionic cases.

First consider the bosonic case. Here $U_j(-1) = 1$, so that

$$U_j(g) = U_j(-g)$$

Lie Groups

for all $g \in \mathrm{SU}(2)$. Since $\rho(-1) = 1$, we also have

$$\rho(g) = \rho(-g)$$

for all $g \in \mathrm{SU}(2)$. This allows us to define a unitary representation V_j of $\mathrm{SO}(3)$ as follows. For each $h \in \mathrm{SO}(3)$, pick $g \in \mathrm{SU}(2)$ with $\rho(g) = h$ — we can do this, since ρ is onto. Then define

$$V_j(h) = U_j(g).$$

Note that this is independent of our choice of g with $\rho(g) = h$! If we chose g with $\rho(g) = h$, we could also have chosen $-g$. We say that both g and $-g$ **cover** h. But $U_j(g) = U_j(-g)$, so the definition of V_j is unaffected by the choice. Note also that V_j is really a representation. If we have elements $h, h' \in \mathrm{SO}(3)$, we can pick $g, g' \in \mathrm{SU}(2)$ covering them, and it follows that $\rho(gg') = hh'$, and

$$V_j(hh') = U_j(gg') = U_j(g)U_j(g') = V_j(g)V_j(g')$$

as desired. It is easy to see that $V_j(1) = 1$, too.

The fermionic case is trickier; here we cannot construct a unitary representation of $\mathrm{SO}(3)$ as in the bosonic case, but only a *projective* one. We define the projective unitary representation V_j of $\mathrm{SO}(3)$ as follows. As before, for each $h \in \mathrm{SO}(3)$, pick an element $g \in \mathrm{SU}(2)$ covering h, and set

$$V_j(h) = U_j(g).$$

Also, if we choose such an element g, we could equally well have chosen $-g$, but this time, $V_j(h)$ is *not* independent of our choice, since $U_j(-1) = -1$, hence

$$U_j(g) = -U_j(-g).$$

Still, we go ahead and randomly make a choice. Then, if $h, h' \in \mathrm{SO}(3)$, we have chosen $g, g' \in \mathrm{SU}(2)$ with $\rho(g) = h$ and $\rho(g') = h'$. This implies $\rho(gg') = hh'$, that is, gg' covers hh'. However, $-gg'$ also covers hh', and we might have picked either gg' or $-gg'$ as our choice of element covering hh'. So all we can say is that

$$V_j(hh') = U_j(\pm gg') = \pm U_j(g)U_j(g') = \pm V_j(h)V_j(h').$$

In other words, V_j is a projective representation, with a cocycle that equals ± 1. One might object that perhaps making the choices *intelligently* would get rid of this irritating cocycle! In other words, one might hope the cocycle is inessential (in the technical sense). Unfortunately, this cocycle *is* essential.

Exercise 27. *Show this. (Hint: show that if the cocycle were inessential we would have $U_j(-1) = 1$, which is not true for j a half-integer.)*

Of course, if one dislikes projective representations, one can instead work directly with the representations U_j. This is what physicists usually do. The price one pays is that one must work not with the rotation group SO(3) but with its double cover SU(2). However, one quickly adapts, because the group SU(2) is mathematically simpler. This is an instance of a very general phenomenon. The group SO(3) is not simply connected, but its double cover SU(2) is, because it is diffeomorphic to S^3, which has no noncontractible loops in it. For a simply connected Lie group all cocycles are inessential. Given any Lie group G, there is a simply-connected 'covering group' $\tilde{G}$ equipped with an onto homomorphism $\rho: \tilde{G} \to G$. Given a projective representation U of G, the projective representation $U \circ \rho$ can always be straightened out to an honest representation of $\tilde{G}$, and we can work with *that*.

Sadly, we must part from the theory of Lie groups now, but not before noting the marvelous fact that everything we have just done generalizes from rotations in Euclidean $\mathbb{R}^3$ to Lorentz transformations. In particular, it turns out that $SL(2, \mathbb{C})$ is a double cover of the identity component of the Lorentz group, $SO_0(3, 1)$. We leave the reader to show this in a series of exercises. The fact that such a basic group, $SL(2, \mathbb{C})$, is so closely related to the structure of spacetime, can only serve as a challenge to our understanding of physics. Is this a coincidence or a clue we have still not fully understood?

Exercise 28. *Suppose that $x \in \mathbb{R}^4$. Show that $x^\mu x_\mu$ as computed using the Minkowski metric,*

$$x^\mu x_\mu = -x_0^2 + x_1^2 + x_2^2 + x_3^2,$$

is equal to minus the determinant of the matrix $x^\mu \sigma_\mu$ (which is to be understood using the Einstein summation convention).

Exercise 29. *Let M denote the space of 2×2 hermitian complex matrices, a 4-dimensional real vector space with basis given by the Pauli matrices σ_μ, $\mu = 0, 1, 2, 3$. Let ρ be the representation of $\mathrm{SL}(2, \mathbb{C})$ on M by*

$$\rho(g)T = gTg^*.$$

Using the identification M with Minkowski space given by

$$\begin{aligned} \mathbb{R}^4 &\to M \\ x &\mapsto x^\mu \sigma_\mu, \end{aligned}$$

show using the previous exercise that ρ preserves the Minkowski metric and hence defines a homomorphism

$$\rho\colon \mathrm{SL}(2, \mathbb{C}) \to \mathrm{O}(3,1).$$

Exercise 30. *Show that the range of $\rho\colon \mathrm{SL}(2, \mathbb{C}) \to \mathrm{O}(3,1)$ lies in $\mathrm{SO}_0(3,1)$.*

Exercise 31. *Show that ρ is two-to-one. In fact, ρ is also onto, so $\mathrm{SL}(2, \mathbb{C})$ is a double cover of the connected Lorentz group $\mathrm{SO}_0(3,1)$.*

Exercise 32. *Investigate the finite-dimensional representations of $\mathrm{SL}(2, \mathbb{C})$ and $\mathrm{SO}(3,1)$, copying the techniques used above for $\mathrm{SU}(2)$ and $\mathrm{SO}(3)$.*

Lie Algebras

Lie algebras are a very powerful tool for studying Lie groups. Recall that a Lie group is a manifold that is also a group, such that the group operations are smooth. It turns out that the group structure is almost completely determined by its behavior near the identity. This, in turn, can be described in terms of an operation on the tangent space of the Lie group, called the 'Lie bracket'.

To be more precise, suppose that G is a Lie group. We define the **Lie algebra** of G, often written $\mathfrak{g}$, to be the tangent space of the identity element of G. This is a vector space with the same dimension as G. A good way to think of Lie algebra elements is as tangent vectors to paths in G that start at the identity. An example of this is the physicists' notion of an 'infinitesimal rotation'. If we let γ be the path in SO(3) such that $\gamma(t)$ corresponds to a rotation by the angle t (counterclockwise) about the z axis:

$$\gamma(t) = \begin{pmatrix} \cos t & -\sin t & 0 \\ \sin t & \cos t & 0 \\ 0 & 0 & 1 \end{pmatrix}.$$

Then the tangent vector to γ as it passes through the identity can be calculated by differentiating the components of $\gamma(t)$ and setting $t = 0$:

$$\gamma'(0) = \begin{pmatrix} 0 & -1 & 0 \\ 1 & 0 & 0 \\ 0 & 0 & 0 \end{pmatrix}.$$

If we do this for rotations about the x and y axes as well, we get three matrices:

$$J_x = \begin{pmatrix} 0 & 0 & 0 \\ 0 & 0 & -1 \\ 0 & 1 & 0 \end{pmatrix}, \; J_y = \begin{pmatrix} 0 & 0 & 1 \\ 0 & 0 & 0 \\ -1 & 0 & 0 \end{pmatrix}, \; J_z = \begin{pmatrix} 0 & -1 & 0 \\ 1 & 0 & 0 \\ 0 & 0 & 0 \end{pmatrix}.$$

These are all in the Lie algebra of SO(3), written $\mathfrak{so}(3)$.

The point of these matrices in the Lie algebra, which describe 'infinitesimal rotations', is that we can obtain matrices describing actual finite rotations by exponentiating them. The **exponential** of an $n \times n$ complex matrix T is defined by the obvious sort of power series

$$\exp(T) = 1 + T + \frac{T^2}{2!} + \frac{T^3}{3!} + \cdots.$$

Exercise 33. *For analysts: show that this sum converges.*

To get the matrix describing a rotation by the angle t around the z axis, for example, we just need to calculate the exponential $\exp(tJ_z)$. To work this out note that

$$J_z^2 = \begin{pmatrix} -1 & 0 & 0 \\ 0 & -1 & 0 \\ 0 & 0 & 0 \end{pmatrix},$$

Lie Algebras

so that $J_z^3 = -J_z$, $J_z^4 = -J_z^2$, and so on. Thus

$$\begin{aligned}
\exp(tJ_z) &= 1 + tJ_z + \frac{t^2}{2!}J_z^2 - \frac{t^3}{3!}J_z - \frac{t^4}{4!}J_z^2 + \cdots \\
&= 1 + (t - \frac{t^3}{3!} + \cdots)J_z + (\frac{t^2}{2!} - \frac{t^4}{4!} - \cdots)J_z^2 \\
&= 1 + \sin t\, J_z + (1 - \cos t)J_z^2 \\
&= \begin{pmatrix} \cos t & -\sin t & 0 \\ \sin t & \cos t & 0 \\ 0 & 0 & 1 \end{pmatrix},
\end{aligned}$$

which is a counterclockwise rotation by the angle t about the z axis. More generally:

Exercise 34. *Show that the matrix describing a counterclockwise rotation of angle t about the unit vector $n = (n^x, n^y, n^z) \in \mathbb{R}^3$ is given by*

$$\exp t(n^x J_x + n^y J_y + n^z J_z).$$

We have already noted earlier that a rotation about the x axis usually does not commute with a rotation about the y axis. This manifests itself 'infinitesimally' in the failure of the matrices J_x and J_y to commute. Suppose, for example, that we rotate by the angle s around the x axis, then t around the y axis. This is typically not the same as rotating by t around the y axis and then s around the x axis. Thus typically

$$\exp(sJ_x)\exp(tJ_y) \neq \exp(tJ_y)\exp(sJ_x).$$

We can visualize this noncommutativity abstractly as follows:

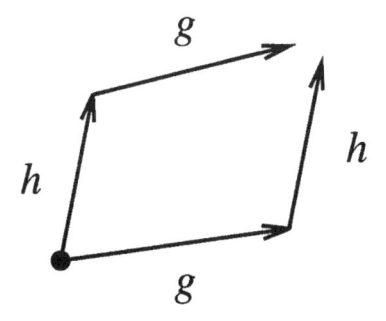

Fig. 1. Noncommutativity: $gh \neq hg$

If we consider the difference

$$\exp(sJ_x)\exp(tJ_y) - \exp(tJ_y)\exp(sJ_x)$$

and expand it as power series in s and t, keeping only the lowest-order terms, we obtain

$$st(J_xJ_y - J_yJ_x) + \text{higher order terms in } s \text{ and } t.$$

Exercise 35. *Check this!*

Note the important fact that the *first-order terms vanish*. Moreover, we did not use any special properties of J_x or J_y to have this happen! Also note that the second-order term measures how much J_x and J_y fail to commute. In general, given two $n \times n$ matrices S and T, or any elements of any algebra, we define their **commutator** or **Lie bracket**, written $[S,T]$, by

$$[S,T] = ST - TS.$$

In other words, the noncommutativity of the group is reflected on the 'infinitesimal' level by the noncommutativity of the Lie algebra. We will explain this in more detail shortly. For now, we simply urge the reader to calculate some of these commutators.

Exercise 36. *Show that*

$$J_x^2 = J_y^2 = J_z^2 = -1$$

and

$$[J_x, J_y] = J_z,\ [J_y, J_z] = J_x,\ [J_z, J_x] = J_y.$$

Note the resemblance to vector cross products and quaternions, but also the differences.

Now let us turn to the problem of describing the Lie algebras of some of the famous matrix groups we introduced in the previous section. Let us start by figuring out what the Lie algebra $\mathfrak{so}(n)$ of the group $SO(n)$ looks like, for any n. Let γ be a path in $SO(n)$ with $\gamma(0) = 1$. Then for any vectors $v, w \in \mathbb{R}^n$,

$$\langle \gamma(t)v, \gamma(t)w \rangle = \langle v, w \rangle$$

Lie Algebras

for all t, where $\langle \cdot, \cdot \rangle$ is the Euclidean inner product. Differentiating this relationship and setting $t = 0$ we get

$$\langle v, \gamma'(0)w \rangle + \langle \gamma'(0)v, w \rangle = 0.$$

If we think of γ as a path in the space of $n \times n$ matrices, so that $T = \gamma'(0)$ is an $n \times n$ matrix, this implies that the matrix entries T_{ij} satisfy

$$T_{ij} + T_{ji} = 0$$

for all i, j. In this situation we say that T is **skew-adjoint**. In short, elements of $\mathfrak{so}(n)$ must be skew-adjoint.

Conversely, if T is a skew-adjoint real $n \times n$ matrix, let

$$\gamma(t) = \exp(tT).$$

It is not too hard to show that $\gamma(t)$ is really a smooth path in the space of $n \times n$ matrices. Moreover, by a power series computation we obtain

$$\begin{aligned}
\gamma'(t) &= \sum_{n=0}^{\infty} \frac{d}{dt} t^n T^n / n! \\
&= \sum_{n=1}^{\infty} t^{n-1} T^n / (n-1)! \\
&= T\gamma(t).
\end{aligned}$$

It follows that for any $v, w \in \mathbb{R}^n$ we have

$$\begin{aligned}
\frac{d}{dt} \langle \gamma(t)v, \gamma(t)w \rangle &= \langle \gamma(t)v, \gamma'(t)w \rangle + \langle \gamma'(t)v, \gamma(t)w \rangle \\
&= \langle \gamma(t)v, T\gamma(t)w \rangle + \langle T\gamma(t)v, \gamma(t)w \rangle \\
&= 0
\end{aligned}$$

where in the last step we use the skew-adjointness of T. It follows that $\gamma(t)$ preserves inner products for all t, that is, $\gamma(t)$ is a path in $O(n)$. Now, when $t = 0$, $\gamma(t)$ is the identity matrix, so its determinant is 1. Moreover, the determinant of $\gamma(t)$ must vary continuously with t, but determinant of a matrix in $O(n)$ must be either 1 or -1, so the determinant of $\gamma(t)$ must equal 1 for all t. Thus $\gamma(t)$ is actually a path in $SO(n)$. This implies that its tangent vector at the identity,

$$\gamma'(0) = T,$$

lies in $\mathfrak{so}(n)$. Thus we have shown that $\mathfrak{so}(n)$ consists precisely of the skew-adjoint $n \times n$ real matrices.

The methods we have used to see this allow us to determine the Lie algebras of the other matrix groups discussed in the previous section. We urge the reader to do the following exercises, or at least to learn the answers by heart:

Exercise 37. *Suppose T is any $n \times n$ complex matrix. Show that*

$$\exp((s+t)T) = \exp(sT)\exp(tT)$$

by a power series calculation. (Hint: use the binomial theorem.) Show that for a fixed T, $\exp(tT)$ is a smooth function from $t \in \mathbb{R}$ to the $n \times n$ matrices. Show that $\exp(tT)$ is the identity when $t = 0$ and that

$$\left.\frac{d}{dt}\exp(tT)\right|_{t=0} = T.$$

Exercise 38. *Show that the Lie algebra $\mathfrak{gl}(n,\mathbb{C})$ of $\mathrm{GL}(n,\mathbb{C})$ consists of all $n \times n$ complex matrices. Show that the Lie algebra $\mathfrak{gl}(n,\mathbb{R})$ of $\mathrm{GL}(n,\mathbb{R})$ consists of all $n \times n$ real matrices.*

Exercise 39. *Show that for any matrix T,*

$$\det(\exp(T)) = e^{\mathrm{tr}(T)}.$$

(Hint: first show it for diagonalizable matrices, then use the fact that these are dense in the space of all matrices.) Use this to show that the Lie algebra $\mathfrak{sl}(n,\mathbb{C})$ of $\mathrm{SL}(n,\mathbb{C})$ consists of all $n \times n$ traceless complex matrices, while the Lie algebra $\mathfrak{sl}(n,\mathbb{R})$ of $\mathrm{SL}(n,\mathbb{R})$ consists of all $n \times n$ traceless real matrices.

Exercise 40. *Let g be a metric of signature (p,q) on $\mathbb{R}^n$, where $p+q = n$. Show that the Lie algebra $\mathfrak{so}(p,q)$ of $\mathrm{SO}(p,q)$ consists of all $n \times n$ real matrices T with*

$$g(Tv, w) = -g(v, Tw)$$

for all $v, w \in \mathbb{R}^n$. Show that the dimension of $\mathfrak{so}(p,q)$, hence that of $\mathrm{SO}(p,q)$, is $n(n-1)/2$. Determine an explicit basis of the Lorentz Lie algebra, $\mathfrak{so}(3,1)$.

Lie Algebras

Exercise 41. *Show that the Lie algebra* $\mathfrak{u}(n)$ *of* $\mathrm{U}(n)$ *consists of all skew-adjoint complex* $n \times n$ *matrices, that is, matrices* T *with*

$$T_{ij} = -\overline{T}_{ji}.$$

In particular, show that $\mathfrak{u}(1)$ *consists of the purely imaginary complex numbers:*

$$\mathfrak{u}(1) = \{ix \colon x \in \mathbb{R}\}.$$

Show that the Lie algebra $\mathfrak{su}(n)$ *of* $\mathrm{SU}(n)$ *consists of all traceless skew-adjoint complex* $n \times n$ *matrices.*

Though we will not prove it here, it is important to know that any Lie group G, not just the matrix groups we have considered here, has an **exponential map**. This is a (smooth) map

$$\exp \colon \mathfrak{g} \to G$$

uniquely determined by the following properties:

1) $\exp(0)$ is the identity element of G.
2) $\exp(sx)\exp(tx) = \exp((s+t)x)$ for all $x \in \mathfrak{g}$ and $s,t \in \mathbb{R}$.
3) $\dfrac{d}{dt}\exp(tx)\Big|_{t=0} = x$.

By these properties and the inverse function theorem, one can show that exp maps any sufficiently small open set containing $0 \in \mathfrak{g}$ onto an open set containing the identity of G. Using some topology, it follows that any element of the identity component of G is the product of elements of the form $\exp(x)$. (In fact, for a compact group every element in the identity component is of the form $\exp(x)$.) This makes it plausible that in some sense most of the structure of a Lie group is encoded in that of its Lie algebra. That is indeed true, and in what follows we will explain how — leaving out all the difficult steps! (See the Notes for references.)

Since the Lie algebra is a tangent space of the Lie group, it is natural to obtain operations on the Lie algebra from those on the group by *differentiation*. Consider first the operation of taking inverses. If one considers a path γ in G with $\gamma(0) = 1$, the inverse operation in G gives a new path $\gamma(t)^{-1}$. We can relate the derivative of $\gamma(t)$ at $t=0$ to that of $\gamma(t)^{-1}$, and the result is

$$\frac{d}{dt}\gamma(t)\Big|_{t=0} = -\frac{d}{dt}\gamma(t)^{-1}\Big|_{t=0}.$$

Exercise 42. *Show this for G a matrix Lie group by differentiating*

$$\gamma(t)\gamma(t)^{-1} = 1$$

with respect to t, using the product rule.

A more concise way of saying this is that the differential of the inverse map $^{-1}: G \to G$ at the identity of G is the linear map

$$\mathfrak{g} \to \mathfrak{g}$$
$$x \mapsto -x$$

In short, inverses in G correspond to negatives in $\mathfrak{g}$. Similarly, the differential of the product map $\cdot : G \times G \to G$ at the identity element of $G \times G$ is just the linear map

$$\mathfrak{g} \oplus \mathfrak{g} \to \mathfrak{g}$$
$$(x, y) \mapsto x + y.$$

That is, multiplication in G corresponds to addition in $\mathfrak{g}$. The reader should check this fact, at least for matrix Lie groups:

Exercise 43. *If G is a matrix Lie group and γ, η are paths in G with $\gamma(0) = \eta(0) = 1$, show that*

$$\frac{d}{dt}\gamma(t)\eta(t)\Big|_{t=0} = \frac{d}{dt}\gamma(t)\Big|_{t=0} + \frac{d}{dt}\eta(t)\Big|_{t=0}.$$

Conclude that the differential of $\cdot : G \times G \to G$ at $(1,1) \in G \times G$ is the addition map from $\mathfrak{g} \oplus \mathfrak{g}$ to $\mathfrak{g}$.

These facts are nice, but they do not really show how the *interesting* aspects of a Lie group are encoded in its Lie algebra! For that, we must turn to the Lie bracket, or commutator. We have already seen in a special case that the Lie bracket in $\mathfrak{g}$ is related to the group commutator

Lie Algebras

in G. Here it is convenient to consider curves $\exp(tv)$ and $\exp(sw)$ in G, where $v, w \in \mathfrak{g}$. As we saw earlier in the special case of $SO(3)$, if G is a matrix Lie group then

$$\exp(sv)\exp(tw) - \exp(tw)\exp(sv) = st[v, w] + \text{higher order terms},$$

so that

$$[v, w] = \frac{d^2}{ds\, dt}\left(\exp(sv)\exp(tw) - \exp(tw)\exp(sv)\right)\Big|_{s,t=0}.$$

In fact, given *any* Lie group G, not necessarily a matrix group, one can define a Lie bracket operation on its Lie algebra $\mathfrak{g}$. In this case, however, it makes no sense to subtract two group elements, so one needs to be more clever; we explain this below.

In the case of matrix Lie groups, where the Lie algebra again consists of matrices and the Lie bracket $[v, w]$ is simply the commutator $vw - wv$, it is easy to check the following identities:

1) $[v, w] = -[w, v]$ for all $v, w \in \mathfrak{g}$.
2) $[u, \alpha v + \beta w] = \alpha[u, v] + \beta[u, w]$ for all $u, v, w \in \mathfrak{g}$ and scalars α, β.
3) The **Jacobi identity**: $[u, [v, w]] + [v, [w, u]] + [w, [u, v]] = 0$ for all $u, v, w \in \mathfrak{g}$.

Exercise 44. *Check these. Note that in 2), the term 'scalars' means real numbers if $\mathfrak{g}$ is a real vector space, but complex numbers if $\mathfrak{g}$ is a complex vector space.*

In fact, we can quite abstractly define a **Lie algebra** to be any vector space $\mathfrak{g}$ equipped with a map $[\cdot, \cdot]\colon \mathfrak{g} \times \mathfrak{g} \to \mathfrak{g}$ such that the identities 1) - 3) hold. Just as with Lie groups, there is a notion of **homomorphism** from one Lie algebra to another; it is simply a linear map f from the Lie algebra $\mathfrak{g}$ to the Lie algebra $\mathfrak{h}$ such that

$$f([v, w]) = [f(v), f(w)]$$

for all $v, w \in \mathfrak{g}$. When f is one-to-one and onto, it is called an **isomorphism**; two isomorphic Lie algebras can be regarded as the same for most practical purposes. Perhaps the most important example of a Lie algebra isomorphism is the following:

Exercise 45. *Show that the Lie algebras* $\mathfrak{su}(2)$ *and* $\mathfrak{so}(3)$ *are isomorphic as follows. First show that* $\mathfrak{su}(2)$ *has as a basis the quaternions* I, J, K, *or in other words, the matrices* $-i\sigma_1, -i\sigma_2, -i\sigma_3$. *Then show that the linear map* $f: \mathfrak{su}(2) \to \mathfrak{so}(3)$ *given by*

$$-\frac{i}{2}\sigma_j \mapsto J_j$$

is a Lie algebra isomorphism.

As we shall see, this isomorphism is the 'infinitesimal' version of the homomorphism $\rho: \mathrm{SU}(2) \to \mathrm{SO}(3)$ discussed in the previous section.

The astute reader will note that everything we are doing is very reminiscent of Chapter 3 of Part I, where we introduced the Lie bracket of vector fields. The space $\mathrm{Vect}(M)$ of all vector fields on the manifold M is indeed a Lie algebra, where we define the Lie bracket by

$$[v, w](f) = v(w(f)) - w(v(f)).$$

It is, however, an *infinite-dimensional* Lie algebra, unlike the ones we have just been considering. The reader should also compare Figure 1 of this chapter to Figure 10 of Chapter 3, Part I. The resemblance is no coincidence! Of course, nothing in mathematics ever is, but here it is worthwhile to make the connection explicit.

We have been paying a lot of attention to the tangent space of the identity of a Lie group, but a Lie group is so symmetrical that every tangent space looks just like every other one. Let us make that more precise. Fixing any element g of a Lie group G, there is a map from G to itself given by

$$h \mapsto gh,$$

called **left multiplication** by g, and written $L_g: G \to G$. Since this map has an inverse, namely left multiplication by g^{-1}, it is a diffeomorphism. This means that we can push forward any vector field on G by L_g:

$$(L_g)_*: \mathrm{Vect}(G) \to \mathrm{Vect}(G).$$

We say that a vector field v on G is **left-invariant** if $(L_g)_* v = v$ for all $g \in G$. Since the pushforward $(L_g)_*$ is linear, the left-invariant vector fields on G form a vector subspace of $\mathrm{Vect}(G)$. More interestingly, they form a **Lie subalgebra**, that is, a subspace closed under the Lie bracket:

Lie Algebras 193

Exercise 46. *Let M be any manifold and $v, w \in \text{Vect}(M)$. Let ϕ be a diffeomorphism of M. Show that*

$$\phi_*[v, w] = [\phi_* v, \phi_* w].$$

Conclude that if v, w are two left-invariant vector fields on a Lie group, so is $[v, w]$.

The wonderful thing is that the space of left-invariant vector fields on G is isomorphic to the Lie algebra $\mathfrak{g}$ of G. First, given any vector $v_1 \in \mathfrak{g}$ — which is just a tangent vector at $1 \in G$ — we can obtain a left-invariant vector field as follows:

$$v_g = (L_g)_* v_1.$$

One can show that v_g depends smoothly on G, so there is really a vector field v on G whose value at any point $g \in G$ is v_g. Moreover, the vector field v is left-invariant! To see this, we need to show that for any $g \in G$

$$(L_g)_* v = v,$$

or in other words, for any $h \in G$,

$$(L_g)_* v_h = v_{L_g h},$$

that is,

$$(L_g)_* v_h = v_{gh}.$$

To see this, we just use the definition of v:

$$\begin{aligned}
(L_g)_* v_h &= (L_g)_* (L_h)_* v_1 \\
&= (L_g L_h)_* v_1 \\
&= (L_{gh})_* v_1 \\
&= v_{gh}.
\end{aligned}$$

Note that in the next-to-last step we have used the fact that $L_g L_h = L_{gh}$, which is easy to check. Conversely, given any left-invariant vector field v on G, we can take its value at the identity of G, obtaining a vector in $\mathfrak{g}$. Thus the space of left-invariant vector fields G is isomorphic as a vector space to $\mathfrak{g}$.

Now we have a way to endow g with a Lie bracket even when G is not a matrix group: simply use the above isomorphism to transfer the Lie bracket of left-invariant vector fields on G over to g. Of course, we should check that when G happens to be a matrix group this definition agrees with our previous one. We leave this to the reader:

Exercise 47. *Let G be a matrix Lie group. Let v be a left-invariant vector field on G and $v_1 \in \mathfrak{g}$ its value at the identity. Let $\phi_t: G \to G$ be given by*

$$\phi_t(g) = g \exp(tv_1).$$

Show that ϕ_t is the flow generated by v, that is, that

$$\left.\frac{d}{dt}\phi_t(g)\right|_{t=0} = v_g$$

for all $g \in G$.

Exercise 48. *Let G be a matrix Lie group and $\mathfrak{g}$ its Lie algebra. Let u_1, v_1, and $w_1 = [u_1, v_1]$ be elements of $\mathfrak{g}$, and let $u, v,$ and w be the corresponding left-invariant vector fields on G. Show that $[u, v] = w$, so that $\mathfrak{g}$ and the left-invariant vector fields on G are isomorphic as Lie algebras. (Hint: use the previous exercise, and if necessary, review the material on flows in Chapter 3 of Part I.)*

Henceforth, we will allow ourselves to define the Lie algebra g of a Lie group G as *either* the tangent space of G at the identity, *or* as the space of left-invariant vector fields on G.

There is much more to say about the relationship between Lie groups and Lie algebras, but we will only touch upon a few more basic points before turning to the real subject of Part II, gauge theory. It is crucial to note that just as every Lie group has a Lie algebra, every homomorphism $\rho: G \to H$ between Lie groups determines a corresponding homomorphism $d\rho: \mathfrak{g} \to \mathfrak{h}$ between their Lie algebras. This simply amounts to pushing forward tangent vectors at the identity of G:

$$d\rho = (\rho)_*: T_1 G \to T_1 H.$$

Exercise 49. *Show that this is a Lie algebra homomorphism.*

Lie Algebras

For example, suppose $\rho: \mathrm{SU}(2) \to SO(3)$ is the two-to-one homomorphism of the previous section. Let us get our hands dirty and work out the corresponding homomorphism $d\rho: \mathfrak{su}(2) \to \mathfrak{so}(3)$ explicitly, and show

$$d\rho(-i\sigma_j/2) = J_j$$

In the process, we will show that ρ is really onto, as claimed.

First we will work out the exponential

$$g_t = \exp(-it\sigma_3/2) \in \mathrm{SU}(2),$$

then we work out $\rho(g_t) \in SO(3)$, and finally we will differentiate with respect to t and set $t = 0$ to get $d\rho(-i\sigma_3/2) \in \mathfrak{so}(3)$, which we claim is the 3×3 matrix J_3. Rolling up our sleeves, we first note that since

$$\sigma_3 = \begin{pmatrix} 1 & 0 \\ 0 & -1 \end{pmatrix},$$

we have

$$g_t = \begin{pmatrix} e^{-it/2} & 0 \\ 0 & e^{it/2} \end{pmatrix}.$$

Now let us work out $\rho(g_t)$. To do this, we need to work out each

$$\rho(g_t)\sigma_j = g_t \sigma_j g_t^{-1} \qquad j = 1, 2, 3,$$

as a linear combination of all three σ's, and summarize the results as a 3×3 matrix. We have

$$\begin{aligned} \rho(g_t)\sigma_1 &= \begin{pmatrix} e^{-it/2} & 0 \\ 0 & e^{it/2} \end{pmatrix} \begin{pmatrix} 0 & 1 \\ 1 & 0 \end{pmatrix} \begin{pmatrix} e^{it/2} & 0 \\ 0 & e^{-it/2} \end{pmatrix} \\ &= \begin{pmatrix} 0 & e^{-it} \\ e^{it} & 0 \end{pmatrix} \\ &= \cos t\, \sigma_1 + \sin t\, \sigma_2 \end{aligned}$$

Similarly — now the reader will need to lend a hand — we have

$$\rho(g_t)\sigma_2 = -\sin t\, \sigma_1 + \cos t\, \sigma_2, \qquad \rho(g_t)\sigma_3 = \sigma_3.$$

Exercise 50. *Do these calculations.*

Thus, if we express $\rho(g_t)$ as a linear transformation of the space spanned by the σ_j, and write it as a matrix in the σ_j basis, we have

$$\rho(g_t) = \begin{pmatrix} \cos t & -\sin t & 0 \\ \sin t & \cos t & 0 \\ 0 & 0 & 1 \end{pmatrix}.$$

In other words, ρ takes $g_t = \exp(-it\sigma_3)$ to a rotation of angle t about the z axis. With more work, one can do the following:

Exercise 51. *Show that $\rho(\exp(-it\sigma_1/2)$ is a rotation of angle t about the x axis, and $\rho(\exp(-it\sigma_2/2)$ is a rotation of angle t about the y axis.*

There are, of course, more conceptual and less computational ways to establish these results, but it is occaisionally good to check that matrix multiplication really works as advertised.

As an immediate consequence we see that ρ is onto, since any rotation in SO(3) can be obtained as a product of rotations about the x, y, and z axes. Moreover, we can differentiate these formulas with respect to t and set $t = 0$ to see that

$$d\rho(-i\sigma_j/2) = J_j,$$

as desired. It is thus a consequence of Exercise 45 that $d\rho$ is a Lie algebra isomorphism, although this also follows from the fact that it is a homomorphism (by general theory) and one-to-one and onto (by inspection).

At this point it is worth mentioning some very powerful theorems relating Lie groups and Lie algebras. Not only does every Lie group G determine a Lie algebra $\mathfrak{g}$, but for every Lie algebra $\mathfrak{g}$ there is a Lie group G having $\mathfrak{g}$ as its Lie algebra. The Lie group G is not unique, but there is a unique such G (up to isomorphism) that is connected and simply connected. All other connected Lie groups having $\mathfrak{g}$ as Lie algebra are **covered** by G, meaning that there is a Lie group homomorphism ρ from G onto them such that $d\rho$ is an isomorphism. For example, if $\mathfrak{g} = \mathfrak{su}(2) \cong \mathfrak{so}(3)$, both SU(2) and SO(3) are connected Lie groups with Lie algebra $\mathfrak{g}$; SU(2) is also simply connected, and covers SO(3).

Lie Algebras

We conclude this chapter by describing how Lie group representations determine Lie algebra representations and vice versa. Recall that a representation of a Lie group G on V is nothing but a homomorphism $\rho\colon G \to \mathrm{GL}(V)$. Similarly, we define a **representation** of a Lie algebra $\mathfrak{g}$ on a vector space V to be a Lie algebra homomorphism $f\colon \mathfrak{g} \to \mathfrak{gl}(V)$, where $\mathfrak{gl}(V)$ is the Lie algebra of all linear operators on V, with the usual commutator. From our earlier results, if we have a representation $\rho\colon G \to \mathrm{GL}(V)$ we can differentiate it to obtain a representation $d\rho\colon \mathfrak{g} \to \mathfrak{gl}(V)$. A deeper result says that if we start with a representation $f\colon \mathfrak{g} \to \mathfrak{gl}(V)$, we can 'exponentiate' it and obtain a representation $\rho\colon G \to \mathrm{GL}(V)$ with $d\rho = f$, given that G is simply connected.

The relation between Lie group and Lie algebra representations is very important in physics. For example, suppose we have a quantum system described by a Hilbert space $\mathcal{H}$. As we have seen, if this system possesses symmetry under rotations in 3-dimensional space, there is a unitary representation U of $\mathrm{SU}(2)$ on $\mathcal{H}$. This gives rise to a representation dU of $\mathfrak{su}(2)$ on $\mathcal{H}$. We can define $dU(ix) = i\,dU(x)$ for any $x \in \mathfrak{su}(2)$, and then the operator $dU(\sigma_z/2)$ on $\mathcal{H}$ called the **angular momentum** about the z axis. What this means is that for any state vector $\psi \in \mathcal{H}$, the expected value — or average observed value — of the z-component of the system's angular momentum about that axis is given by
$$\langle \psi, dU(\sigma_z/2)\psi \rangle.$$
In a similar manner, the angular momentum about some unit vector $v \in \mathbb{R}^3$ corresponds to the operator
$$dU(v^i \sigma_i/2).$$

Other symmetry groups give other observables in a similar manner: translation in space gives momentum, translation in time gives energy, and so on.

Exercise 52. *Show that in the spin-1/2 representation of* $\mathrm{SU}(2)$, *the expected value of the angular momentum about the z axis in the so-called* **spin-up state**,
$$\uparrow = \begin{pmatrix} 1 \\ 0 \end{pmatrix},$$

is 1/2, while in the **spin-down state**,

$$\downarrow = \begin{pmatrix} 0 \\ 1 \end{pmatrix},$$

it is $-1/2$. Similarly, compute the expected value of the angular momentum about the y and z axes in these states.

The relation between Lie group and Lie algebra representations is also very useful in classifying Lie group representations, basically because vector spaces are easier to deal with than manifolds. The 20th century has seen a tremendous amount of work on Lie algebra representation theory, starting with Lie, Killing, Cartan, and Weyl, and continued by Wigner, Harish-Chandra, and many other researchers. Most of the work has gone into the study of **semisimple** Lie algebras, that is, those such that every element is a linear combination of the Lie brackets of other elements. This includes many of the cases of greatest interest in physics; for example, $\mathfrak{sl}(n, \mathbb{R})$, $\mathfrak{sl}(n, \mathbb{C})$, $\mathfrak{so}(p, q)$, and $\mathfrak{su}(n)$ are all semisimple (except for a few low-dimensional cases that are easy to handle separately). This theory is part of the toolkit of every good mathematician or theoretical physicist, but we will not go into it further here, instead referring the reader to the Notes.

Exercise 53. *Show that $\mathfrak{sl}(n, \mathbb{R})$, $\mathfrak{sl}(n, \mathbb{C})$, $\mathfrak{so}(p, q)$, and $\mathfrak{su}(n)$ are semisimple, except for certain low-dimensional cases, which you should determine.*

Exercise 54. *Show that if $\mathfrak{g}$ and $\mathfrak{h}$ are Lie algebras, so is the direct sum $\mathfrak{g} \oplus \mathfrak{h}$, with bracket given by*

$$[(x, x'), (y, y')] = ([x, y], [x', y']).$$

Show that if G and H are Lie groups with Lie algebras $\mathfrak{g}$ and $\mathfrak{h}$, the Lie algebra of $G \times H$ is isomorphic to $\mathfrak{g} \oplus \mathfrak{h}$. Show that if $\mathfrak{g}$ and $\mathfrak{h}$ are semisimple, so is $\mathfrak{g} \oplus \mathfrak{h}$

Chapter 2
Bundles and Connections

Maxwell's equations and the principles of quantum mechanics led to the idea of gauge invariance. Attempts to generalize this idea, motivated by physical concepts of phases, symmetry, and conservation laws, led to the theory of non-Abelian gauge fields. That non-Abelian gauge fields are conceptually identical to ideas in the beautiful theory of fiber bundles, developed by mathematicians without reference to the physical world, was a great marvel to me. In 1975 I discussed my feelings with Chern, and said "this is both thrilling and puzzling, since you mathematicians dreamed up these concepts out of nowhere." He immediately protested: "No, no. These concepts were not dreamed up. They were natural and real." — C. N. Yang

Bundles

In Part I we dealt with the electromagnetic field, which is a 2-form on the spacetime manifold M. Gauge theory deals with more general fields on spacetime, so we need to say what these more general fields are. This requires the language of vector bundles. The simplest sort of 'field' on a manifold M is a function

$$f \colon M \to V$$

from M to some fixed vector space V. However, the case of electromagnetism already shows that this conception of fields is inadequate; a 2-form *cannot* generally be regarded as a function from M to some vector space V, except when working locally in a chart. A vector field is a simpler example. A vector field v on M assigns to each point $p \in M$ a vector — but not a vector in a fixed vector space, rather, a vector v_p in the tangent space $T_p M$. This has tremendous consequences. It means that we cannot easily compare the values of a vector field at different points in M, since they live in different spaces (see Figure 4 in Chapter 3 of Part I). This, in turn, means that we cannot naively *differentiate* a vector field the same way as we do a function, since derivatives involve comparing values at different points. Gauge theory is all about this sort of situation, where rather than a fixed vector space, we have a 'bundle' of vector spaces, one for each point in M. In gauge theory, fields are 'sections' of vector bundles, which assign to each point in spacetime a vector in the vector space *for that point*. To write down differential equations in gauge theory, like the Yang-Mills equations, we need a mechanism whereby we can compare vectors in these different vector spaces. This is called a 'connection'. The connection is a field in its own right, and the Yang-Mills equations are particularly elegant because they are equations for the connection itself! This is the subject of the next chapter. We must begin with the definition of a vector bundle.

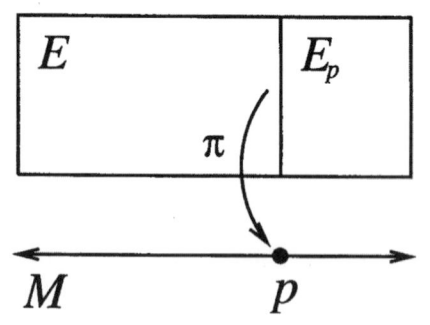

Fig. 1. A bundle $\pi: E \to M$

First, a **bundle** is a structure consisting of a manifold E, a manifold M, and an onto map $\pi: E \to M$. A simple bundle is shown in Figure 1, where M is the real line, E is the plane $\mathbb{R} \times \mathbb{R}$, and $\pi: E \to M$ is the standard projection from the plane down to the x axis. In general, the manifold E is called the **total space**, the manifold M is called the **base space**, and π is called the **projection** map. For each point

Bundles

$p \in M$, the space
$$E_p = \{q \in E: \pi(q) = p\}$$
is called the **fiber over** p. In Figure 1 we have indicated a point p in the base space and the fiber over p, which is literally sitting right over it. Note that the total space E is the union of all the fibers:
$$E = \bigcup_{p \in M} E_p.$$
This is why we speak of a 'bundle' of fibers. Sometimes we will call such a bundle '$\pi: E \to M$', and sometimes we will be lazy and simply call the bundle 'E'. If we want to remind the reader about the base space, we will say that E is a bundle **over** M.

In the cases we are interested in, M is physical space or spacetime, and each fiber E_p is a vector space. The **tangent bundle** of a manifold M is a good example. Here, the total space, denoted by TM, is simply the union of all the tangent spaces of M:
$$TM = \bigcup_{p \in M} T_p M.$$
The projection $\pi: TM \to M$ maps each tangent vector $v \in T_p M$ to the point $p \in M$. The fiber over any point $p \in M$ is thus the tangent space $T_p M$. However, we must give TM the structure of a manifold in such a way that π is a smooth map. There is a standard good way of doing this, which we leave as an exercise to the reader. The *idea* is as follows: if M is an n-manifold, it locally looks like $\mathbb{R}^n$, and the tangent space at each point looks like $\mathbb{R}^n$. To specify a point in TM is the same as specifying a point p in M together with a vector $v \in T_p M$. Thus, locally TM looks like $\mathbb{R}^n \times \mathbb{R}^n$. Thus we should make TM into a manifold of dimension $2n$.

Exercise 55. *Given a manifold M, define charts for TM starting from charts $\varphi_\alpha: U_\alpha \to \mathbb{R}^n$ for M as follows. Let V_α be the subset of TM given by*
$$V_\alpha = \{v \in TM: \pi(v) \in U_\alpha\}.$$
Show that every point in TM lies in some set V_α. Define maps $\psi_\alpha: V_\alpha \to \mathbb{R}^n \times \mathbb{R}^n$ by
$$\psi_\alpha(v) = (\varphi_\alpha(\pi(v)), (\varphi_\alpha)_* v),$$

where we think of $(\varphi_\alpha)_* v$, which is really a tangent vector to $\mathbb{R}^n$, as a vector in $\mathbb{R}^n$. Give TM the topology in which open sets are the unions of sets of the form $O \subseteq V_\alpha$ such that $\psi_a(O) \subseteq \mathbb{R}^n \times \mathbb{R}^n$ is open. Check that ψ_a are charts, so that TM is a manifold. Check that $\pi: TM \to M$ is smooth.

In general there is no need for all the fibers of a bundle to look alike. However, for the bundles we will be interested in, they do. In particular, given manifolds M and F, the **trivial bundle** over M with **standard fiber** F is simply the Cartesian product $E = M \times F$, with the projection map given by

$$\pi(p, f) = p$$

for all $(p, f) \in M \times F$. The example in Figure 1 is a trivial bundle. In the trivial bundle $E = M \times F$, the fiber over p is

$$E_p = \{p\} \times F.$$

Thus, not only are all the fibers diffeomorphic to the standard fiber F, there is an obvious 'best' — or in math jargon, 'canonical' — diffeomorphism between each fiber and F, sending $(p, f) \in E_p$ to $f \in F$. This property is special to trivial bundles.

While trivial bundles are a very narrow class of bundles, the most interesting bundles are those that are *locally* trivial. Roughly speaking, a locally trivial bundle is one that looks trivial if we examine it over a small enough neighborhood of any point of the base space. Globally, however, it may have 'twists'. For example, the cylinder is a trivial bundle over S^1 with standard fiber $\mathbb{R}$, while the Möbius strip is not a trivial bundle, as we can see in Figure 2. Nonetheless, if we look at the piece of a Möbius strip that sits over a small portion of S^1, it looks just like the cylinder. Topologically speaking, only if we take a trip all the way around the circle do we notice the twist in the Möbius strip. To make the concept of a 'locally trivial' precise, we will need to introduce the notion of 'isomorphic' bundles and the notion of the 'restriction' of a bundle to a submanifold of the base space. Having done so, we will say that a bundle is locally trivial if each point in the base space has a neighborhood for which the restriction of the bundle to that neighborhood is isomorphic to a trivial bundle.

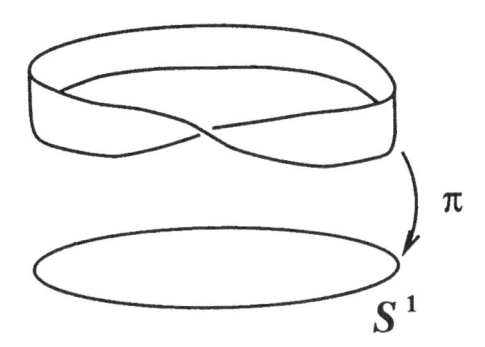

Fig. 2. The Möbius strip as a nontrivial bundle over S^1

Suppose, then, that we have two bundles, $\pi : E \to M$ and $\pi' : E' \to M'$. A **morphism** from the first to the second is a map $\psi : E \to E'$ together with a map $\phi : M \to M'$ such that ψ maps each fiber E_p into the fiber $E'_{\phi(p)}$:

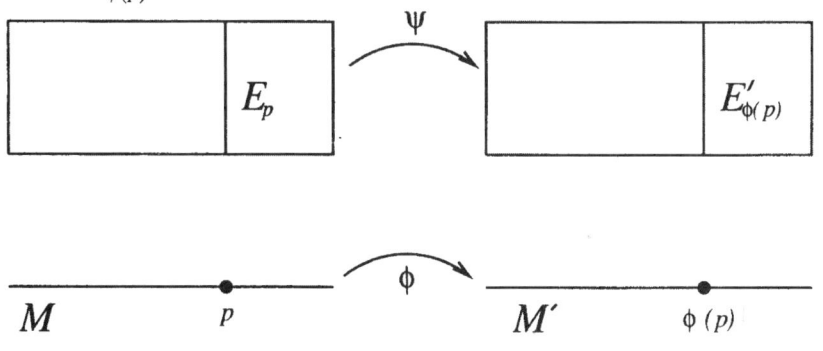

Fig. 3. A bundle morphism

We say this morphism is an **isomorphism** if ϕ and ψ are both diffeomorphisms.

Exercise 56. *Given bundles $\pi : E \to M$ and $\pi' : E' \to M'$, show that the maps $\psi : E \to E'$ and $\phi : M \to M'$ are a bundle morphism if and only if $\pi' \circ \psi = \phi \circ \pi$. This condition is shown in Figure 4, where we have drawn the total spaces E and E' over the corresponding base spaces M and M'. Show that ψ uniquely determines ϕ.*

Since $\psi : E \to E'$ uniquely determines $\phi : M \to M'$ when we have a bundle morphism, we feel free to call ψ the bundle morphism.

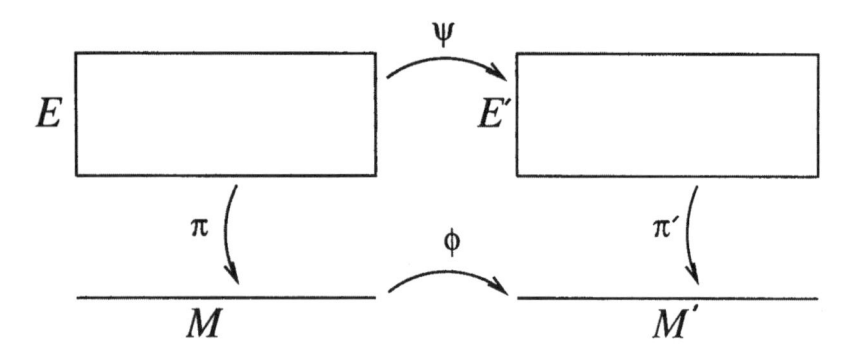

Fig. 4. The condition $\pi' \circ \psi = \phi \circ \pi$

The best example of a bundle morphism comes from the pushforward of tangent vectors. If $\phi: M \to M'$ is any map between manifolds, we can use it to push forward tangent vectors from the tangent space T_pM to the tangent space $T_{\phi(p)}M'$:

$$\phi_*: T_pM \to T_{\phi(p)}M'.$$

Since TM is just the union of the tangent space T_pM, and similarly for TM', we really have a map

$$\phi_*: TM \to TM'$$

One can check that ϕ_* is smooth. It follows that $\phi_*: TM \to TM'$ is a bundle morphism.

Exercise 57. *Check that ϕ_* is smooth when we make the tangent bundle into a manifold as in the previous exercise.*

Exercise 58. *Show that if $\phi: M \to M'$ is a diffeomorphism, then $\phi_*: TM \to TM'$ is a bundle isomorphism.*

Given a bundle $\pi: E \to M$ and a submanifold $S \subseteq M$, we define its **restriction** to S as follows. Take as the total space

$$E|_S = \{q \in E: \pi(q) \in S\},$$

take S as the base space, and use π (restricted to $E|_S$) as the projection. See Figure 5 for a picture of the Möbius strip, which is a bundle over the

Vector Bundles

circle, and its restriction to a small open set. Note that its restriction to a small open set looks just like the trivial bundle with standard fiber $\mathbb{R}$. In general, we say a bundle $\pi: E \to M$ is **locally trivial** with **standard fiber** F if for each point $p \in M$, there is a neighborhood U of p and a bundle isomorphism

$$\phi: E|_U \to U \times F$$

sending each fiber E_p to the fiber $\{p\} \times F$. We call ϕ a **local trivialization**. The idea of a local trivialization of a bundle is very much like the idea of a chart on a manifold: it allows us to assume locally that the situation is a simple, standard one.

Exercise 59. *Show that for any manifold M, the tangent bundle $\pi: TM \to M$ is locally trivial.*

Exercise 60. *Describe a bundle that is not locally trivial.*

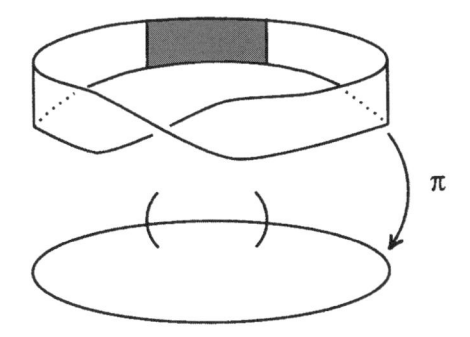

Fig. 5. The Möbius strip as a locally trivial bundle

Vector Bundles

Now we are finally ready to define a vector bundle, which is the kind of bundle we are really interested in! An n-dimensional **real vector bundle** is a locally trivial bundle $\pi: E \to M$ such that each fiber E_p is a n-dimensional vector space. Furthermore, we require that, for each point $p \in M$, there is a neighborhood U of p and a local trivialization

$$\phi: E|_U \to U \times \mathbb{R}^n$$

that maps each fiber E_q to the fiber $\{q\} \times \mathbb{R}^n$ *linearly*. In short, we require the trivialization to be **fiberwise linear**. We can similarly define a **complex vector bundle**, with $\mathbb{C}^n$ taking the role of $\mathbb{R}^n$ above. The real and complex cases are very similar, so in various definitions below we will only mention the real case, with the implicit understanding that the complex case goes the same way.

Exercise 61. *Check that the tangent bundle of a manifold is a vector bundle.*

Exercise 62. *A 1-dimensional bundle is called a (real or complex)* **line bundle.** *Check that the Möbius strip is a real line bundle if we regard the standard fiber as being* $\mathbb{R}$.

Note that while each fiber of a vector bundle is isomorphic to $\mathbb{R}^n$ (or $\mathbb{C}^n$), there need not be a canonical isomorphism. The tangent bundle of S^2 is a good example. You might wish to identify the tangent space to the north pole with $\mathbb{R}^2$ in one way, but I, viewing the sphere from a different angle, might prefer to do so in some other way — and neither of us would be 'right'. This is of extreme importance in physics, where we use bundles to describe fields! It means that, just as the laws of physics should be expressed in a way that looks equally nice in any local coordinate system on spacetime, they should look equally nice in any local trivialization of whatever vector bundles are involved. We will try to make this principle, the principle of 'gauge invariance', more and more clear as we proceed.

Along with our definition of vector bundles we should give a definition of the sort of maps between vector bundles that we are interested in. Suppose that $\pi: E \to M$ and $\pi': E' \to M'$ are vector bundles. Then a **vector bundle morphism** from the first to the second is a bundle morphism $\psi: E \to E'$ whose restriction to each fiber E_p of E is linear.

Exercise 63. *Show that if a vector bundle morphism is a diffeomorphism, its inverse is a vector bundle morphism.*

Vector Bundles

We said at the beginning of this section that fields in physics are often described as 'sections' of vector bundles. Let us now define this term precisely. A **section** of a bundle $\pi: E \to M$ is a function $s: M \to E$ such that for any $p \in M$,

$$s(p) \in E_p.$$

In other words, the section assigns to each point in the base space a vector in the fiber over that point. Another way of saying this is that $\pi \circ s$ is the identity map. We can draw a section as follows:

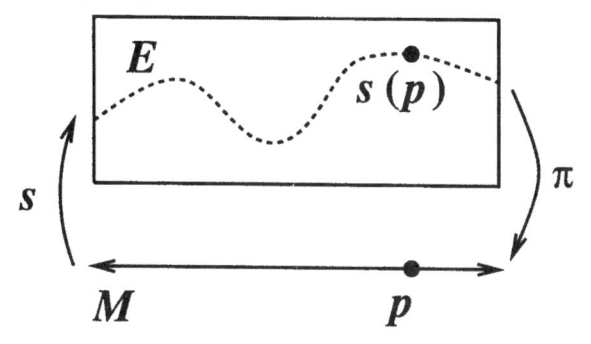

Fig. 6. A section s of a bundle $\pi: E \to M$

Note that if $E = M \times F$ is a trivial bundle with standard fiber F, a section of E is really just a complicated way of talking about a function from M to F! In other words, if we have a section $s: M \to E$, there is function $f: M \to F$ such that

$$s(p) = (p, f(p)) \in E_p.$$

Conversely, if we have a function $f: M \to F$, the formula above defines a section. Indeed, the bundle we have shown in Figure 6 is trivial, so the picture of the section s is really just the graph of a function. The concept of a section becomes really useful when considering nontrivial bundles. For example, a vector field on M is nothing other than a section of the tangent bundle of M.

Exercise 64. *Show that a section of the tangent bundle is a vector field.*

A section of a nontrivial vector bundle, namely the Möbius strip, appears in Figure 7.

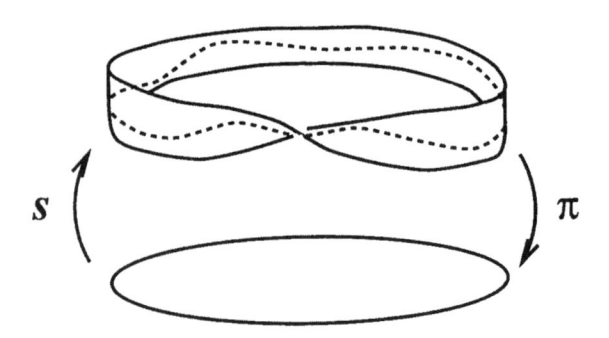

Fig. 7. A section of a nontrivial bundle

Suppose E is a vector bundle over M. We can add sections of E and multiply them by functions on M as follows. If s and s' are sections of E, and $f \in C^\infty(M)$, we define the section $s + s'$ by

$$(s + s')(p) = s(p) + s'(p),$$

and define the section fs by

$$(fs)(p) = f(p)s(p).$$

We denote the set of all sections of E by $\Gamma(E)$. We can summarize the properties of addition and multiplication by functions as follows:

Exercise 65. *Show that $\Gamma(E)$ is a module over $C^\infty(M)$.*

When we do computations with vector bundles, it is very handy to work with a 'basis of sections'. Given a vector bundle E, we say that the sections $e_1, \ldots, e_n$ of E form a **basis of sections** of E if any section $s \in \Gamma(E)$ can be written uniquely as a sum

$$s = s^i e_i$$

(using the Einstein summation convention), where $s^i \in C^\infty(M)$. Unfortunately, a vector bundle has a basis of sections if and only if it is isomorphic to a trivial bundle! To see this, note that a basis of sections determines a vector bundle isomorphism

$$\psi \colon M \times \mathbb{R}^n \to E$$

by
$$\psi(p,v) = v^i e_i(p)$$
where $p \in M$, $v \in \mathbb{R}^n$. Conversely, if we have a vector bundle isomorphism $\psi \colon M \times \mathbb{R}^n \to E$, we can use it to transfer the obvious basis of sections for $M \times \mathbb{R}^n$ to a basis of sections for E. Luckily, the definition of a vector bundle says that every vector bundle is *locally* isomorphic to a trivial bundle. In local computations, therefore, we can always work with a basis of sections over some neighborhood of any point in the base space.

Exercise 66. *Show that every section of the Möbius strip (viewed as real line bundle over S^1) vanishes somewhere. Conclude that the Möbius strip has no basis of sections, hence is not trivial.*

Vector Bundle Constructions

Now let us describe a few ways to get new vector bundles from old ones. The basic thing to keep in mind is that any natural operation on vector spaces can also be done with vector bundles. (Indeed, a vector space is really just a vector bundle with base space equal to a single point!) For example, just as we can define duals, direct sums, and tensor products of vector spaces, we can define them for vector bundles. Starting with the tangent bundle, these constructions will allow us to define the 'cotangent bundle' and various 'tensor bundles' that are very important in general relativity.

First, given a vector bundle E over M, we can define the **dual vector bundle** E^* over M as follows. Each fiber E_p is a vector space, and thus has a dual space E_p^*. We define the total space E^* to be the union of the spaces E_p^* for all $p \in M$, and let the projection $\pi \colon E^* \to M$ map each E_p^* to the corresponding point p. Thus the fiber over $p \in M$ is E_p^*. The only thing to check is that we can make E^* into a manifold so that there is local trivialization of E^* that is fiberwise linear. One can do this starting with the manifold structure and local trivialization of E.

Exercise 67. *Check the above statement. Also, show that given a basis of sections e_i of a vector bundle E, there is a unique **dual basis** e^i of sections* of E^* such that for each point $p \in M$, $e^i(p)$ is the basis of E_p^* dual to the basis $e_i(p)$ of E_p.

Exercise 68. *Show that if s is a section of a vector bundle E over M and λ is a section of E^*, there is a smooth function $\lambda(s)$ on M given by*

$$\lambda(s)(p) = \lambda(p)(s(p))$$

for all $p \in M$. Show that $\lambda(s)$ depends $C^\infty(M)$-linearly on λ and s.

A good example of a dual bundle is the **cotangent bundle** T^*M of a manifold. Here the fiber T_p^*M is just the cotangent space at p as we have earlier defined it! Furthermore, a section of T^*M is nothing but a 1-form on M.

Exercise 69. *Show that a section of the cotangent bundle is the same as a 1-form.*

Similarly, given two vector bundles E and E' over M, we can define the **direct sum vector bundle** $E \oplus E'$ over M. This bundle has fiber over p equal to the vector space $E_p \oplus E'_p$. Likewise, the **tensor product vector bundle** $E \otimes E'$ over M has fiber over p equal to $E_p \otimes E'_p$. Again, one can check that these are really vector bundles.

Exercise 70. *Check this fact.*

Exercise 71. *Suppose that E and E' are vector bundles over M, s is a section of E, and s' is a section of E'. Show that there is a unique section (s,s') of $E \oplus E'$ such that for each point $p \in M$, $(s,s')(p) = (s(p), s'(p))$. Show that there is a unique section $s \otimes s'$ of $E \otimes E'$ such that for each $p \in M$, $(s \otimes s')(p) = s(p) \otimes s'(p)$.*

Exercise 72. *Suppose that E and E' are vector bundles over M. Show that any section of $E \otimes E'$ can be written, not necessarily uniquely, as a locally finite sum of sections of the form $s \otimes s'$, where $s \in \Gamma(E)$ and $s' \in \Gamma(E')$.*

Given a vector bundle E over M, we can also define an 'exterior algebra bundle' ΛE over M. We show the reader how to do this in Exercise 73. This construction is another way of thinking about differential forms: starting with the tangent bundle TM, one forms its dual, the cotangent bundle T^*M, and then the exterior algebra bundle of that, ΛT^*M, called the **form bundle**; a differential form on M is just a section of the form bundle.

Vector Bundle Constructions

Exercise 73. *Suppose that E is a vector bundle over M. Define the exterior algebra bundle ΛE over M to have total space equal to the union of the vector spaces ΛE_p and projection map π sending ΛE_p to p. Show how to make ΛE into a manifold such that $\pi\colon \Lambda E \to M$ is a vector bundle.*

Exercise 74. *Show that ΛE is the direct sum of bundles $\Lambda^i E$,*

$$\Lambda E = \bigoplus_{i=0}^{n} \Lambda^i E$$

where n is the dimension of the fibers of E, and the vector bundle $\Lambda^i E$ has fiber over $p \in M$ given by $\Lambda^i E_p$. Show that sections of $\Lambda^0 E$ are in natural one-to-one correspondence with functions on M and sections of $\Lambda^1 E$ are in natural one-to-one correspondence with sections of E.

Exercise 75. *Show that for any sections ω, μ of ΛE there is a section $\omega \wedge \mu$ given by*

$$(\omega \wedge \mu)(p) = \omega(p) \wedge \mu(p).$$

Show that the sections of ΛE form an algebra. Show that that the sections of $\Lambda^i E$ form a subspace of the sections of ΛE, and that the sections of $\Lambda^i E$ are all locally finite sums of wedge products of sections of E.

Exercise 76. *Show that sections of $\Lambda^i T^* M$ are in natural one-to-one correspondence with i-forms on M.*

Another very important way to construct vector bundles is by gluing together trivial vector bundles. This is where the group theory we discussed in the previous chapter comes in. Before we describe the general construction let us describe it with a simple example. We can make the Möbius strip bundle over S^1 (see Figure 2) by gluing together trivial line bundles over three open sets U_1, U_2, U_3 that cover the circle:

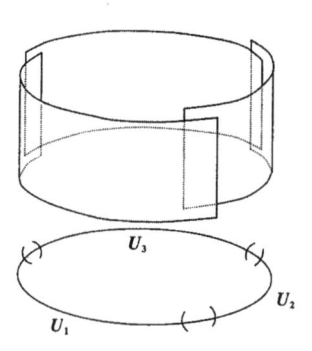

Fig. 8. Building the Möbius strip out of trivial bundles

We get the 'twist' as follows: if $p \in U_1 \cap U_2$, then we glue the point $(p,v) \in U_1 \times \mathbb{R}$ to the point $(p,-v) \in U_2 \times \mathbb{R}$. The minus sign here is a special case of a 'transition function'.

In general, let M be a manifold and $\{U_\alpha\}$ a cover of M by open sets. Let V be a vector space and let ρ be a representation of some group G on V. We will glue together the trivial bundles $U_\alpha \times V$ to get a vector bundle $\pi \colon E \to M$ using **transition functions** $g_{\alpha\beta} \colon U_\alpha \cap U_\beta \to G$. To get E, we start with the disjoint union

$$\bigcup_\alpha U_\alpha \times V,$$

and then we regard any two points $(p,v) \in U_\alpha \times V$ and $(p,v') \in U_\beta \times V$ as equal — in math jargon, we 'identify' them — if

$$v = \rho(g_{\alpha\beta}(p))v'.$$

To save space (and time), we will usually just write this as

$$v = g_{\alpha\beta}v'.$$

However, this procedure will only give a vector bundle if the transition functions satisfy a couple of consistency conditions. Suppose that $p \in U_\alpha$. Then by the above recipe we have to identify $(p,v) \in U_\alpha \times V$ with $(p, g_{\alpha\alpha}v) \in U_\alpha \times V$. However, we do not want to identify two different points in the same trivial bundle $U_\alpha \times V$, so we require

$$g_{\alpha\alpha} = 1 \quad \text{on } U_\alpha.$$

Vector Bundle Constructions

There is also a more subtle condition along these lines if $p \in U_\alpha \cap U_\beta \cap U_\gamma$. In this case we will identify

$$(p, v) \in U_\alpha \times V$$

with

$$(p, g_{\gamma\alpha} v) \in U_\gamma \times V,$$

which we identify with

$$(p, g_{\beta\gamma} g_{\gamma\alpha} v) \in U_\beta \times V,$$

which in turn we identify with

$$(p, g_{\alpha\beta} g_{\beta\gamma} g_{\gamma\alpha} v) \in U_\alpha \times V.$$

Again, so that we do not identify two different points in $U_\alpha \times V$, we will require the **cocycle condition**:

$$g_{\alpha\beta} g_{\beta\gamma} g_{\gamma\alpha} = 1 \quad \text{on } U_\alpha \cap U_\beta \cap U_\gamma.$$

The reader can easily think up many more consistency conditions of this general sort, but the wonderful thing is that they all follow from these two.

Exercise 77. *Show that these conditions imply $g_{\alpha\beta} = g_{\beta\alpha}^{-1}$. Show that for any sequences $\alpha_1, \ldots, \alpha_n$ and $\beta_1, \ldots, \beta_m$ with $\alpha_1 = \beta_1, \alpha_n = \beta_m$, they imply*

$$g_{\alpha_1 \alpha_2} \cdots g_{\alpha_{n-1} \alpha_n} = g_{\beta_1 \beta_2} \cdots g_{\beta_{m-1} \beta_m}.$$

Let us write $[p, v]_\alpha$ for the point of E corresponding to $(p, v) \in U_\alpha \times V$. (The bracket here has nothing to do with a Lie bracket!) Because of the identifications we have made, we have $[p, v]_\alpha = [p, g_{\beta\alpha} v]_\beta$. We define the projection $\pi \colon E \to M$ by

$$\pi[p, v]_\alpha = p.$$

The fiber E_p is thus the set of all points in E of the form $[p, v]_\alpha$.

Now, one can show that if the consistency conditions on the transition functions hold, $\pi \colon E \to M$ is really a vector bundle.

Exercise 78. *Prove that if $g_{\alpha\alpha} = 1$ and $g_{\alpha\beta}g_{\beta\gamma}g_{\gamma\alpha} = 1$ where defined, $\pi\colon E \to M$ is a vector bundle. (Hint: first show how to give each fiber E_p the structure of a vector space, and then show that E is trivial over each set U_α, with a fiberwise linear local trivialization.)*

We call this sort of vector bundle a **G-bundle**. The group G is called the **gauge group** of the bundle and V is called the **standard fiber**. In gauge theory, fields are described by sections of G-bundles, with different choices of the gauge group being used for different forces, as mentioned in Chapter 1.

The trivial bundles $U_\alpha \times V$ give rise to local trivializations of E, say

$$\phi_\alpha\colon E|_{U_\alpha} \to U_\alpha \times V,$$

given by

$$\phi_\alpha[p,v]_\alpha = (p,v)$$

for $p \in U_\alpha$, $v \in V$. To compare two of these trivializations on the overlap $U_\alpha \cap U_\beta$ we use the transition function $g_{\alpha\beta}$; in other words,

$$\phi_\alpha \circ \phi_\beta^{-1} = \rho(g_{\alpha\beta})$$

on any fiber $\{p\} \times V$ with $p \in U_\alpha \cap U_\beta$. The idea is that I can do local calculations over U_α by working in $U_\alpha \times V$ using the trivialization ϕ_α, while you do local calculations over U_β by working in $U_\beta \times V$, but to compare notes about what is going on in the overlap, we need to use the transition function $g_{\alpha\beta}$.

Since each fiber E_p really 'looks like V', but not canonically, and G has a representation ρ on V, it makes sense to ask if a linear transformation $T\colon E_p \to E_p$ is of the form $\rho(g)$ — but not to ask *which* $g \in G$! More precisely, suppose $p \in U_\alpha$. Then we say T **lives in** G if it is of the form

$$[p,v]_\alpha \mapsto [p,gv]_\alpha$$

for some $g \in G$. The point is that this definition is independent of the choice of α. For suppose $p \in U_\beta$ as well: then

$$[p,v]_\alpha = [p, g_{\beta\alpha}v]_\beta$$

since we have identified them, and similarly

$$[p,gv]_\alpha = [p, g_{\beta\alpha}gv]_\beta$$

so T is also given on $U_\alpha \cap U_\beta$ by

$$[p, g_{\beta\alpha}v]_\beta \mapsto [p, g_{\beta\alpha}gv]_\beta$$

or in other words

$$[p, v']_\beta \mapsto [p, g'v']_\beta$$

where we have made the change of variables $v' = g_{\beta\alpha}v$, $g' = g_{\beta\alpha}gg_{\beta\alpha}^{-1}$.

The same thing holds for the Lie algebra of G. Namely, we say $T: E_p \to E_p$ **lives in** $\mathfrak{g}$ if it is of the form

$$[p, v]_\alpha \mapsto [p, d\rho(x)v]_\alpha$$

for some $x \in \mathfrak{g}$.

Exercise 79. *Show that if the above condition holds, and $p \in U_\alpha \cap U_\beta$, then T is also of the form*

$$[p, v']_\beta \mapsto [p, d\rho(x')v']_\beta$$

for some $v' \in \mathfrak{g}$.

Gauge Transformations

The conservation of isotopic spin is identical with the requirement of invariance of all interactions under isotopic spin rotation. This means that when electromagnetic interactions can be neglected, as we shall hereafter assume to be the case, the orientation of the isotopic spin is of no physical significance. The differentiation between a neutron and a proton is then a purely arbitrary process. As usually conceived, however, once one chooses what to call a proton, and what to call a neutron, at one space-time point, one is then not free to make any choices at other space-time points.

It seems that this is not consistent with the localized field concept that underlies the usual physical theories. In the present paper we wish to explore the possibility of requiring all interactions to be invariant under independent *rotations of the isotopic spin at all space-time points, so that the relative orientation of the isotopic spin becomes a physically meaningless quantity....*
— *C. N. Yang and R. L. Mills*

Having introduced vector bundles, it is now time to introduce the natural groups of *symmetries* of vector bundles. The most basic kind of symmetry is called a 'gauge transformation'. Yang and Mills became interested in finding equations that were invariant under gauge transformations when they were thinking about the physics of 'hadrons' — particles that interact via the strong force.

The most common hadrons in everyday matter are neutrons and protons. These two particles, which are the constituents of atomic nuclei, are in a funny way both very similar and quite different. Both are fermions with spin 1/2. The mass of the proton is about 1836 times that of the electron, while the mass of the neutron is about 1839 times the electron mass. The proton is positively charged, while the neutron is neutral, so they interact very differently with respect to the electromagnetic force, but they interact in a very similar way with respect to the strong nuclear force. The proton appears to be absolutely stable, despite careful experiments seeking to observe proton decay. An isolated neutron is not stable; it has a mean lifetime of about 15 minutes and decays into a proton, an electron and an electron anti-neutrino, in what is called 'beta-decay'. In the 1930s Heisenberg introduced the concept of 'isospin' to try to account for some of these facts. The idea was that, if we could ignore all interactions apart from the strong force, we could treat the proton and neutron as two states of a single particle, the 'nucleon'. By analogy with spin, Heisenberg described the internal degrees of freedom of a nucleon by a unit vector in $\mathbb{C}^2$, with the proton being arbitrarily assigned the 'isospin-up' state

$$p = \begin{pmatrix} 1 \\ 0 \end{pmatrix},$$

and the neutron being the down state

$$n = \begin{pmatrix} 0 \\ 1 \end{pmatrix}.$$

One says, for short, that the nucleon is an isospin **doublet**. Moreover, he described the symmetry between protons and neutrons by the spin-1/2 representation of SU(2). This kind of symmetry hypothesis provides a useful guiding principle, because it limits ones search for the correct laws governing the strong force to those admitting an SU(2) symmetry in which the nucleon transforms in this way.

Gauge Transformations

Of course, such a hypothesis has to be tested empirically. For example, in Heisenberg's time it was believed that the strong force was carried by particles called pions. There are three pions, positive, negative and neutral in their electric charge, written $\pi^\pm$ and π^0, and, like the proton and neutron, these are very similar to each other. The π^+ and π^- have a mass about 140 times that of the electron, while the π^0 has mass 135 times the electron mass. All are unstable and decay quite rapidly. The π^- has a half-life of about 2.6×10^{-8} seconds and almost always decays into a muon (μ^-) and muon antineutrino ($\bar{\nu}_\mu$), while the π^+, its antiparticle, has the same half-life but decays into an antimuon (μ^+) and muon neutrino (ν_μ). The π^0 has a half-life of about 8×10^{-17} seconds and almost always decays into a pair of photons. Ignoring the rather violent differences in half-lives and decay modes, which we can suppose are due to interactions other than the strong force, we can hypothesize that all three pions are states of a single particle, the pion, which transforms according to the spin-1 representation of SU(2), making the following assignments:

$$\pi^+ = \begin{pmatrix} 1 \\ 0 \\ 0 \end{pmatrix}, \quad \pi^0 = \begin{pmatrix} 0 \\ 1 \\ 0 \end{pmatrix}, \quad \pi^- = \begin{pmatrix} 0 \\ 0 \\ 1 \end{pmatrix}.$$

In other words, the pions form an isospin **triplet**.

So far we have only been thinking of the internal degrees of freedom of the nucleon or pion, rather than their motion in spacetime, but to describe actual physical processes we need to treat the nucleon as, roughly speaking, a *field*

$$\psi\colon \mathbb{R}^4 \to \mathbb{C}^2$$

on Minkowski spacetime, and the pion as a field

$$\phi\colon \mathbb{R}^4 \to \mathbb{C}^3.$$

The idea is then to write down nonlinear wave equations describing the behavior of these fields, making sure that if we act on any solution by an element of SU(2) as follows:

$$\psi'(x) = g\psi(x), \qquad \phi'(x) = g\phi(x),$$

we again obtain a solution. Here $g \in \mathrm{SU}(2)$ acts on $\psi(x)$ and $\phi(x)$ according to the spin-1/2 and spin-1 representations, respectively, so a more explicit formula would be

$$\psi'(x) = U_{1/2}(g)\psi(x), \qquad \phi'(x) = U_1(g)\phi(x)$$

where U_j is the spin-j representation described in Chapter 1. The simplest example of such a nonlinear equation would be the following equation for a self-interacting pion field:

$$(\partial^\mu \partial_\mu + m^2)\phi + \lambda \phi^i \phi_i \phi = 0$$

where $\phi^i \phi_i$ denotes the inner product in $\mathbb{C}^3$ and $\lambda > 0$ is a real number, the 'coupling constant', which adjusts the strength of the nonlinear term.

Exercise 80. *Check that if $\phi: \mathbb{R}^4 \to \mathbb{C}^3$ is a solution of this equation, so is $U_1(g)\phi$ for any $g \in \mathrm{SU}(2)$.*

Actually, we have written down this equation just to give the slightest flavor of the real physics, and we should immediately emphasize that we are simplifying things a lot. A more realistic model would involve both the nucleon and the pion fields, with a nonlinear term describing a nucleon-pion interaction. Moreover, we have neglected the fact that the nucleon has ordinary spin as well as isospin, so that it is really a 'spinor field'. Even worse, particles are really described, not by a 'classical field theory', but by a *quantum* field theory in which ψ and ϕ are replaced by *operators*. Quantum field theory is an enormous subject which this book merely skirts; for introductory references on it, see the notes at the end of Part II.

The great idea of Yang and Mills was to look for equations possessing much more symmetry, namely, symmetries under transformations of the form

$$\psi'(x) = g(x)\psi(x), \qquad \phi'(x) = g(x)\phi(x),$$

where

$$g: \mathbb{R}^4 \to \mathrm{SU}(2)$$

Gauge Transformations

is any SU(2)-valued *function* on spacetime. This is called a 'gauge transformation'. It is much more difficult to invent equations possessing so much symmetry, but they succeeded by copying Maxwell's equations! Their work led to an SU(2) 'gauge theory' of protons, neutrons, pions and other hadrons. Various groups of hadrons could be classified into **multiplets**, families corresponding to different representations of SU(2), and many qualitative and approximate quantitative properties were explained.

At about the same time, Gell-Mann realized that more patterns in the data about hadrons could be explained if one thought of SU(2) as a subgroup of SU(3): various multiplets could be lumped together into still larger families. A stunning triumph of this approach was Gell-Mann's prediction of the existence and properties of the Ω particle, a hadron about 3272 times as massive as the electron. Gell-Mann also originated the notion of elementary constituents of the hadrons, transforming according to the fundamental (3-dimensional) representation of SU(3). These were called the up, down, and strange quarks, since the former two form an isospin doublet, while the first evidence for the third was the anomalously long decay of certain 'strange' particles. At first it was not at all clear if the quarks were really particles or simply mathematical conveniences, because they were never observed in isolation.

Eventually a good theory of the strong force predicting 'quark confinement' was devised to account for the fact that quarks are never seen alone. In a very curious twist, the key idea here was 'color' symmetry, a gauge symmetry with gauge group SU(3), utterly unrelated to the SU(3) symmetry between the various 'flavors' of quarks — a mixed metaphor referring to up, down, and strange. Recall that the symmetry between up and down was never proposed as anything more than an approximation suitable when the electromagnetic field could be neglected. Indeed, in the quark theory, the up has charge 2/3, while the down has charge $-1/3$. In addition, the strange quark is considerably more massive than the up or down, so SU(3) flavor symmetry is only a rough approximation. On the other hand SU(3) color symmetry is intended as an *exact* symmetry. The gauge theory associated to this symmetry is called quantum chromodynamics. Later, other flavors of quarks were discovered — the charm and bottom quarks — and for reasons too subtle to go into here, a further, the top quark is expected as well. But the SU(3) theory of the strong force is still regarded as correct, and is part of the standard model. Similarly, the electroweak force is described by a $SU(2)\times U(1)$ gauge theory in the standard model. There is, of course, far more to the standard model than we can treat in this book, and the reader is urged to read the Notes for some good introductory references to this wonderful theory of particles and forces. While the standard model is unlikely to be the last word on physics — in particular, it does not treat gravity — it is likely to be a useful basis for future developments.

In the rest of this section, we will make the notion of a gauge transformation precise for general vector bundles. Suppose we have a vector bundle $\pi\colon E \to M$. A gauge transformation is essentially just a one-to-one and onto linear transformation of the fiber E_p which varies smoothly with the point $p \in M$. However, if E is a G-bundle, as defined in the previous section, we will demand that the linear transformation lives in G.

Given a vector space V, the linear functions from V to itself are sometimes called **endomorphisms** by those who wish to show off their Greek, and the set of all endomorphisms of V is denoted $\text{End}(V)$. This is a vector space in its own right, where we define

$$(\alpha T)(v) = \alpha T(v)$$

and

$$(S+T)(v) = S(v) + T(v)$$

for any scalar α, any $S,T \in \text{End}(V)$, and any $v \in V$. Note that $\text{End}(V)$ is also an algebra, with the product defined by

$$(ST)(v) = S(T(v)).$$

For example, $\text{End}(\mathbb{R}^n)$ is simply the algebra of $n \times n$ real matrices, while $\text{End}(\mathbb{C}^n)$ is the algebra of $n \times n$ complex matrices, with matrix multiplication as the product.

In general, if we take a basis e_i of V and let e^j denote the dual basis of V^*, a convenient basis of $\text{End}(V)$ is given by the elements e^i_j, where

$$e^i_j\, e_k = \delta^i_k e_j,$$

Gauge Transformations

δ denoting the Kronecker delta. When V is $\mathbb{R}^n$ and e_i is the standard basis, the matrices e^i_j are called **matrix units**. In this case, e^i_j is the matrix with a 1 in the ith row and jth column and all other entries zero.

Note that there is an isomorphism

$$V \otimes V^* \cong \operatorname{End}(V)$$

taking the element $v \otimes f \in V \otimes V^*$ to the linear function given by

$$x \mapsto f(x)v$$

for all $x \in V$. In terms of bases, this isomorphism is given by

$$e^i \otimes e_j \mapsto e^i_j.$$

In fact, there is no harm in *defining* $\operatorname{End}(V)$ to be $V \otimes V^*$. The advantage is that, since we already know about duals and tensor products of vector bundles, this immediately lets us generalize from vector spaces to vector bundles. Thus: given a vector bundle E over a manifold M, let $\operatorname{End}(E)$, the **endomorphism bundle** of E, denote the bundle $E \otimes E^*$. This name is a good one, since sections of $\operatorname{End}(E)$ really do determine vector bundle morphisms from E to itself, as follows. The fiber of $\operatorname{End}(E)$ over any point $p \in M$ is just the same as $\operatorname{End}(E_p)$, the endomorphisms of E_p. As a result, any section T of $\operatorname{End}(E)$ defines a map from E to itself sending $v \in E_p$ to $T(p)v \in E_p$, which is a vector bundle morphism.

As a result, any section T of $\operatorname{End}(E)$ acts on any section s of E pointwise, giving a new section Ts of E, as follows:

$$(Ts)(p) = T(p)s(p).$$

Thus T determines a function

$$T \colon \Gamma(E) \to \Gamma(E),$$

where $\Gamma(E)$ is the set of all sections of E. This function is $C^\infty(M)$-linear, i.e.,

$$T(fs) = fT(s)$$

for all functions f and sections s of E. With some work, one can show that all $C^\infty(M)$-linear maps
$$T\colon \Gamma(E) \to \Gamma(E)$$
correspond to sections of $\text{End}(E)$ in this way.

Exercise 81. *Show this. (Hint: use a local trivialization and the 'partition of unity' trick described in Chapter 6 of Part I to reduce this to the case of a trivial bundle.)*

Now suppose that $\pi\colon E \to M$ is a G-bundle, where G is some Lie group. Remember that this means there is an open cover $\{U_\alpha\}$ of M such that E is built by gluing together trivial bundles $U_\alpha \times V$, where V is a vector space on which G has a representation ρ. Suppose also that $T \in \text{End}(E)$. We saw at the end of the previous section that there is a well-defined notion of $T(p) \in \text{End}(E_p)$ living in the gauge group G or the Lie algebra $\mathfrak{g}$. If $T(p)$ lives in $\mathfrak{g}$ for all $p \in M$, we say simply that T **lives in $\mathfrak{g}$**. If $T(p)$ lives in G for all $p \in M$, we say that T is a **gauge transformation**. The set of all gauge transformations, which we call $\mathcal{G}$, is actually a group, with products and inverses given by
$$\begin{aligned}(gh)(p) &= g(p)h(p),\\ g^{-1}(p) &= g(p)^{-1}.\end{aligned}$$

Exercise 82. *Show that the product or inverse of gauge transformations is a gauge transformation, and that the identity is a gauge transformation.*

Beware: sometimes physicists call *this* group the 'gauge group', while others reserve that term for G.

The principle of gauge theory is that fields should be sections of G-bundles, and that the laws of physics should be differential equations such that if the section s is a solution, so is gs for any $g \in \mathcal{G}$. Such differential equations are said to be **gauge invariant**. For example, in the isospin theory, the pion field is a section of an SU(2)-bundle with standard fiber given by the spin-1 representation of SU(2). It is not so easy to write down differential equations that are gauge invariant! The problem is that the derivatives of gs can be very different than those of s, since $g(p)$ depends on the point p. It took Yang and Mills quite a while to figure out the trick. However, the trick had already been figured out by mathematicians — it goes by the name of a 'connection'.

Connections

It should be clear by now that differentiating a section of a vector bundle is a nontrivial affair. The usual derivative of a function on the real line, for example:

$$\frac{df}{dx} = \lim_{\epsilon \to 0} \frac{f(x+\epsilon) - f(x)}{\epsilon}$$

involves subtracting the value of f at two different points, x and $x+\epsilon$. A section of a bundle, however, assigns to each point in the base a vector in the fiber *over that point*. There is no canonical way to add or subtract vectors in different fibers. As a consequence, there is usually not a single 'best' way to differentiate sections of a vector bundle; instead, there are many different ways. A way to differentiate sections is called a 'connection'.

To be precise, let E be a vector bundle over the manifold M. Recall that $\Gamma(E)$ denotes the space of sections of E. A **connection** D on M assigns to each vector field v on M a function D_v from $\Gamma(E)$ to $\Gamma(E)$ satisfying the following properties:

$$\begin{aligned} D_v(\alpha s) &= \alpha D_v s \\ D_v(s+t) &= D_v s + D_v t \\ D_v(fs) &= v(f)s + f D_v s \\ D_{v+w} s &= D_v s + D_w s \\ D_{fv} s &= f D_v s \end{aligned}$$

for all $v, w \in \text{Vect}(M)$, $s, t \in \Gamma(E)$, $f \in C^\infty(M)$ and all scalars α. (Here 'scalars' are real or complex numbers depending on whether E is a real or complex vector bundle.) Note that this definition is very similar to the definition of a vector field! In particular, the third property, the Leibniz law, is what makes D_v act like differentiation. Given any section s and vector field v, we call $D_v s$ the **covariant derivative** of s in the direction v.

This definition of a connection may seem abstract, so let us see what it amounts to in terms of local coordinates and a local basis of sections of the bundle E. Let x^μ be coordinates on an open set $U \subseteq M$, let ∂_μ

be the corresponding basis of coordinate vector fields, and let e_i be a basis of sections of E over U. We use the abbreviation

$$D_\mu = D_{\partial_\mu}.$$

Note that for any μ, j we can express $D_\mu e_j$ uniquely as a linear combination of the sections e_i, with functions on U as the coefficients. Thus we can define functions $A^i_{\mu j}$ on U by

$$D_\mu e_j = A^i_{\mu j} e_i.$$

These functions are called the components of the **vector potential**; in a bit, we will give a more coordinate-independent description of the vector potential. The point of the vector potential is that we can use it to work out the covariant derivative $D_v s$ of any section s of E over U, in the direction of any vector field v on U. Namely, we have

$$\begin{aligned} D_v s &= D_{v^\mu \partial_\mu} s \\ &= v^\mu D_\mu s \\ &= v^\mu D_\mu(s^i e_i) \\ &= v^\mu((\partial_\mu s^i) e_i + A^j_{\mu i} s^i e_j) \\ &= v^\mu(\partial_\mu s^i + A^i_{\mu j} s^j) e_i \end{aligned}$$

where in the third step we used the Leibniz law, and in the fourth we switched the names of the indices i and j in the second term. Alternatively, if we define functions $(D_\mu s)^i$ by $D_\mu s = (D_\mu s)^i e_i$, the above equation gives

$$(D_\mu s)^i = \partial_\mu s^i + A^i_{\mu j} s^j.$$

Physicists often use this sort of notation, except that they usually just write $D_\mu s^i$ instead of $(D_\mu s)^i$. This is a bit confusing, because we are *not* taking the covariant derivative of the function s^i (which is undefined, after all); instead, we are taking the covariant derivative of s, and *then* looking at one of its components in the basis e_i. Be forewarned!

Let us figure out the geometrical meaning of the vector potential. In other words, let us try to think of it, not as a big batch of functions $A^i_{\mu j}$ labeled by indices μ, i, j and depending on local coordinates and a local trivialization of the bundle E, but as a section of some bundle.

To do this, we should figure out what job the vector potential actually does in the formulas above. We started with a vector field v and a section s of E over U, and when we calculated $D_v s$, we got a term of the form
$$A^j_{\mu i} v^\mu s^i e_j,$$
which is a new section of E over U. Note that this expression is linear in both v and s; in particular, it is linear over $C^\infty(U)$, meaning that if we multiply either v or s by a *function* the expression is multiplied by that function, since no derivatives appear in it. So the real role of the vector potential is to eat a vector field and a section of E over U and spit out a new section of E over U in a $C^\infty(U)$-linear way.

Pondering along these lines, one is led to think of the vector potential as an **End(**E**)-valued 1-form** on U, that is, a section of the bundle
$$\mathrm{End}(E|_U) \otimes T^*U.$$
The reason is that if we define the **vector potential** A this way:
$$A = A^j_{\mu i}\, e_j \otimes e^i \otimes dx^\mu,$$
the 1-form part eats any vector field v on U and spits out a section of $\mathrm{End}(E)$ over U,
$$\begin{aligned} A(v) &= A^j_{\mu i}(e_j \otimes e^i)\, dx^\mu(v) \\ &= A^j_{\mu i} v^\mu (e_j \otimes e^i) \end{aligned}$$
which in turn eats any section s of E over U and spits out
$$\begin{aligned} A(v)s &= A^j_{\mu i} v^\mu (e_j \otimes e^i) s \\ &= A^j_{\mu i} v^\mu s^i e_j \end{aligned}$$
which is the section of E over U that we are interested in. In these terms, we have
$$(D_v s)^i = v s^i + (A(v)s)^i.$$

It is sometime nice to suppress the **internal indices** i, j, and so on, which are associated to the basis of sections e_i, and write the vector potential in terms of components
$$A_\mu = A^j_{\mu i}\, e_j \otimes e^i.$$

Each of the components A_μ is a section of $\text{End}(E)$ over U. The reader should be wondering about the precise relationship between the 'vector potential' we are talking about here and the electromagnetic vector potential discussed in Part I. A simple way of thinking about gauge theory is that it generalizes electromagnetism by letting the vector potential be, not merely a 1-form with components A_μ, but a matrix-valued 1-form with components $A^j_{\mu i}$. Of course, we need a local trivialization of $\text{End}(E)$ to define the components $A^j_{\mu i}$, so a more invariant way to think of the vector potential is as an $\text{End}(E)$-valued 1-form with components A_μ.

One advantage of thinking about the vector potential as an $\text{End}(E)$-valued 1-form is that we can do so, not just locally over some open set U, but globally, throughout M. Let us describe how. Suppose that A is an $\text{End}(E)$-valued 1-form. By Exercise 72, we can write

$$A = \sum_i T_i \otimes \omega_i$$

where T_i are sections of $\text{End}(E)$ and ω_i are 1-forms on M. Thus for any vector field v on M we can define the section $A(v)$ of $\text{End}(E)$ by

$$A(v) = \sum_i \omega_i(v) T_i.$$

Exercise 83. *Check that $A(v)$ is well-defined, i.e., independent of how we write A as a sum $\sum T_i \otimes \omega_i$.*

This section $A(v)$ acts on any section s of E to give a new section $A(v)s$.

Now, we claim that if D^0 is any connection on E, so is $D = D^0 + A$, by which we mean the connection such that

$$D_v s = D^0_v s + A(v) s.$$

Moreover, we claim that *any* connection D can be written as $D^0 + A$. In short, once we have chosen a single connection, we can write all others as that one plus some vector potential. This is why when people are being sloppy they do not worry much about the difference between a connection and a vector potential, particularly when there is an obvious 'best' connection D^0 around. Indeed, this was the case back when we

Connections

were working using local coordinates and a local trivialization of E. There, we were implicitly making use of the connection

$$D^0_v s = v(s^j) e_j$$

to write any connection D as $D^0 + A$:

$$\begin{aligned} D_v s &= \left(v(s^i) + A^i_{\mu j} v^\mu s^j\right) e_i \\ &= D^0_v s + A(v) s. \end{aligned}$$

This connection D^0 is called the **standard flat connection** on $E|_U$. The standard flat connection depends on the *choice* of local trivialization of E; it is not canonical.

Let us sketch the proofs of the above claims. First we need to check that if D^0 is a connection on E and A is an $\text{End}(E)$-valued 1-form, $D = D^0 + A$ is also a connection on E. The only nontrivial part of checking this is the Leibniz law:

$$\begin{aligned} D_v(fs) &= D^0_v(fs) + A(v)(fs) \\ &= v(f)s + f D^0_v s + f A(v) s \\ &= v(f)s + f D_v(s). \end{aligned}$$

Second, we need to check that if D is any other connection on E, we have $D = D^0 + A$ for some $\text{End}(E)$-valued 1-form A. The main thing is to show that $A = D - D^0$ really has the property that $A(v)s$ depends $C^\infty(M)$-linearly on v and s. Linearity in v is obvious, since $D_v s$ and $D^0_v s$ depend linearly on v. Linearity in s is a bit less obvious, but not hard to show:

$$\begin{aligned} D_v(fs) - D^0_v(fs) &= v(f)s + f D_v s - v(f)s - f D^0_v s \\ &= f(D_v s - D^0_v s) \end{aligned}$$

for any function f on M. Then, using the fact that $A(v)s$ has these linearity properties, we can cook up the desired $\text{End}(E)$-valued 1-form A. For example, for any local coordinates x^μ and local basis of sections e_i of E, we can define

$$A = A^j_{\mu i} e_j \otimes e^i \otimes dx^\mu$$

where

$$A(\partial_\mu) e_j = A^i_{\mu j} e_i.$$

Exercise 84. *Work out the details of the proof we have sketched here.*

When our vector bundle E has some extra structure, the connections that are compatible with this structure (in some sense or other) become especially important. In general relativity, for example, the crucial vector bundle is the tangent bundle of spacetime. Since spacetime has a metric on it which allows one to measure lengths of tangent vectors, the most important connections for physics are those that 'preserve' the metric — in a sense that we define in Chapter 1 of Part III. In Yang-Mills theory, on the other hand, the vector bundles at hand are always G-bundles, where G is the gauge group of the theory. Here the physically relevant connections are those whose vector potentials look locally like a $\mathfrak{g}$-valued 1-form. To make this more precise, we will use the notion of a section of $\text{End}(E)_p$ that lives in $\mathfrak{g}$. Suppose that E is a G-bundle with standard fiber given by some vector space V on which G has a representation ρ. Then there are local trivializations

$$\phi_\alpha \colon E|_{U_\alpha} \to U_\alpha \times V$$

such that the transition functions $\phi_\alpha \circ \phi_\beta^{-1}$ are of the form $\rho(g_{\alpha\beta})$ for some G-valued function $g_{\alpha\beta}$ on the overlap. Now suppose that D is a connection on E. Over any one of the U_α we can write D as the standard flat connection D^0 plus a vector potential A. We say that D is a *G*-**connection** if in local coordinates the components $A_\mu \in \text{End}(E)$ live in $\mathfrak{g}$. This definition may appear to depend on the local coordinates x^μ used to define the components $\text{End}(E)$, but it does not really: if we change to new coordinates x'^ν, we get new components A'_ν given by

$$A'_\nu = \frac{\partial x'^\mu}{\partial x^\nu} A_\mu,$$

which will live in $\mathfrak{g}$ if the original components A_μ did.

Just as we can apply a gauge transformation to a section of a G-bundle, we can apply it to a G-connection. Let D be a G-connection on E and let $g \in \mathcal{G}$ be a gauge transformation. Then we claim that there is a new G-connection D' on E such that

$$D'_\nu(gs) = gD_\nu s$$

Connections

for all vector fields v on M and sections s of E. The explicit formula for D' is
$$D'_v(s) = gD_v(g^{-1}s),$$
so we just need to check that D' is a G-connection. We check that D' satisfies the Leibniz law, leaving the rest to the reader. Suppose f is any function on M. Then
$$\begin{aligned}D'_v(fs) &= gD_v(g^{-1}fs)\\ &= gD_v(fg^{-1}s)\\ &= gv(f)g^{-1}s + fgD_v(g^{-1}s)\\ &= v(f)s + fD'_v s.\end{aligned}$$

Exercise 85. *Check that D' has the rest of the properties of a connection.*

Exercise 86. *Using a local trivialization of E over $U_\alpha \subseteq M$ write the G-connection D as the standard flat connection plus a vector potential: $D = D^0 + A$. Show that the vector potential A' for D' is given in local coordinates by*
$$A'_\mu = gA_\mu g^{-1} + g\partial_\mu g^{-1}.$$
Show that since A_μ lives in $\mathfrak{g}$, so does A'_μ. (Hint: show that if A_μ lives in $\mathfrak{g}$ and $g \in G$, then $gA_\mu g^{-1}$ lives in $\mathfrak{g}$. Also show that if $g \in \mathcal{G}$, $g\partial_\mu g^{-1}$ lives in $\mathfrak{g}$.) Conclude that D' is a G-connection.

If the G-connection D' is obtained from the G-connection D by a gauge transformation, we say that D and D' are **gauge-equivalent**. In gauge theory, two connections are regarded as describing the same physical field if they are gauge-equivalent. Sometimes people denote the space of all G-connections on a G-bundle as $\mathcal{A}$, and write $\mathcal{A}/\mathcal{G}$ for the space of gauge equivalence classes of connections, also known as the space of **connections modulo gauge transformations**. The geometry of $\mathcal{A}/\mathcal{G}$ plays a crucial role in gauge theory.

Let us see what gauge transformations do to connections in the case $G = \mathrm{U}(1)$, that is, the case of electromagnetism. For simplicity, let us assume E is a trivial complex line bundle over M. In other words, we assume $E = M \times \mathbb{C}$, so that the fiber E_p over any point $p \in M$ equals $\mathbb{C}$. A connection D on E can then be described by its vector potential A, which is an $\text{End}(E)$-valued 1-form, but since $\text{End}(\mathbb{C}) = \mathbb{C}$ (more precisely, they are canonically isomorphic) this is the same as a complex-valued 1-form. Note that E becomes a $U(1)$-bundle if we think of its standard fiber, $\mathbb{C}$, as the fundamental representation of the group $U(1)$. If the connection D is a $U(1)$-connection, the components A_μ of the vector potential must live in $\mathfrak{u}(1)$. Since

$$\mathfrak{u}(1) = \{ix\colon x \in \mathbb{R}\},$$

this means that the components A_μ are purely imaginary functions, or in other words, A equals i times a real-valued 1-form. This may puzzle the reader, since in Part I we said that the vector potential was a real-valued 1-form! The point is that the vector potential we are dealing with here is really equal to i times the one used in Part I. This actually turns out to be rather handy, but one has to be careful to keep things straight.

Now suppose we apply a gauge transformation g to the vector potential A. Since E is trivial we can think of g as a $U(1)$-valued function, and by Exercise 86 we obtain

$$A'_\mu = g A_\mu g^{-1} + g \partial_\mu g^{-1}.$$

Since $U(1)$ is abelian, this simplifies to

$$A'_\mu = A_\mu + g \partial_\mu g^{-1},$$

and if we can write

$$g = e^{-f}$$

for some imaginary-valued function f, we have $g \partial_\mu g^{-1} = \partial_\mu f$, hence

$$A'_\mu = A_\mu + \partial_\mu f$$

or simply

$$A' = A + df.$$

In other words, in this special case, applying a gauge transformation to A simply amounts to adding the exact 1-form df. This was the definition of a gauge transformation in electromagnetism that we gave in Part I.

We will not go too much deeper into the study of gauge freedom for G-connections, but leave off with a useful exercise. If we have a G-bundle over a spacetime M that we can split into a product $\mathbb{R} \times S$ of space and time, we can define the notion of a G-connection in 'temporal gauge', just as in electromagnetism, and we can show that any G-connection is gauge equivalent to one in temporal gauge:

Exercise 87. *Suppose that E is a trivial G-bundle over the spacetime $\mathbb{R} \times S$, where S is any manifold. Given an $\mathrm{End}(E)$-valued 1-form A on M, let $A_0 = A(\partial_t)$, where t is the usual time coordinate on $\mathbb{R} \times S$. We say that a G-connection D on E is in **temporal gauge** if $D = D^0 + A$ where $A_0 = 0$. Modify the argument given in the section on gauge freedom in Chapter 6 of Part I to show that any G-connection on E is gauge-equivalent to one in temporal gauge.*

Holonomy

A connection on a vector bundle actually plays two closely related roles. First, as we have seen, it allows us to differentiate sections. Second, it allows us to perform 'parallel transport'. As we have described, there is no canonical way to compare vectors lying in the fibers of a vector bundle over two different points p and q. If we pick a connection and a path γ from p to q, however, there is a canonical way to drag a vector in the fiber over p along the path γ, winding up with a vector in the fiber over q. Except in certain special circumstances, the result will depend on the path γ.

A good example is the tangent bundle of the sphere, TS^2. As we shall see in Part III, the usual Riemannian metric on S^2 determines a connection on its tangent bundle. Not worrying about the formula for this connection for now, let us simply describe what happens when we use it to parallel transport a vector tangent to the north pole down to the south pole. There are, of course, many paths from the north pole to the south pole. In Figure 9 we have chosen one of these and shown how a particular vector tangent to the north pole is parallel transported to the south pole. Note that this corresponds to what one would do if, standing at the north pole with a big arrow in ones hands, one were told to carry it to the south pole while keeping it tangent to the sphere and keeping its direction as constant as possible each step of the way, never rotating it unnecessarily. Of course, keeping it tangent to the sphere and never letting it change direction are contradictory if we think of the vector as a vector in $\mathbb{R}^3$, but we are thinking of it as a tangent vector to S^2, and the orders for the traveler are simply to keep its direction *as constant as possible* each step of the way.

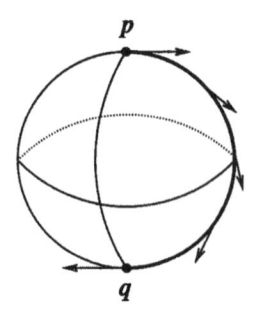

Fig. 9. Parallel translating a tangent vector along a path

In Figure 10 we show the result of parallel transporting the same vector to the south pole along a different path. Note that the result is different! Alternatively, we can parallel transport a vector around a loop that begins and ends at the north pole, as shown in Figure 11; typically, it does not wind up where it began. Parallel transport around a loop defines a linear map from a given fiber of a vector bundle to itself, called the 'holonomy' of the connection around the loop.

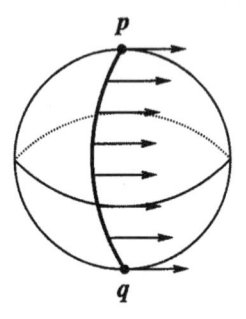

Fig. 10. Parallel translating a tangent vector along another path

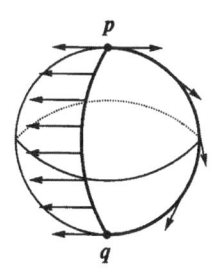

Fig. 11. Parallel translating a tangent vector around a loop

The notion of holonomy plays a basic role in gauge theories of physics. An example that we have already seen in electromagnetism is the *phase* a charged particle's wavefunction is multiplied by when the particle is moved around a loop. The phase is an element of the group U(1), the gauge group of electromagnetism. Recently there has been a lot of work on formulating gauge theories completely (or as much as possible) in terms of holonomies around loops. This approach, called the 'loop representation' of a gauge theory, has been used in quantum chromodynamics starting in the late 1970s. In 1990, Rovelli and Smolin published a paper in which they used Ashtekar's formulation of general relativity as an $SL(2, \mathbb{C})$ gauge theory to construct a loop representation of quantum gravity. We hope this book will help prepare the reader to understand their paper and its many interesting spinoffs!

Let us make the concept of parallel translation precise. Let E be a vector bundle over M equipped with a connection D. Let $\gamma\colon [0, T] \to M$ be a smooth path from the point p to the point q, and suppose that for $t \in [0, T]$, $u(t)$ is a vector in the fiber of E over $\gamma(t)$. We want to write an equation saying that $u(t)$ is 'parallel transported' along γ. This should somehow say that the covariant derivative of $u(t)$ in the direction γ is going, namely $\gamma'(t)$, is equal to zero. Formally we can write this as

$$D_{\gamma'(t)} u(t) = 0.$$

The problem is that so far we only know how to define the covariant derivative of a section of E defined on some open set, while u lives only on the curve γ. However, if we use a local trivialization

$$E|_U \cong U \times V$$

of E over a neighborhood U of $\gamma(t)$, we can think of sections s of E as V-valued functions and think of the connection D as the standard flat connection plus a vector potential A:

$$D_\mu s = \partial_\mu s + A_\mu s$$

where A_μ is an $\text{End}(V)$-valued function. Thus, by analogy, we define the **covariant derivative**

$$D_{\gamma'(t)} u(t) = \frac{d}{dt} u(t) + A(\gamma'(t)) u(t).$$

Exercise 88. *Show that $D_{\gamma'(t)} u(t)$ defined in this manner is actually independent of the choice of local trivialization.*

We then say that $u(t)$ is **parallel transported** (or **parallel translated**) along γ if $D_{\gamma'(t)} u(t) = 0$ for all t.

In fact, starting with any vector $u \in E_p$, we can parallel translate it along γ, that is, we can find $u(t) \in E_{\gamma(t)}$ such that

$$u(0) = u, \qquad D_{\gamma'(t)} u(t) = 0.$$

To see this, it suffices to work locally and show that we can solve the differential equation

$$\frac{d}{dt} u(t) + A(\gamma'(t)) u(t) = 0.$$

This follows from the basic existence result on linear differential equations, but it is also very nice to have an explicit formula for the solution. This is easiest to describe if we assume the whole curve γ lies in an open set over which we have trivialized E; otherwise, we can cut γ up into pieces, each of which lies in such an open set, and treat them separately. Rewriting the equation as

$$\frac{d}{dt} u(t) = -A(\gamma'(t)) u(t),$$

we see that the solution, if it exists, satisfies

$$u(t) = u - \int_0^t A(\gamma'(t_1)) u(t_1) \, dt_1.$$

Holonomy

Of course, we cannot say yet that we have 'solved for' $u(t)$ in this equation, since the quantity to be solved for also appears in the right hand side. Nonetheless, we can use a sneaky recursive trick to really solve for $u(t)$. Simply plug the left hand side into the right hand side! If we do it once we get:

$$u(t) = u - \int_0^t A(\gamma'(t_1))u\,dt_1 + \int_0^t \int_0^{t_1} A(\gamma'(t_1))A(\gamma'(t_2))u(t_2)\,dt_2 dt_1.$$

If we do it again, we get

$$\begin{aligned} u(t) &= u - \int_0^t A(\gamma'(t_1))u\,dt_1 + \\ &\quad \int_0^t \int_0^{t_1} A(\gamma'(t_1))A(\gamma'(t_2))u\,dt_2 dt_1 - \\ &\quad \int_0^t \int_0^{t_1} \int_0^{t_2} A(\gamma'(t_1))A(\gamma'(t_2))A(\gamma'(t_3))u(t_3)\,dt_3 dt_2 dt_1. \end{aligned}$$

It seems that we are getting nowhere fast! However, if we repeat this process *infinitely* many times we get a formula for $u(t)$ as the following infinite sum:

$$u(t) = \sum_{n=0}^\infty \left((-1)^n \int_{t \geq t_1 \geq \cdots \geq t_n \geq 0} A(\gamma'(t_1)) \cdots A(\gamma'(t_n))\,dt_n \cdots dt_1 \right) u.$$

The wonderful thing is that this sum actually converges to the right answer! This is a little exercise in analysis:

Exercise 89. *Put a norm on the vector space V and give $\mathrm{End}(V)$ the norm*

$$\|T\| = \sup_{\|u\|=1} \|Tu\|.$$

Let

$$K = \sup_{t \in [0,t]} \|A(\gamma'(t))\|.$$

Show that the nth term in the sum above has norm $\leq t^n K^n \|u\|/n!$, so that the sum converges. Show using similar estimates that $u(t)$ is differentiable (in fact, smooth), and that $u(0) = u$ and $\frac{d}{dt}u(t) = -A(\gamma'(t))u(t)$.

Physicists have a nice way of thinking about this sum. Let us define the **path-ordered product**

$$P\, A(\gamma'(t_1)) \cdots A(\gamma'(t_n))$$

to be the product with the factors permuted

$$A(\gamma'(t_{\sigma(1)})) \cdots A(\gamma'(t_{\sigma(n)}))$$

so that the larger values of t_i appear first:

$$t_{\sigma(1)} \geq \cdots \geq t_{\sigma(n)}.$$

The integral

$$\int_{t \geq t_1 \geq \cdots \geq t_n \geq 0} A(\gamma'(t_1)) \cdots A(\gamma'(t_n))\, dt_n \cdots dt_1$$

is equal to

$$\frac{1}{n!} \int_{t_i \in [0,t]} P\, A(\gamma'(t_1)) \cdots A(\gamma'(t_n))\, dt_n \cdots dt_1,$$

or, for short,

$$\frac{1}{n!} P \left(\int_0^t A(\gamma'(s))ds \right)^n.$$

Thus if we define the **path-ordered exponential** by

$$P e^{-\int_0^t A(\gamma'(s))ds} = \sum_{n=0}^{\infty} \frac{(-1)^n}{n!} P \left(\int_0^t A(\gamma'(s))ds \right)^n$$

we have

$$u(t) = P e^{-\int_0^t A(\gamma'(s))ds}\, u.$$

Note that the path-ordered exponential reduces to the ordinary exponential in various cases. If $A(\gamma'(t))$ is independent of t, so that it equals a fixed element $A \in \text{End}(V)$, our original differential equation has constant coefficients:

$$\frac{d}{dt} u(t) = -A u(t)$$

Holonomy

and the solution is the ordinary exponential

$$u(t) = e^{-tA}u,$$

defined by the power series

$$e^{-tA} = 1 - tA + \frac{(tA)^2}{2!} - \cdots.$$

Also, if the $A(\gamma'(t))$ for different values of t commute, the path-ordering process has no effect:

$$P\, A(\gamma'(t_1)) \cdots A(\gamma'(t_n)) = A(\gamma'(t_1)) \cdots A(\gamma'(t_n)),$$

so we have

$$u(t) = e^{-\int_\gamma A} u.$$

This occurs whenever the gauge group is abelian. Indeed, we saw a similar expression in Chapter 6 of Part I, namely

$$e^{-\frac{i}{\hbar} q \int_\gamma A},$$

which describes the phase acquired by a charged particle as it moves along a path through a magnetic field. In our new setup, the U(1)-connection A automatically includes the factor of i that we had to write down explicitly in Part I. Also, the factor of $q/\hbar$ we saw in Part I is now being set equal to 1 by an appropriate choice of units. More precisely, if choose units so that $q/\hbar = 1$ for q the fundamental unit of charge in nature, then the wavefunction of a particle with this charge will be a section of a U(1)-bundle with standard fiber given by the fundamental representation of U(1). Similarly, the wavefunction of a particle of n times this charge will correspond to a section of a U(1)-bundle with standard fiber given by the representation ρ_n of U(1).

Suppose now that $\gamma \colon [0, T] \to M$ is a smooth path from p to q in the manifold M and E is a vector bundle with connection D. Given $u \in E_p$, let $H(\gamma, D)u$ denote the result of parallel transporting u to q along the path γ. Since the differential equation defining parallel transport is linear, the map

$$H(\gamma, D) \colon E_p \to E_q$$

is linear. We call this the **holonomy** along the path γ. More generally, if γ is piecewise smooth, we can find the points at which it is not smooth, break it up into maximal smooth pieces $\gamma_i: [t_i, t_{i+1}] \to M$, where $1 \le i \le n$, and define the holonomy by

$$H(\gamma, D) = H(\gamma_n, D) \cdots H(\gamma_1, D).$$

In other words, we parallel translate a vector along a piecewise smooth path by parallel translating it along one piece at a time.

This leads us to the notion of the 'product' of paths. Suppose that we have a path α in M from p to q, and a path β in M from q to r. We say that α and β are **composable**, since we can 'compose' or stick them together to get a path $\beta\alpha$ from p to r, as in Figure 12. Note the unfortunate fact that we have decided to call this path $\beta\alpha$ instead of the seemingly more sensible name $\alpha\beta$. This is because we wish to be compatible with the traditional notation for composites of functions, where $g \circ f$ denotes the result of *first* doing f and *then* g. Some people have gotten so annoyed by this business that they have proposed writing the composite of functions the other way around. We will stick with the traditional way.

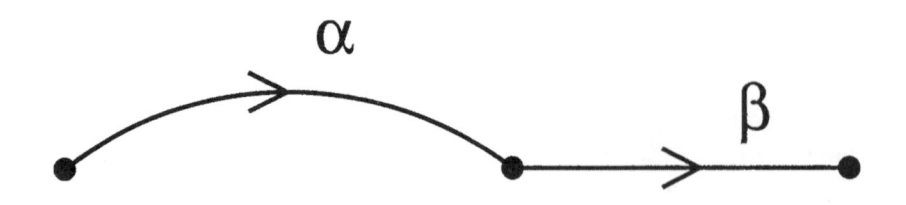

Fig. 12. The product of paths α and β

More precisely, if we have a path $\alpha: [0, S] \to M$ and $\beta: [0, T] \to M$ with $\beta(0) = \alpha(S)$ we define the **product**

$$\beta\alpha: [0, S+T] \to M$$

by

$$(\beta\alpha)(t) = \begin{cases} \alpha(t) & \text{if } 0 \le t \le S \\ \beta(t-S) & \text{if } S \le t \le S+T \end{cases}$$

Holonomy

Note that even if α and β are smooth, $\beta\alpha$ need only be piecewise smooth; this is the main reason for working with piecewise smooth paths. The parametrization that we have chosen for $\beta\alpha$ is somewhat arbitrary, but luckily the holonomy along a path does not depend upon its parametrization:

Exercise 90. *Let $\alpha: [0, T] \to M$ be a piecewise smooth path and let $f: [0, S] \to [0, T]$ be any piecewise smooth function with $f(0) = 0$, $f(S) = T$. Let β be the reparametrized path given by $\beta(t) = \alpha(f(t))$. Show that for any connection D on a vector bundle $\pi: E \to M$, $H(\alpha, D) = H(\beta, D)$.*

The notion of a product of paths is nice because it is not hard to show that if α and β are composable paths,

$$H(\beta\alpha, D) = H(\beta, D)H(\alpha, D).$$

Also, for any path $\alpha: [0, T] \to M$ from p to q there is an **inverse** path α^{-1} from q to p given by

$$\alpha^{-1}(t) = \alpha(T - t),$$

and one can show that

$$H(\alpha^{-1}, D) = H(\alpha, D)^{-1},$$

Moreover, for any point p we can define the **identity loop** $1_p: [0, 1] \to M$ which is the path that simply stays at p:

$$1_p(t) = p.$$

If α is a path from p to q we *do not* have identities like $1_q \alpha = \alpha$, $\alpha 1_p = \alpha$, or $\alpha^{-1}\alpha = 1_p$, but we *do* have

$$\begin{align} H(1_q\alpha, D) &= H(\alpha, D) \\ H(\alpha 1_p, D) &= H(\alpha, D) \\ H(1_p, D) &= 1. \end{align}$$

Exercise 91. *Check these identities.*

The holonomy $H(\gamma, D)$ is affected in a very simple way when we apply a gauge transformation g to the connection D. This is easy to see in a local trivialization. Suppose that $u(t) \in E_{\gamma(t)}$ satisfies the parallel transport equation

$$D_{\gamma'(t)} u(t) = 0$$

If A is the vector potential of D, this equation says

$$\frac{d}{dt} u(t) = -A(\gamma'(t)) u(t)$$

or

$$\frac{d}{dt} u(t) = -\gamma'^{\mu}(t) A_\mu \, u(t)$$

where we write just A_μ for $A_\mu(\gamma(t))$. Now apply a gauge transformation g to $u(t)$, defining $w(t)$ by

$$w(t) = g(\gamma(t)) u(t).$$

We can differentiate w and obtain

$$\begin{aligned}\frac{d}{dt} w(t) &= \left(\frac{d}{dt} g(\gamma(t))\right) u(t) + g(\gamma(t)) \frac{d}{dt} u(t) \\ &= \gamma'^{\mu}(t)(\partial_\mu g) u(t) - g\gamma'^{\mu}(t) A_\mu u(t) \\ &= \gamma'^{\mu}(t)(\partial_\mu g) g^{-1} w(t) - \gamma'^{\mu}(t) g A_\mu g^{-1} w(t).\end{aligned}$$

Since gg^{-1} is constant, we have

$$(\partial_\mu g) g^{-1} = -g \partial_\mu g^{-1}$$

so

$$\begin{aligned}\frac{d}{dt} w(t) &= -\gamma'^{\mu}(t) g (\partial_\mu g^{-1}) w(t) - \gamma'^{\mu}(t) g A_\mu g^{-1} w(t) \\ &= -\gamma'^{\mu}(t) A'_\mu w(t)\end{aligned}$$

where A' is the vector potential obtained from A by the gauge transformation g:

$$A'_\mu = g A_\mu g^{-1} + g \partial_\mu g^{-1}.$$

Holonomy

Thus $w(t)$ satisfies the parallel transport equation

$$D'_{\gamma'(t)} w(t) = 0$$

where D' is the result of applying the gauge transformation g to D. Since the holonomy $H(\gamma, D)$ is the linear map sending $u(0)$ to $u(T)$, and similarly $H(\gamma, D')$ sends $w(0) = g(\gamma(0))u(0)$ to $w(T) = g(\gamma(T))w(T)$, it follows that

$$H(\gamma, D') = g(\gamma(T)) \, H(\gamma, D) \, g(\gamma(0))^{-1}.$$

This is the formula for how holonomies transform under gauge transformations.

Exercise 92. *Check that this formula holds even when the path γ does not stay within an open set over which we have trivialized the G-bundle E, by breaking up γ into smaller paths.*

Something very special happens when we consider the holonomy around a *loop*. If γ is loop based at $p \in M$, the holonomy $H(\gamma, D)$ is a linear map from E_p to itself. In other words,

$$H(\gamma, D) \in \mathrm{End}(E)_p.$$

In this case, when we apply a gauge transformation g to D, we have

$$H(\gamma, D') = g(p) H(\gamma, D) g(p)^{-1}$$

so if we take the trace of $H(\gamma, D)$, we obtain a number that does not change under gauge transformations:

$$\begin{aligned}\mathrm{tr}(H(\gamma, D')) &= \mathrm{tr}(g(p) H(\gamma, D) g(p)^{-1}) \\ &= \mathrm{tr}(H(\gamma, D)).\end{aligned}$$

We therefore say that $\mathrm{tr}(H(\gamma, D))$ is **gauge invariant**.

Recall that two connections are regarded as physically the same if they differ by a gauge transformation. Physically observable quantities in a gauge theory should not change under gauge transformations. For this reason, gauge-invariant quantities are very precious in physics, and the trace of the holonomy around a loop has a special name; we call it a **Wilson loop** and write it as

$$W(\gamma, D) = \text{tr}(H(\gamma, D)).$$

These are named after Kenneth Wilson, who introduced them in the context of statistical mechanics. In the case of a U(1) gauge theory the Wilson loop is simply the phase acquired by a charged particle as it moves around γ in the vector potential corresponding to the U(1) connection D. As we saw in Chapter 6 of Part I, it is this sort of phase that gives rise to interference effects like the Bohm-Aharonov effect.

In general, we can think of the Wilson loop as measuring the self-interference of a particle moving in a loop through the gauge field. The marvelous thing about a G-connection is that its holonomy around any loop lives in the gauge group G. However, unless G is abelian, the holonomy around the loop changes when we change the connection by a gauge transformation. (See the formula above for how the holonomy transforms under gauge transformations.) We take the trace of the holonomy and form the Wilson loop in order to extract some gauge-invariant information out of the holonomy.

Exercise 93. *Show that if D is a G-connection on a G-bundle and γ is a loop, the holonomy $H(\gamma, D)$ lives in G. (Hint: first work in a local trivialization and use the fact that $\mathfrak{g}$ is the tangent space of the identity element of G.)*

We cannot resist noting at this point that Wilson loops are the way *knots* become involved in recent work on quantum gravity and other field theories. We will discuss this a bit more in Chapter 5. First, however, we need to talk about the grand generalization of Maxwell's equations that arises when we generalize the vector potential to an arbitrary G-connection as we have. This is the Yang-Mills equation.

Chapter 3

Curvature and the Yang-Mills Equation

What Mills and I were doing in 1954 was generalizing Maxwell's theory. We knew of no geometrical meaning of Maxwell's theory, and we were not looking in that direction. To a physicist, gauge potential is a concept rooted in our description of the electromagnetic field. Connection is a geometrical concept which I only learned around 1970. — C. N. Yang

Curvature

Suppose that E is a vector bundle over M with connection D. The 'curvature' of a connection D measures the failure of covariant derivatives to commute, or if you like, the failure of equality of mixed 'covariant partial derivatives'. Given two vector fields v and w on M, we define the **curvature** $F(v,w)$ to be the operator on sections of E given by

$$F(v,w)s = D_v D_w s - D_w D_v s - D_{[v,w]}s.$$

The first two terms are a rather obvious way to measure the failure of the derivatives D_v and D_w to commute, but the third term calls for some comment. The point is that even in the best of situations, when E is a trivial bundle and D is the standard flat connection, the covariant derivatives may not commute, solely because the vector fields v and w may have nonvanishing Lie bracket. (See the end of Chapter 3 in Part

I where we discuss this.) The term $-D_{[v,w]}$ is meant to correct for this effect. For example, in the case of the standard flat connection on a trivial bundle with fiber V, where a section is really just a function $f: M \to V$, we have

$$F(v,w)s = vwf - wvf - [v,w]f = 0$$

thanks to the third term. A connection with vanishing curvature, that is, one with $F(v,w)s = 0$ for all vector fields v and w and sections s, is said to be **flat**. We have just seen that the standard flat connection on a trivial bundle is really flat.

We can think of the curvature $F(v,w)$ as an entity in its own right, the operator upon sections given by

$$F(v,w) = D_v D_w - D_w D_v - D_{[v,w]},$$

or if one wants to be still more terse,

$$F(v,w) = [D_v, D_w] - D_{[v,w]}.$$

The simplest property of the curvature is that it is antisymmetric:

$$F(v,w) = -F(w,v).$$

A subtler but very important property is that it is also linear over $C^\infty(M)$ in each argument:

$$F(fv,w)s = F(v,fw)s = F(v,w)(fs) = fF(v,w)s$$

for all functions f and vector fields v, w. Sometimes people express this property by saying that the curvature is a 'tensor', which we make clearer later. This property is rather remarkable, because one might expect *derivatives* of f or s to show up.

First let us show that $F(v, fw) = fF(v,w)$. By definition,

$$F(v, fw) = D_v D_{fw} - D_{fw} D_v - D_{[v,fw]}.$$

Now, the Lie bracket of vector fields satisfies

$$[v, fw] = f[v,w] + v(f)w.$$

Curvature

Exercise 94. *Check the above identity.*

Thus we have

$$\begin{aligned}
F(v, fw) &= D_v D_{fw} - D_{fw} D_v - D_{f[v,w]+v(f)w} \\
&= D_v f D_w - f D_w D_v - f D_{[v,w]} - v(f) D_w \\
&= f D_v D_w + v(f) D_w - f D_w D_v - f D_{[v,w]} - v(f) D_w \\
&= f F(v, w).
\end{aligned}$$

Note that the third term in the definition of the curvature is crucial here! To see that $F(fv, w) = fF(v, w)$, we just use the previous calculation and the antisymmetry of F:

$$F(fv, w) = -F(w, fv) = -fF(w, v) = fF(v, w).$$

Finally to show that $F(v, w)fs = fF(v, w)s$, we calculate:

$$\begin{aligned}
F(v, w)(fs) &= D_v D_w(fs) - D_w D_v(fs) - D_{[v,w]}(fs) \\
&= D_v(fD_w s + w(f)s) - D_w(fD_v s + v(f)s) - \\
&\quad f D_{[v,w]} s - ([v,w](f))s \\
&= f D_v D_w s + v(f) D_w s + w(f) D_v s + v(w(f))s - \\
&\quad f D_w D_v s - w(f) D_v s - v(f) D_w s - w(v(f))s - \\
&\quad f D_{[v,w]} s - ([v,w](f))s \\
&= f[D_v, D_w]s - f D_{[v,w]} s \\
&= f F(v, w)s.
\end{aligned}$$

This last result shows that for any fixed v and w, $F(v, w)$ defines a $C^\infty(M)$-linear map from $\Gamma(E)$ to itself. By a result of the previous chapter, this means that $F(v, w)$ corresponds to a section of $\text{End}(E)$.

Let us think a bit about the curvature in terms of local coordinates x^μ on some open set $U \subset M$. Define $F_{\mu\nu}$ to be the section of $\text{End}(E)$ given by

$$F_{\mu\nu} = F(\partial_\mu, \partial_\nu).$$

Note that since $[\partial_\mu, \partial_\nu] = 0$, we also have

$$F_{\mu\nu} = [D_\mu, D_\nu].$$

By the linearity properties of the curvature, we can write $F(v,w)$ for any vector fields v, w on U as

$$F(v,w) = v^\mu w^\nu F_{\mu\nu}.$$

If we also have a local basis of sections e_i for E over U, we can define the components $A^j_{\mu i}$ of the vector potential by

$$D_\mu e_j = A^i_{\mu j} e_i.$$

The covariant derivative of any section s of E over U is then given by

$$(D_\mu s)^i = \partial_\mu s^i + A^i_{\mu j} s^j.$$

We can use these facts to work out a formula for the curvature in terms of the vector potential:

$$\begin{aligned} F_{\mu\nu} e_i &= D_\mu D_\nu e_i - D_\nu D_\mu e_i \\ &= D_\mu(A^j_{\nu i} e_j) - D_\nu(A^j_{\mu i} e_j) \\ &= (\partial_\mu A^j_{\nu i}) e_j + A^k_{\mu j} A^j_{\nu i} e_k - (\partial_\nu A^j_{\mu i}) e_j - A^k_{\nu j} A^j_{\mu i} e_k \end{aligned}$$

or relabeling indices

$$F_{\mu\nu} e_i = ((\partial_\mu A^j_{\nu i}) - (\partial_\nu A^j_{\mu i}) + A^j_{\mu k} A^k_{\nu i} - A^j_{\nu k} A^k_{\mu i}) e_j$$

Since the sections $e_j \otimes e^i$ form a local basis of sections for $\text{End}(E)$, we can also write

$$F_{\mu\nu} = F^j_{\mu\nu i} \, e_j \otimes e^i$$

for some set of functions $F^j_{\mu\nu i}$, the 'components' of the curvature. In particular,

$$F_{\mu\nu} e_i = F^j_{\mu\nu i} e_j$$

so the result of the previous paragraph gives

$$F^j_{\mu\nu i} = \partial_\mu A^j_{\nu i} - \partial_\nu A^j_{\mu i} + A^j_{\mu k} A^k_{\nu i} - A^j_{\nu k} A^k_{\mu i}.$$

This ugly formula can be very useful for certain calculations, but like a fire extinguisher, it should be pulled out only in an emergency. Physicists often suppress the internal indices i, j, k associated to the basis of

Curvature

sections of E over U. This allows them to write the following prettier formula for the curvature in terms of the vector potential:

$$F_{\mu\nu} = \partial_\mu A_\nu - \partial_\nu A_\mu + [A_\mu, A_\nu].$$

Lest the reader think that curvature is a rather abstract thing, defined using either sophisticated mathematics or masses of indices, we hasten to explain what it has to do with parallel transport. We have already seen that parallel transporting a vector from one point to another yields a result that depends on the path taken — see Figures 9 and 10 in the previous chapter. In fact, this effect is often due to curvature. Let us work in local coordinates with a given point $p \in M$ as the origin. Take a vector $v \in E_p$ and parallel transport it around a small square in the x^μ-x^ν plane whose sides are both of 'length' ϵ in this coordinate system:

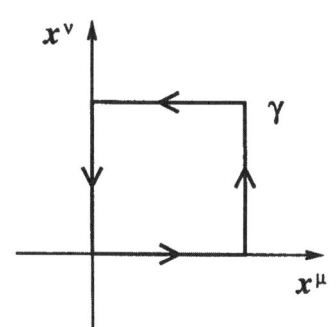

Fig. 1. The holonomy around a small square

The result is a vector $v' \in E_p$ slightly different from v. In fact, up to terms of order ϵ^2,

$$v - v' = \epsilon^2 F_{\mu\nu} v,$$

or in other words, if γ denotes the loop going around the square,

$$H(\gamma, D) = 1 - \epsilon^2 F_{\mu\nu}.$$

In short, the curvature really measures the holonomy around 'infinitesimal loops'!

Exercise 95. *Prove this formula for the holonomy around a small square. (Hint: use the path-ordered exponential formula for parallel transport and keep only terms of order ϵ^2 or less.)*

This result makes it not too surprising that if the connection D is flat — that is, if its curvature vanishes — the holonomies along any two homotopic paths from some point p to another point q are the same.

Exercise 96. *Show this. (Hint: choose a homotopy γ_s between two paths γ_0 and γ_1 from p to q, express the parallel transport map $H(\gamma_s, D)$ using the path-ordered exponential, and show*

$$\frac{d}{ds} H(\gamma_s, D) = 0$$

if D is flat.

This implies that if M is simply connected and D is flat, the holonomy around any loop in M is the identity. If M is not simply connected, there may be nontrivial holonomies even when D is flat. The simplest example is the Möbius strip — see Figure 2.

Exercise 97. *Show that every connection on a vector bundle $\pi\colon E \to M$ is flat if M is 1-dimensional.*

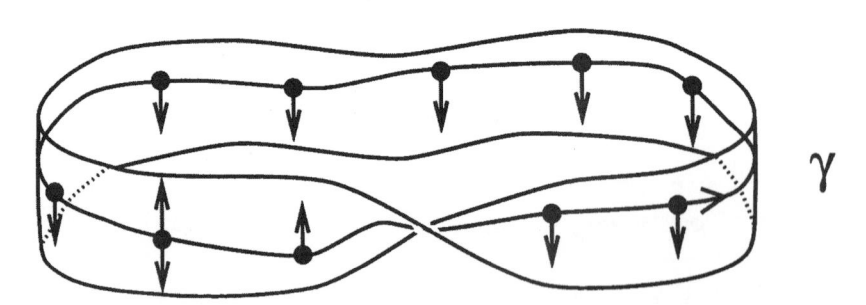

Fig. 2. Nontrivial holonomy around a noncontractible loop

The calculus of connections and curvature can be clarified using **End(E)-valued differential forms**, that is, sections of the bundle

$$\mathrm{End}(E) \otimes \Lambda T^*M$$

or its restriction to some open set in M. We saw already in Chapter 2 that if we work in an open set $U \subseteq M$ where we have introduced local coordinates and a local trivialization of E, we can write any connection

Curvature

D as $D^0 + A$, where D^0 is the standard flat connection and the vector potential A is an End(E)-valued 1-form. We then have

$$A = A_\mu \otimes dx^\mu$$

where each A_μ is a section of End($E|_U$).

Similarly, there is a nice way to view the curvature of the connection D on E as an End(E)-valued 2-form F. As we have already seen, for any vector fields v and w, $F(v,w)$ is a section of End(E), so that working with coordinates on an open set U, the 'components'

$$F_{\mu\nu} = F(\partial_\mu, \partial_\nu)$$

are sections of End(E) over U. We can then define the **curvature 2-form** F, an End(E)-valued 2-form, by

$$F = \frac{1}{2} F_{\mu\nu} dx^\mu \wedge dx^\nu.$$

The factor of 1/2 is introduced for the same reasons as in Chapter 4 of Part I, namely to correct for the double-counting that occurs because

$$F_{\mu\nu} = -F_{\nu\mu}, \qquad dx^\mu \wedge dx^\nu = -dx^\nu \wedge dx^\mu.$$

Now, while our definition of the curvature 2-form F involves coordinates, it is actually coordinate-independent and defined on all of M. The most elegant way to see this, and to really understand the meaning of F, involves a generalization of the usual exterior derivative called the 'covariant exterior derivative'. This reduces to the usual exterior derivative d in the special case of the standard flat connection on a trivial line bundle, where sections are just *functions* on M. As we have seen, one of the most important properties of the usual exterior derivative is that $d^2 = 0$. It turns out that the square of the exterior covariant derivative is *not* zero; in fact, it is proportional to F. This makes sense, because the reason why d^2 vanishes is that partial derivatives ∂_μ commute, and the curvature measures the failure of 'covariant partials' D_μ to commute.

Suppose that E is a vector bundle over a manifold M equipped with a connection D. We define an E-**valued** p-**form** to be a section of

$$E \otimes \Lambda^p T^* M.$$

An E-valued 0-form is really nothing but a section of E. Moreover, just as a 1-form on M is just a $C^\infty(M)$-linear function from Vect(M) to $C^\infty(M)$, a E-valued 1-form is really nothing but a $C^\infty(M)$-linear function from Vect(M) to $\Gamma(E)$.

Exercise 98. *Use Exercise 72 to show that any E-valued differential form can be written — not necessarily uniquely — as a sum of those of the form $s \otimes \omega$, where s is a section of E and ω is an ordinary differential form on M.*

Exercise 99. *Using the previous exercise, show that there is a unique way to define the wedge product of an E-valued form and an ordinary form such that the wedge of the E-valued form $s \otimes \omega$ and the ordinary form μ is given by*
$$(s \otimes \omega) \wedge \mu = s \otimes (\omega \wedge \mu).$$
and such that the wedge product depends $C^\infty(M)$-linearly on each factor.

We define the **exterior covariant derivative** d_D of E-valued differential forms as follows. First, we define d_D of a section s of E to be the E-valued 1-form $d_D s$ such that
$$d_D s(v) = D_v s$$
for any vector field v on M. This is just a generalization of the old formula
$$df(v) = v(f).$$
Alternatively, in local coordinates x^μ on some open set $U \subseteq M$, we have
$$d_D s = D_\mu s \otimes dx^\mu.$$

Exercise 100. *Check that these definitions are equivalent.*

To define d_D on arbitrary E-valued differential forms, it suffices, by Exercise 98, to define it on those of the form $s \otimes \omega$ where s is a section of E and ω is an ordinary differential form. We do this as follows:
$$d_D(s \otimes \omega) = d_D s \wedge \omega + s \otimes d\omega,$$

Curvature

where wedge product $d_D s \wedge \omega$ is defined as in Exercise 99. Strictly speaking, we should check that d_D is well-defined, since an E-valued form may be written as a sum of those of the form $s \otimes \omega$ in many different ways. A concrete way to take care of this is to obtain an unambiguous formula for d_D in local coordinates x^μ on an open set U. In this situation there is a basis of differential forms dx^I, where I ranges over multi-indices. Then we can write any E-valued differential form on U uniquely as

$$s_I \otimes dx^I,$$

for some sections s_I of $E|_U$. We then obtain

$$\begin{aligned} d_D(s_I \otimes dx^I) &= d_D s_I \wedge dx^I + s_I \otimes d(dx^I) \\ &= (D_\mu s_I \otimes dx^\mu) \wedge dx^I \\ &= D_\mu s_I \otimes dx^\mu \wedge dx^I. \end{aligned}$$

This is just a generalization of the following formula for the exterior derivative:

$$d(\omega_I dx^I) = (\partial_\mu \omega_I) \, dx^\mu \wedge dx^I.$$

Now let us show that d_D^2 is proportional to the curvature of D. For this, we need to define the wedge product of an $\mathrm{End}(E)$-valued form and an E-valued form, as follows. By linearity it suffices to consider the product of $T \otimes \omega$ with $s \otimes \mu$, where T is a section of $\mathrm{End}(E)$, s is a section of E, and ω, μ are ordinary differential forms. We define

$$(T \otimes \omega) \wedge (s \otimes \mu) = T(s) \otimes (\omega \wedge \mu).$$

Exercise 101. *Check that the definition above extends uniquely to a wedge product of arbitrary $\mathrm{End}(E)$-valued forms and E-valued forms that is $C^\infty(M)$-linear in each argument.*

Now we claim that

$$d_D^2 \eta = F \wedge \eta$$

for any E-valued form η. Working in local coordinates, we write $\eta = s_I \otimes \omega^I$ for some sections s_I, and compute:

$$\begin{aligned} d_D^2 \eta &= d_D(D_\nu s_I \otimes dx^\nu \wedge dx^I) \\ &= D_\mu D_\nu s_I \otimes dx^\mu \wedge dx^\nu \wedge dx^I \end{aligned}$$

$$= \frac{1}{2}[D_\mu, D_\nu]s_I \otimes dx^\mu \wedge dx^\nu \wedge dx^I$$
$$= \frac{1}{2}F_{\mu\nu}s_I \otimes dx^\mu \wedge dx^\nu \wedge dx^I$$
$$= F \wedge \eta,$$

where in the third line we used the antisymmetry of the wedge product. Not only does this formula show how curvature arises naturally from trying to generalize differential calculus to arbitrary bundles, it also proves that the curvature 2-form F is defined independent of any choice of coordinates — since d_D is!

To wrap up this section, let us see what happens to the curvature when E is a G-bundle and the connection D is a G-connection. We can then work in a local trivialization of E we then have $D = D^0 + A$ where D^0 is the standard flat connection and the components A_μ of the vector potential live in $\mathfrak{g}$. Since

$$F_{\mu\nu} = \partial_\mu A_\nu - \partial_\nu A_\mu + [A_\mu, A_\nu]$$

and the bracket of two sections of $\text{End}(E)$ that live in $\mathfrak{g}$ again lives in $\mathfrak{g}$ (since $\mathfrak{g}$ is a Lie algebra) it follows that the components of $F_{\mu\nu}$ live in $\mathfrak{g}$. If $G = \text{U}(1)$ and the standard fiber of E is the fundamental representation of $\text{U}(1)$ on $\mathbb{C}$, this means that in a local trivialization we can think of F as i times a real-valued 2-form. Since $\text{U}(1)$ is abelian, we have

$$F_{\mu\nu} = \partial_\mu A_\nu - \partial_\nu A_\mu$$

in this case. Apart from the factor of i that is now included in the definition of A and F, this is just the formula for the electromagnetic field in terms of the vector potential! It is for this reason that the curvature plays a role in gauge theory that generalizes the role of the electromagnetic field in Maxwell's equations.

In the next section, we will see that the first pair of Maxwell equations, which are simply a tautology in terms of the vector potential, are a special case of a tautology called the Bianchi identity. In the section after that, we will see how Yang and Mills generalized the second pair of Maxwell equations to obtain the Yang-Mills equation.

The Bianchi identity

The Bianchi identity has many forms and many applications. For example, we will see that in electromagnetism it is simply the equation $dF = 0$, which implies conservation of charge, while in general relativity it is equivalent to (local) conservation of energy and momentum. Perhaps the simplest way to start is by thinking of it as a special case of the Jacobi identity. Recall that for any three linear operators $X, Y, Z: V \to V$ on a vector space V the Jacobi identity holds:

$$[X, [Y, Z]] + [Y, [Z, X]] + [Z, [X, Y]] = 0.$$

Exercise 102. *Check this identity.*

In particular, if u, v, and w are vector fields on M, the **Bianchi identity** simply says that

$$[D_u, [D_v, D_w]] + [D_v, [D_w, D_u]] + [D_w, [D_u, D_v]] = 0.$$

This may not seem to have much to do with the curvature. Suppose, however that we take u, v, and w to be the coordinate vector fields $\partial_\mu, \partial_\nu, \partial_\lambda$. Using the fact that $[D_\mu, D_\nu] = F_{\mu\nu}$, we obtain another form of the Bianchi identity:

$$[D_\mu, F_{\nu\lambda}] + [D_\nu, F_{\lambda\mu}] + [D_\lambda, F_{\mu\nu}] = 0.$$

In short, the Bianchi identity says that a certain combination of derivatives of the curvature vanishes. It takes more work to see the real *meaning* of this combination of derivatives. In order to do this, and also to familiarize the reader with various commonly used notations in gauge theory, we will derive the Bianchi identity in two other ways.

For these, we need to define an exterior covariant derivative for $\text{End}(E)$-valued differential forms — this will let us see the Bianchi identity in its most beautiful form. We could define d_D just as we did for E, if only we had a connection on $\text{End}(E)$. Thus we will describe a recipe for getting a connection on $\text{End}(E)$ from the connection D on E. The subject of getting new connections from old ones is actually interesting in its own right. Let us start with the connection D on a vector bundle E over M. Then there is a unique connection on the dual vector bundle E^*, which we could call D^*, such that for any vector field v on M and sections $s \in \Gamma(E)$, $\lambda \in \Gamma(E^*)$,

$$v(\lambda(s)) = (D_v^*\lambda)(s) + \lambda D_v(s).$$

(Think of this as a version of the product rule, or Leibniz law.) To obtain this, we simply *define* D^* on E^* by

$$(D_v^*\lambda)(s) = v(\lambda(s)) - \lambda D_v(s).$$

Exercise 103. *Check that D^* is a connection on E^*.*

Next, suppose that we also have a connection D' on another vector bundle E' over M. Then there is a connection on the vector bundle $E \oplus E'$, which we could call $D \oplus D'$, such that, for any vector field v and sections $s \in \Gamma(E)$, $s' \in \Gamma(E')$,

$$(D \oplus D')_v(s, s') = (D_v s, D'_v s').$$

Exercise 104. *Check that $D \oplus D'$ is a connection.*

Similarly, there is a connection on $E \otimes E'$, which we could call $D \otimes D'$, such that

$$(D \otimes D')_v(s \otimes s') = (D_v s) \otimes s' + s \otimes (D'_v s').$$

(Again, this is a form of the Leibniz law.)

Exercise 105. *Check that $D \otimes D'$ is a connection.*

Since $\text{End}(E) = E \otimes E^*$, we can get a connection on $\text{End}(E)$ from a connection on E using the above constructions. Abusing language a bit, we will use D to denote both a connection on E and the resulting connection on $\text{End}(E)$ — the context should make things clear. More explicitly, we have the following formula for the connection on $\text{End}(E)$:

Exercise 106. *Starting with a connection D on E, and using the above constructions to define a connection D on $\text{End}(E)$, show that*

$$(D_v T)(s) = D_v(Ts) - T(D_v s)$$

for all vector fields v on M, sections T of $\text{End}(E)$, and sections s of E.

The Bianchi Identity

Now that we have a connection D on $\text{End}(E)$, we can use it to define the exterior covariant derivative d_D of $\text{End}(E)$-valued differential forms. In these terms, we claim that the **Bianchi identity** takes the elegant form

$$d_D F = 0.$$

This is really just a generalization of the first pair of Maxwell's equations, $dF = 0$, which are an *identity* if we assume $F = dA$. The nontrivial Yang-Mills equations, which we discuss in the next chapter, are the generalization of the second pair of Maxwell's equations, $\star d \star F = J$. To get ahead of ourselves a little bit, these are simply

$$\star d_D \star F = J!$$

So Yang-Mills theory is very much like electromagnetism.

To derive the above form of the Bianchi identity, we need a fact about the covariant exterior derivative of a wedge product:

Exercise 107. *Show that if D is a connection on E, ω is an $\text{End}(E)$-valued p-form, and μ is an E-valued form, we have*

$$d_D(\omega \wedge \mu) = d_D\omega \wedge \mu + (-1)^p \omega \wedge d_D\mu.$$

(Hint: do the calculation in local coordinates.)

Now let us look at the action of d_D^3 on any E-valued form η on M. On the one hand,

$$d_D^3 \eta = d_D(d_D^2 \eta) = d_D(F \wedge \eta) = d_D F \wedge \eta + F \wedge d_D \eta,$$

but on the other hand,

$$d_D^3 \eta = d_D^2(d_D \eta) = F \wedge d_D \eta.$$

Since this is true for any η, we must have the **Bianchi identity**,

$$d_D F = 0.$$

To see that this really *is* just the Bianchi identity, the reader should work out $d_D F$ in local coordinates:

Exercise 108. *Writing*

$$F = \frac{1}{2}F_{\mu\nu}dx^\mu \wedge dx^\nu$$

in local coordinates, show from the definition of d_D on $\mathrm{End}(E)$-valued 1-forms that

$$d_D F = \frac{1}{3!}(D_\mu F_{\nu\lambda} + D_\nu F_{\lambda\mu} + D_\lambda F_{\mu\nu}) \otimes dx^\mu \wedge dx^\nu \wedge dx^\lambda.$$

Thus the vanishing of $d_D F$ is equivalent to the equation

$$D_\mu F_{\nu\lambda} + D_\nu F_{\lambda\mu} + D_\lambda F_{\mu\nu} = 0.$$

Using Exercise 106, this is equivalent to

$$[D_\mu, [D_\nu, D_\lambda]] + [D_\nu, [D_\lambda, D_\mu]] + [D_\lambda, [D_\mu, D_\nu]] = 0,$$

which is a version of the Bianchi identity that we have already seen.

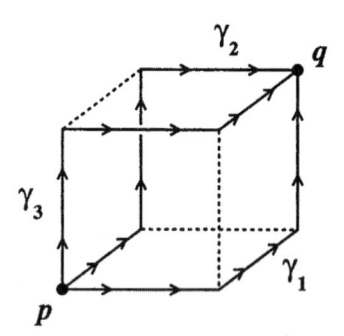

Fig. 3. A cube in x^μ-x^ν-x^λ space

The reader may think the Bianchi identity is a pretty formula but still wonder about its geometrical significance! For this, let us briefly mention what Bianchi identity is really saying about *holonomies*. We have already seen that, working in local coordinates, the holonomy of the connection D around a square loop γ of size ϵ in the x^μ-x^ν plane is, to second order in ϵ,

$$1 - \epsilon^2 F_{\mu\nu}.$$

With more work along these lines, one can show that the Bianchi identity really concerns the holonomies around faces of a *cube* in x^μ-x^ν-x^λ

The Bianchi Identity

space. On the cube above we have marked three paths γ_i from one vertex, p, to the opposite vertex, q.

Starting with this result, and working a bit more, one can show that the holonomies around the three loops shown in Figure 4 are, up to third order in ϵ,

$$H(\gamma_2^{-1}\gamma_1, D) \sim 1 - \epsilon^2(F_{\mu\nu} - F_{\lambda\mu}) + \epsilon^3 D_\nu F_{\lambda\mu}$$
$$H(\gamma_3^{-1}\gamma_2, D) \sim 1 - \epsilon^2(F_{\nu\lambda} - F_{\mu\nu}) + \epsilon^3 D_\lambda F_{\mu\nu}$$
$$H(\gamma_1^{-1}\gamma_3, D) \sim 1 - \epsilon^2(F_{\lambda\mu} - F_{\nu\lambda}) + \epsilon^3 D_\mu F_{\nu\lambda}.$$

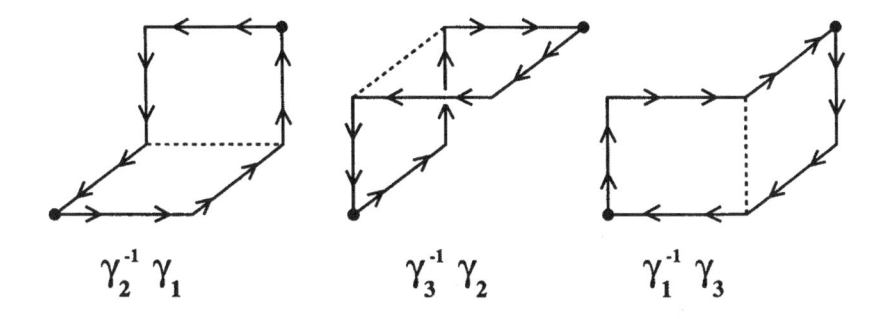

Fig. 4. Proof of the Bianchi identity

If we compose these three loops, using the facts about holonomies that we described in the previous chapter we get

$$H(\gamma_1^{-1}\gamma_3\gamma_3^{-1}\gamma_2\gamma_2^{-1}\gamma_1, D) = H(1_p, D) = 1.$$

On the other hand, up to third order in ϵ we have

$$H(\gamma_1^{-1}\gamma_3\gamma_3^{-1}\gamma_2\gamma_2^{-1}\gamma_1, D) =$$
$$H(\gamma_1^{-1}\gamma_3, D) H(\gamma_3^{-1}\gamma_2, D) H(\gamma_2^{-1}\gamma_1, D) =$$
$$1 + \epsilon^3(D_\mu F_{\nu\lambda} + D_\nu F_{\lambda\mu} + D_\lambda F_{\mu\nu}).$$

Thus we obtain the Bianchi identity,

$$D_\mu F_{\nu\lambda} + D_\nu F_{\lambda\mu} + D_\lambda F_{\mu\nu} = 0$$

in a form we have just seen.

Exercise 109. *Prove the above formulas for the holomies around $\gamma_1^{-1}\gamma_3$, $\gamma_3^{-1}\gamma_2$, and $\gamma_2^{-1}\gamma_1$. (Hint: use the path-ordered exponential and keep only terms of order ϵ^3 or less.)*

Before we finish this section, we would like to derive the Bianchi identity one more time in a slightly different way. This makes use of some more of the algebra of E-valued and $\text{End}(E)$-valued forms, which we present as a series of exercises below. The point is that one can define the wedge product of $\text{End}(E)$-valued forms in an obvious sort of way, as well as a kind of commutator of such forms.

Exercise 110. *Show that there is a unique way to define the wedge product of two $\text{End}(E)$-valued forms such that the wedge of the $\text{End}(E)$-valued forms $S \otimes \omega$ and $T \otimes \mu$ is given by*

$$(S \otimes \omega) \wedge (T \otimes \mu) = ST \otimes (\omega \wedge \mu).$$

and such that the wedge product depends $C^\infty(M)$-linearly on each factor.

Exercise 111. *Show that if D is a connection on E, ω is an $\text{End}(E)$-valued p-form, and μ is an $\text{End}(E)$-valued form, we have*

$$d_D(\omega \wedge \mu) = d_D\omega \wedge \mu + (-1)^p \omega \wedge d_D\mu.$$

Exercise 112. *Given an $\text{End}(E)$-valued p-form ω and an $\text{End}(E)$-valued q-form μ, define the* **graded commutator** *by*

$$[\omega, \mu] = \omega \wedge \mu - (-1)^{pq} \mu \wedge \omega.$$

(The factor of $(-1)^{pq}$ is to correct for the antisymmetry of the wedge product of ordinary differential forms.) Show that

$$[\omega, \mu] = -(-1)^{pq}[\mu, \omega].$$

Also show the **graded Jacobi identity***: if ω, μ, η are $\text{End}(E)$-valued p-, q-, and r-forms, respectively, then*

$$[\omega, [\mu, \eta]] + (-1)^{p(q+r)}[\mu, [\eta, \omega]] + (-1)^{r(p+q)}[\eta, [\omega, \mu]] = 0.$$

Show that if A is an $\text{End}(E)$-valued form, we need not have $A \wedge A = 0$, but we do have $[A, A \wedge A] = 0$.

The Bianchi Identity

Now let us assume that the bundle E admits a flat connection D^0. This is true for all trivial bundles, so it is always true if we work *locally*. Let us write
$$d = d_{D^0}$$
for the exterior covariant derivative of E-valued or $\text{End}(E)$-valued forms with respect to the connection D^0. This abuse of notation is somewhat reasonable, since, as D^0 is flat, this d satisfies
$$d^2 = 0$$
just as the exterior derivative of ordinary differential forms does. Given any other connection D on E, we can write it as $D^0 + A$, where A is an $\text{End}(E)$-valued 1-form. We claim that then
$$d_D \omega = d\omega + A \wedge \omega,$$
for any E-valued form ω, while
$$d_D \eta = d\eta + [A, \eta]$$
for any $\text{End}(E)$-valued form η. To show the first, work in local coordinates and write $\omega = \omega_I \otimes dx^I$ where the ω_I are sections of E, and note:
$$\begin{aligned} d_D \omega &= D_\mu \omega_I \otimes dx^\mu \wedge dx^I \\ &= (D^0_\mu + A_\mu) \omega_I \otimes dx^\mu \wedge dx^I \\ &= d\omega + A \wedge \omega \end{aligned}$$

For the second, work in local coordinates and write the $\text{End}(E)$-valued p-form η as $\eta_I \otimes dx^I$, where ω_I are sections of $\text{End}(E)$:
$$\begin{aligned} d_D \eta &= [D_\mu, \eta_I] \otimes dx^\mu \wedge dx^I \\ &= [D^0_\mu + A_\mu, \eta_I] \otimes dx^\mu \wedge dx^I \\ &= d\eta + A \wedge \eta - (-1)^p \eta \wedge A \\ &= d\eta + [A, \eta] \end{aligned}$$
using the definition of the graded commutator.

Now, we have already seen that if ω is any E-valued form we have

$$d_D^2 \omega = F \wedge \omega$$

where F is the curvature of D. We can also compute

$$\begin{aligned} d_D^2 \omega &= d_D(d\omega + A \wedge \omega) \\ &= d^2\omega + A \wedge d\omega + d(A \wedge \omega) + A \wedge A \wedge \omega \\ &= A \wedge d\omega + dA \wedge \omega - A \wedge d\omega + A \wedge A \wedge \omega \\ &= (dA + A \wedge A) \wedge \omega. \end{aligned}$$

This strongly suggests that

$$F = dA + A \wedge A.$$

Indeed, this is true; it is simply a compressed form of our earlier formula

$$F_{\mu\nu} = \partial_\mu A_\nu - \partial_\nu A_\mu + [A_\mu, A_\nu].$$

In this context we can prove the Bianchi identity

$$d_D F = 0$$

by the following nice computation:

$$\begin{aligned} d_D F &= dF + [A, F] \\ &= d(dA + A \wedge A) + [A, (dA + A \wedge A)] \\ &= dA \wedge A - A \wedge dA + [A, dA] + [A, A \wedge A] \\ &= [dA, A] + [A, dA] \\ &= 0 \end{aligned}$$

using the identities in Exercise 112. This approach is particularly handy in studying Chern-Simons theory, as we will do in the next chapter. Chern-Simons theory is a gauge theory defined for trivial vector bundles, so we can use this formalism *globally*, not just locally.

The Yang-Mills Equation

Recall Maxwell's equations:

$$dF = 0, \qquad \star d \star F = J.$$

In the key case when the electromagnetic field is the exterior derivative of a vector potential

$$F = dA$$

the first equation becomes a tautology, so all the physics is concentrated in the second. As we have seen, the tautologous half of Maxwell's equations has a far-reaching generalization, the Bianchi identity: for any connection D we have

$$d_D F = 0.$$

To generalize the other half of Maxwell's equations we need to define a Hodge star operator for endomorphism-valued differential forms.

To do this, let $\pi \colon E \to M$ be a vector bundle over an oriented semi-Riemannian manifold M. Then let the **Hodge star operator** $\star$ acting on $\mathrm{End}(E)$-valued differential forms be the unique $C^\infty(M)$-linear operator such that for any section T of $\mathrm{End}(E)$ and any differential form ω,

$$\star(T \otimes \omega) = T \otimes \star\omega.$$

On the right side, $\star$ represents the Hodge star operator on ordinary differential forms. Then for any $\mathrm{End}(E)$-valued 1-form J on M, called the **current**, the **Yang-Mills equation** is

$$\star d_D \star F = J.$$

In particular, when E is a trivial U(1) bundle with standard fiber given by the fundamental representation of U(1), the Yang-Mills equation reduces to Maxwell's equation $\star d \star F = J$.

To see the Bianchi identity and the Yang-Mills equation in a more concrete physical form, let us split spacetime into space and time. Let M be a 4-dimensional static spacetime of the form $\mathbb{R} \times S$. Then M has a metric of the form

$$g = -dt^2 + {}^3g$$

as in Chapter 5 of Part I. We can write

$$F = B + E \wedge dt$$

where B, the Yang-Mills version of the **magnetic field**, is an $\mathrm{End}(E)$-valued 2-form on space, while the **electric field** E is an $\mathrm{End}(E)$-valued 1-form on space. (We will have to live with the annoying fact that E stands both for the bundle and the electric field!) Similarly we write

$$J = j - \rho dt$$

where j is an $\mathrm{End}(E)$-valued 1-form on space and ρ is a section of $\mathrm{End}(E)$, both depending on t. We can split the exterior covariant derivative into space and time parts:

$$d_D \omega = dt \wedge D_t \omega + d_S \omega.$$

Then the Bianchi identity becomes

$$d_D F = d_S B + dt \wedge (D_t B + d_S E) = 0,$$

hence

$$d_S B = 0, \qquad D_t B + d_S E = 0.$$

Let $\star_S$ denote the Hodge star operator on $\mathrm{End}(E)$-valued forms on S. Then

$$\star F = \star_S E - \star_S B \wedge dt$$

and the Yang-Mills equation becomes

$$\star d \star F = -D_t E - \star_S d_S \star_S E \wedge dt + \star_S d_S \star_S B = j - \rho dt,$$

hence

$$\star_S d_S \star_S E = \rho, \qquad -D_t E + \star_S d_S \star_S B = j.$$

Let us show that the Yang-Mills equation is **gauge-invariant**, that is, that if D is any connection satisfying the Yang-Mills equation, so is any gauge transformation of D. To show this, we need to figure out how the curvature transforms under gauge transformations. So fix a

The Yang-Mills Equation

connection D on E, let g be a gauge transformation, and let D' be the gauge transform of D by g, defined by

$$D'_v s = g D_v(g^{-1} s).$$

Let F be the curvature of D and let F' be the curvature of D'. Then

$$\begin{aligned} F'(u,v)s &= D'_u D'_v s - D'_v D'_u s - D'_{[u,v]} s \\ &= g D_u D_v(g^{-1}s) - g D_v D_u(g^{-1}s) - g D_{[u,v]}(g^{-1}s) \\ &= g F(u,v)(g^{-1}s) \end{aligned}$$

or simply

$$F' = g F g^{-1}$$

where we are taking the product of the sections g, g^{-1} of $\mathrm{End}(E)$ with the $\mathrm{End}(E)$-valued form F. (This is a special case of the wedge product of $\mathrm{End}(E)$-valued forms, thinking of g as an $\mathrm{End}(E)$-valued 0-form.) In local coordinates, this means simply that

$$F'_{\mu\nu} = g F_{\mu\nu} g^{-1}.$$

We can then show directly using local coordinates that

$$\star d_D \star F = J \implies \star d_{D'} \star F' = J'$$

where $J' = g J g^{-1}$. Suppose that $\star d_D \star F = J$, and recall that

$$\star d_D \star F = \frac{1}{2}[D_\mu, F_{\nu\lambda}] \otimes \star(dx^\mu \wedge \star(dx^\nu \wedge dx^\lambda)),$$

where on the right hand side D_μ denotes the covariant derivative on sections of E, *not* on sections of $\mathrm{End}(E)$. (This notation may be a bit confusing here, so we advise the reader to review the relevant parts of the previous section.) It follows that

$$\begin{aligned} \star d_{D'} \star F' &= \frac{1}{2}[D'_\mu, F'_{\nu\lambda}] \otimes \star(dx^\mu \wedge \star(dx^\nu \wedge dx^\lambda)) \\ &= \frac{1}{2}[g D_\mu g^{-1}, g F_{\nu\lambda} g^{-1}] \otimes \star(dx^\mu \wedge \star(dx^\nu \wedge dx^\lambda)) \\ &= \frac{1}{2} g [D_\mu, F_{\nu\lambda}] g^{-1} \otimes \star(dx^\mu \wedge \star(dx^\nu \wedge dx^\lambda)) \\ &= g(\star d_D \star F) g^{-1} \\ &= J'. \end{aligned}$$

Let us take advantage of the gauge-invariance to write the Yang-Mills equations in an even more explicit form on Minkowski space. Every vector bundle over $\mathbb{R}^n$ is trivial, so we will take advantage of this and write any connection D as the standard flat connection plus a vector potential, or in component form,

$$D_\mu = \partial_\mu + A_\mu.$$

Moreover, by Exercise 87 we can work in temporal gauge, so that $A_0 = 0$ and

$$D_t = \partial_t, \quad D_i = \partial_i + A_i$$

for $i = 1, 2, 3$. Then the Yang-Mills electric and magnetic fields are given by

$$E = E_i dx^i, \quad B = \frac{1}{2}\epsilon_{ijk} B^i dx^j \wedge dx^k,$$

where

$$E_i = -\partial_t A_i$$

and

$$B^i = \epsilon^{ijk}\left(\partial_j A_k - \partial_k A_j + [A_j, A_k]\right).$$

If we think of A as an $\mathrm{End}(E)$-valued 1-form on space (depending on time), the Bianchi identity becomes

$$\partial^i B_i + [A^i, B_i] = 0, \quad \partial_t B^i + \epsilon^{ijk}(\partial_j E_k + [A_j, E_k]) = 0,$$

and the Yang-Mills equation becomes

$$\partial^i E_i + [A^i, E_i] = \rho, \quad -\partial_t E^i + \epsilon^{ijk}(\partial_j B_k + [A_j, B_k]) = j^i.$$

Note that the only real difference between these and Maxwell's equations are the nonlinear terms that arise when the commutators do not vanish. There is thus a direct link between the *nonabelian* nature of the gauge group and the *nonlinear* nature of the corresponding gauge theory. Note that, as with Maxwell's equations, the Yang-Mills equation consists of one constraint on the Cauchy data (A, E), called the **Gauss law**, together with three evolutionary equations.

In what follows we let the reader work out the transformation properties of exterior covariant derivatives, and use this to prove the gauge invariance of the Yang-Mills equations in a somewhat more conceptual manner.

The Yang-Mills Equation

Exercise 113. *Let $\pi: E \to M$ be a vector bundle with a connection D, and let D' be the gauge transform of D given by $D'_v s = gD_v(g^{-1}s)$. Show that the exterior covariant derivative of E-valued forms transforms as follows: if η is any E-valued form, then*

$$d_{D'}\eta = g d_D(g^{-1}\eta).$$

Exercise 114. *Using the same notation as in the previous exercise, show that the covariant derivative of any section T of $\mathrm{End}(E)$ transforms as follows:*

$$D'_v T = \mathrm{Ad}(g) D_v(\mathrm{Ad}(g^{-1})T),$$

where $\mathrm{Ad}(g)T = gTg^{-1}$.

Exercise 115. *Show that the exterior covariant derivative of any $\mathrm{End}(E)$-valued form η transforms as follows:*

$$d_{D'}\eta = \mathrm{Ad}(g) d_D(\mathrm{Ad}(g^{-1})\eta)$$

where $\mathrm{Ad}(g)\eta = g\eta g^{-1}$.

To see the gauge invariance of the Yang-Mills equations, we now note that
$$\star d_{D'} \star F' = g(\star d_D \star F)g^{-1},$$
so
$$\star d_D \star F = J \implies \star d_{D'} \star F' = J'.$$

Chapter 4
Chern-Simons Theory

On a manifold it is necessary to use covariant differentiation; curvature measures its noncommutativity. Its combination as a characteristic form measures the nontriviality of the underlying bundle. This train of ideas is so simple and natural that its importance can hardly be exaggerated. — Shiing-shen Chern

The Action Principle

In the previous chapter we came to the Yang-Mills equation by following an analogy with Maxwell's equations. In modern physics, however, one rarely *starts* with differential equations for fields; rather, one *derives* them from a 'Lagrangian'. The advantages of this approach are that symmetries and other important properties of the equations can easily seen by looking at the Lagrangian, and, more importantly, the Lagrangian gives one information on how to 'quantize' the theory. We emphasize that everything we have been doing so far in Part II is classical field theory, where a field is simply a section of a bundle! At the end of this section, we will touch upon the uses of Lagrangians in *quantum* field theory. This is a vast subject of its own.

Action principles arose in physics with Fermat's discovery that light tended to take the path of least time. More generally, classical systems often tend to follow a path that minimizes a quantity called the action, which is the integral of a quantity called the Lagrangian. In fact, the common term 'principle of least action' is a bit of a misnomer, since the action is not always minimized. For example, light bouncing off a mirror does not take the path of least time, but merely a path such that small variations do not change the action to first order. Thus we prefer to speak simply of an action principle.

Let us begin by illustrating this principle for the motion of a classical point particle. Recall Newton's law of motion,

$$F = ma.$$

Here we are studying the motion of a particle in $\mathbb{R}^n$ as time passes; for each time $t \in \mathbb{R}$, the particle has some **position** $q(t) \in \mathbb{R}^n$. The **velocity** of the particle is given by

$$v = \dot{q}(t),$$

where, following Newton, we use a dot to denote differentiation with respect to t. Similarly, **acceleration** of the particle is

$$a = \ddot{q}(t).$$

The **mass** of the particle is simply a positive constant m. In the simplest case, the **force** F on the particle is a vector that depends only on the particle's position; that is, F is a fixed vector field on $\mathbb{R}^n$, and Newton's law says

$$m\ddot{q}(t) = F(q(t)).$$

As it turns out, the force is often the gradient of some function on $\mathbb{R}^n$. For reasons we will soon explain, it is handy to include a minus sign, and call the function V on $\mathbb{R}^n$ with

$$F = -\nabla V$$

the **potential energy**. The potential energy of the particle at time t is $V(q(t))$. We can also define a quantity called the **kinetic energy** T by

$$T = \frac{1}{2}mv^2,$$

meaning that the kinetic energy of the particle at time t is $\frac{1}{2}m\dot{q}(t)^2$. The main point of these quantities is that their sum, the total energy or **Hamiltonian**,

$$H = T + V$$

The Action Principle

is conserved:

$$\begin{aligned} \frac{dH}{dt} &= \frac{dT}{dt} + \frac{dV}{dt} \\ &= \frac{1}{2}\frac{d}{dt}m\dot{q}(t)^2 + \nabla V(q(t)) \cdot \dot{q}(t) \\ &= m\ddot{q}(t) \cdot \dot{q}(t) - F(q(t)) \cdot \dot{q}(t) \\ &= 0. \end{aligned}$$

However, it is also profitable to introduce a quantity called the **Lagrangian**, the difference of the kinetic and potential energy:

$$L = T - V.$$

The particle's path can be determined by the **action principle**, which says that the path must be a critical point of the **action**, which is the integral over time of the Lagrangian. More precisely, suppose $q\colon [0, T] \to \mathbb{R}^n$ is any path. Then we define the action $S(q)$ by

$$\begin{aligned} S(q) &= \int_0^T L\,dt \\ &= \int_0^T \left(\frac{1}{2}m\dot{q}(t)^2 - V(q(t))\right) dt. \end{aligned}$$

Now suppose q is a path that begins at some point a and ends at some point b:

$$q(0) = a, \quad q(T) = b.$$

We can vary the path q to a nearby path with the same endpoints by taking any function $f\colon [0,T] \to \mathbb{R}^n$ with $f(0) = f(T) = 0$, and considering

$$q_s(t) = q(t) + sf(t)$$

for some small value of $s \in \mathbb{R}$.

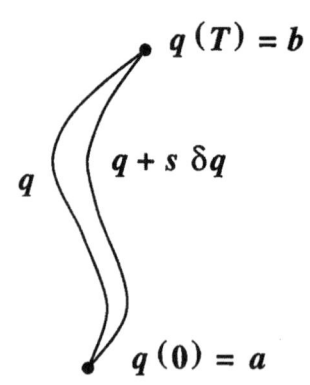

Fig. 1. Varying a path with fixed endpoints

We call this function f a **variation** of the path q, and write it as δq from now on! Note that

$$\delta q(t) = \frac{d}{ds} q_s(t)\Big|_{s=0}.$$

More generally, given any function G of paths from a to b, we define its **variation** by

$$\delta G = \frac{d}{ds} G(q_s)\Big|_{s=0}.$$

For example, we can calculate the variation of the action as follows:

$$\begin{aligned}
\delta S &= \delta \int_0^T \left(\frac{1}{2} m\dot{q}(t)^2 - V(q(t))\right) dt \\
&= \int_0^T \frac{d}{ds}\left(\frac{1}{2} m\dot{q}_s(t)^2 - V(q_s(t))\right)\Big|_{s=0} dt \\
&= \int_0^T \left(m\dot{q}(t) \cdot \dot{f}(t) - \nabla V(q(t)) \cdot f(t)\right) dt \\
&= -\int_0^T \left(m\ddot{q}(t) - F(q(t))\right) \cdot f(t)\, dt
\end{aligned}$$

where in the last step we did an integration by parts in the first term, and the boundary terms vanished because $f(0) = f(T) = 0$. (This sort of thing almost *always* happens when calculating the variation of the action.) Now, note that the variation of the action is zero for *all* functions f vanishing at $t = 0, T$ if and only if

$$m\ddot{q}(t) - F(q(t)) = 0,$$

The Action Principle

that is,
$$F = ma!$$
In other words, Newton's law is equivalent to demanding
$$\delta S = 0$$
for all variations f that vanish at the endpoints $t = 0, T$.

Let us repeat the calculation of δS is a more terse sort of way, to show how physicists usually do it:

$$\begin{aligned}\delta S &= \delta \int (\frac{1}{2}m\dot{q}^2 - V(q))\, dt \\ &= \int (m\dot{q} \cdot \delta \dot{q} - \nabla V(q) \cdot \delta q)\, dt \\ &= -\int (m\ddot{q} + \nabla V(q)) \cdot \delta q\, dt.\end{aligned}$$

More generally, the Lagrangian could be any function of the particle's position and velocity,
$$L = L(q(t), \dot{q}(t)),$$
and we obtain

$$\begin{aligned}\delta S &= \delta \int L\, dt \\ &= \int (\frac{\partial L}{\partial q^i}\delta q^i + \frac{\partial L}{\partial \dot{q}^i}\delta \dot{q}^i)\, dt \\ &= \int (\frac{\partial L}{\partial q^i} - \frac{d}{dt}\frac{\partial L}{\partial \dot{q}^i})\delta q^i\, dt\end{aligned}$$

where we are using the Einstein summation convention. It follows that $\delta S = 0$ for all variations δq vanishing at the endpoints if and only if the **Euler-Lagrange equations**

$$\frac{\partial L}{\partial q^i} = \frac{d}{dt}\frac{\partial L}{\partial \dot{q}^i}$$

hold for $i = 1, \ldots, n$.

At this point we should interject a remark about the 'calculus of variations' we are using here. We are taking a quick and dirty approach to the subject, but to delve into it more deeply one should take a more sophisticated approach. In particular, one should think of the space of all (smooth) paths from a to b,

$$\mathcal{P} = \{q\colon [0,T] \to \mathbb{R}^n\colon q(0) = a,\ q(T) = b\}$$

as a kind of 'infinite-dimensional manifold'. In fact, one can develop a very nice theory of infinite-dimensional manifolds and make this precise. We have defined the variation of any function G on $\mathcal{P}$ by

$$\delta G = \frac{d}{ds}G(q+sf)|_{s=0}$$

where the variation $\delta q = f$ is a fixed function vanishing at the endpoints of $[0,T]$. With this definition, δG is just the *directional derivative* of G in the direction f. However, it is better to define δG to be the differential dG, which summarizes all the directional derivatives of G. Then δG is a 1-form on $\mathcal{P}$. The action principle then says that a particle always travels along paths that are 'critical points' of the action, that is, points of $\mathcal{P}$ at which the 1-form δS vanishes. The only reason we will *not* take this more sophisticated approach to the calculus of variations is that we do not want to take the time to develop the theory of differential forms on infinite-dimensional manifolds! We urge the reader to read some of the references in the Notes that explore this subject.

Now let us show how to derive the Yang-Mills equation — hence, as a special case, Maxwell's equations — from a Lagrangian. Recall that to set up the Yang-Mills equation, we need a vector bundle E over a semi-Riemannian oriented manifold M. Suppose that M is n-dimensional. Then the Yang-Mills Lagrangian will be an n-form that we integrate over M to get the action.

To define the Yang-Mills Lagrangian, we need to define the 'trace' of an $\text{End}(E)$-valued form. Recall that the trace of a matrix is the sum of its diagonal entries. Thus if we have a real vector space V and pick a basis for V, we can write elements of $\text{End}(V)$ as matrices and define their trace this way. Actually, however, the trace is an invariant notion that is independent of the choice of basis. A definition of the trace that makes this clear is as follows. As we saw in the previous chapter, we have $\text{End}(V) \cong V \otimes V^*$ — an isomorphism that does not depend on

The Action Principle

any choice of basis — so the pairing between V and V^* defines a linear map

$$\text{tr}: \text{End}(V) \to \mathbb{R}$$
$$v \otimes f \mapsto f(v).$$

To see that this is really the usual trace, pick a basis e_i of V and let e^j be a dual basis of V^*. Writing $T \in \text{End}(V)$ as

$$T = T^i_j\, e_i \otimes e^j,$$

we have

$$\text{tr}(T) = T^i_j e^j(e_i) = T^i_j \delta^j_i = T^i_i,$$

the sum of the diagonal entries! (Similar remarks apply if V is a complex vector space.)

This implies that if we have a section T of $\text{End}(E)$, we can define a function $\text{tr}(T)$ on the base manifold M whose value at $p \in M$ is the trace of the endomorphism $T(p)$ of the fiber E_p:

$$\text{tr}(T)(p) = \text{tr}(T(p)).$$

We can similarly define the trace of an $\text{End}(E)$-valued form, which is an ordinary differential form. If T is a section of $\text{End}(E)$ and ω is a differential form, we define

$$\text{tr}(T \otimes \omega) = \text{tr}(T)\,\omega.$$

Now we can write down the **Yang-Mills Lagrangian**: if D is a connection on E, this is the n-form given by

$$\mathcal{L}_{YM} = \frac{1}{2}\text{tr}(F \wedge \star F)$$

where F is the curvature of D. Note that by the definition of the Hodge star operator, we can write this in local coordinates as

$$\mathcal{L}_{YM} = \frac{1}{4}\text{tr}(F_{\mu\nu} F^{\mu\nu})\,\text{vol}.$$

If we integrate $\mathcal{L}_{YM}$ over M, we obtain the **Yang-Mills action**

$$S_{YM} = \frac{1}{2} \int_M \mathrm{tr}(F \wedge \star F).$$

This integral may not converge if M is not compact, so we will assume for the moment that M is compact. In Exercise 120 we show how to remove this assumption. We warn the reader, by the way, that it is customary to put a minus sign in front of the Yang-Mills action, so that it comes out positive when D is a G-connection and the gauge group G is compact.

We want to show that the Yang-Mills equations are equivalent to the action principle

$$\delta S_{YM} = 0.$$

First we need to make clear what this equation means! The Yang-Mills action depends on a connection D, but we can fix a connection D^0 on E and write

$$D = D^0 + A,$$

where the vector potential A is an $\mathrm{End}(E)$-valued 1-form. This allows us to think of the Yang-Mills action as a function on the vector space of $\mathrm{End}(E)$-valued 1-forms:

$$S_{YM}(A) = \frac{1}{2} \int_M \mathrm{tr}(F \wedge \star F).$$

We can then consider varying the vector potential by adding to it s times any $\mathrm{End}(E)$-valued 1-form δA:

$$A_s = A + s\delta A.$$

The variation of any function G of A is then defined by

$$\delta G = \frac{d}{ds} G(A_s)\Big|_{s=0}$$

and when we write $\delta G = 0$, we mean that this variation vanishes for *all* variations δA.

To calculate the variation of the Yang-Mills action, we need to see how F varies when we vary A. We have already seen that if D^0 is flat, the curvature F of D is given by

$$F = dA + A \wedge A,$$

The Action Principle

where d is short for the exterior covariant derivative d_{D^0}. However, as we shall see, not every vector bundle admits a flat connection, so we need a formula for F that applies when D^0 is not flat. Note that when D^0 is not flat, the square of d_{D^0} does not vanish, so it is somewhat dangerous to write d_{D^0} as d — but being hardy souls, we will risk it! We then have

$$d^2\omega = F_0 \wedge \omega$$

for any E-valued form ω, where F_0 is the curvature of D^0. Similarly, we have

$$d_D^2 \omega = F \wedge \omega,$$

but also

$$\begin{aligned} d_D^2 \omega &= d_D(d\omega + A \wedge \omega) \\ &= d(d\omega + A \wedge \omega) + A \wedge (d\omega + A \wedge \omega) \\ &= F_0 \wedge \omega + dA \wedge \omega + A \wedge A \wedge \omega, \end{aligned}$$

so, while we have not quite proved it, it is pretty clear that

$$F = F_0 + dA + A \wedge A.$$

Exercise 116. *Check this by a calculation using local coordinates and a local trivialization of E.*

This allows us to calculate the variation of F:

$$\begin{aligned} \delta F &= \left. \frac{d}{ds}(F_0 + dA_s + A_s \wedge A_s) \right|_{s=0} \\ &= \left. d\left(\frac{d}{ds}A_s\right) + \left(\frac{d}{ds}A_s\right) \wedge A + A \wedge \left(\frac{d}{ds}A_s\right) \right|_{s=0} \\ &= d\delta A + \delta A \wedge A + A \wedge \delta A \\ &= d\delta A + [A, \delta A] \\ &= d_D \delta A. \end{aligned}$$

Here $[\cdot, \cdot]$ is the graded commutator introduced in the previous chapter. The result is quite simple and beautiful: the variation of the curvature equals the exterior covariant derivative of the variation of the vector potential!

Now let us compute the variation of the Yang-Mills action. For this, we will use a few properties of the trace of $\text{End}(E)$-valued forms, which we let the reader work out:

Exercise 117. *Suppose $\pi\colon E \to M$ is a vector bundle. Show that if ω is a $\text{End}(E)$-valued p-form and μ is an $\text{End}(E)$-valued q-form, then*

$$\text{tr}(\omega \wedge \mu) = (-1)^{pq}\text{tr}(\mu \wedge \omega).$$

*We call this the **graded cyclic property** of the trace, as it generalizes the usual cyclic property of the trace, namely that $\text{tr}(ST) = \text{tr}(TS)$ for any two $n \times n$ matrices S, T. Show that this implies*

$$\text{tr}([\omega, \mu]) = 0.$$

Exercise 118. *Now let D be a connection on E. Show that if ω is an $\text{End}(E)$-valued p-form then*

$$\text{tr}(d_D\omega) = d\,\text{tr}(\omega).$$

Exercise 119. *Now suppose that M is oriented and n-dimensional. Suppose that ω is an $\text{End}(E)$-valued p-form and μ is an $\text{End}(E)$-valued q-form on M. Using Exercise 118, show that if M is compact and $p + q = n - 1$, then*

$$\int_M \text{tr}(d_D\omega \wedge \mu) = (-1)^{p+1} \int_M \text{tr}(\omega \wedge d_D\mu).$$

Show that if M has a semi-Riemannian metric and $p = q$, then

$$\int_M \text{tr}(\omega \wedge \star\mu) = \int_M \text{tr}(\mu \wedge \star\omega).$$

With these identities in hand, computing the variation of the Yang-Mills action is a snap, at least if we assume M is compact:

$$\begin{aligned}
\delta S_{YM} &= \frac{1}{2}\delta \int_M \text{tr}(F \wedge \star F) \\
&= \frac{1}{2} \int_M \text{tr}(\delta F \wedge \star F + F \wedge \star \delta F) \\
&= \int_M \text{tr}(\delta F \wedge \star F)
\end{aligned}$$

The Action Principle

by Exercise 119; so using the formula for δF together with Exercise 119 again,

$$\delta S_{YM} = \int_M \text{tr}(d_D \delta A \wedge \star F) = \int_M \text{tr}(\delta A \wedge d_D \star F).$$

The integrand vanishes for an arbitary variation δA if and only if the Yang-Mills equation

$$d_D \star F = 0$$

holds!

Exercise 120. *Show how to derive the Yang-Mills equation from an action principle when M is not compact. (Hint: In this case note that, while the integral in $S_{YM}(A)$ may not converge, if we define*

$$\delta S_{YM}(A) = \int_M \delta \mathcal{L}_{YM}(A)$$

we get an integral that converges when δA vanishes outside some compact subset of M. Restricting ourselves to variations of this kind, we can show $\delta S_{YM}(A) = 0$ if and only if the Yang-Mills equations hold.)

Exercise 121. *Derive Maxwell's equations directly from the action*

$$S(A) = -\frac{1}{2} \int_M F \wedge \star F$$

where $F = dA$, A being a 1-form on the oriented semi-Riemannian manifold M. (This is easier than the full-fledged Yang-Mills case and sort of fun in its own right.) Show that when $M = \mathbb{R} \times S$ with the metric $dt^2 - {}^3g$,

$$-F \wedge \star F = \big(\langle E, E\rangle - \langle B, B\rangle\big) \text{vol}$$

in this case (see Exercise 58 of Part I). Generalize this to a formula for the Yang-Mills Lagrangian in terms of the Yang-Mills analogs of the electric and magnetic fields.

Now suppose that E is a G-bundle and we are restricting our attention to G-connections. Since the solutions of the Yang-Mills equation are precisely the critical points of the action, the gauge-invariance of the Yang-Mills equations will follow from the gauge-invariance of the

Lagrangian. Of course, we have already shown the gauge-invariance of the Yang-Mills equations directly, but it is nice to see how to prove invariance properties starting with the Lagrangian. It is very easy in this case; if A' is the transform of the vector potential A by the gauge transformation g, the curvature F' of A' is related to that of A by

$$F' = gFg^{-1},$$

so

$$\begin{aligned}\mathcal{L}_{YM}(A') &= \frac{1}{2}\mathrm{tr}(F' \wedge \star F') \\ &= \frac{1}{2}\mathrm{tr}(gFg^{-1} \wedge g(\star F)g^{-1}) \\ &= \frac{1}{2}\mathrm{tr}(F \wedge \star F) \\ &= \mathcal{L}_{YM}(A)\end{aligned}$$

using the graded cyclic property of the trace proved in Exercise 117.

In addition to its gauge invariance, the Yang-Mills equation has other symmetries that follow from symmetries of the Lagrangian. A nice way to think of these symmetries is in terms of **bundle automorphisms**, that is, bundle isomorphisms $\psi\colon E \to E$ mapping E to itself. Any bundle automorphism gives rise to a diffeomorphism $\phi\colon M \to M$ of spacetime such that $\phi \circ \pi = \pi \circ \psi$, where $\pi\colon E \to M$ is the projection. It turns out that a bundle automorphism ψ preserves the Yang-Mills equation whenever the corresponding diffeomorphism ϕ of spacetime preserves the metric, so that it preserves the Hodge star operator up to a sign. For example, the Yang-Mills equation for a trivial bundle over Minkowski space is invariant under the Poincaré group.

In physics it is always worth looking for laws that have as much symmetry as possible. In the next sections we will discuss some Lagrangians that are invariant under almost *all* bundle automorphisms. These give rather boring equations when we find critical points of the action, so that the classical field theories with these Lagrangians are not very interesting. However, these Lagrangians are interesting in quantum field theory and also in topology.

Chern Classes

The Yang-Mills equation depends on the metric on spacetime, since the Yang-Mills action

$$S_{YM}(A) = \frac{1}{2}\int_M \text{tr}(F \wedge \star F)$$

involves the Hodge star operator. This metric is what physicists would call a 'fixed background structure', since it plays a crucial role, but rather than being a solution of some equations, it is simply postulated to have a particular form, such as the Minkowski metric on $\mathbb{R}^4$. Starting with Einstein's theories of special and general relativity, the philosophy of much modern physics has been that such fixed background structures are undesirable. Of course, this philosophy may or may not turn out to be correct. For more discussion, see the references in the Notes — but remember that the last word on the laws of physics belongs to Nature. In any event, as we shall see in Part III, the theory of general relativity puts an end to thinking of the metric as a a fixed background structure by saying that *any* metric is allowed as long as it satisfies Einstein's equation. There is a more naive approach worth thinking about, however. We could attempt to write down an action for a gauge theory that does not involve the metric at all!

For example, in 4 dimensions we could try the action

$$S(A) = \int_M \text{tr}(F \wedge F)$$

as an alternative to the Yang-Mills action. It is worth noting that this action is proportional to the Yang-Mills action when A is **self-dual**, that is, when $F = \star F$. When A is self-dual (or anti-self-dual), the Bianchi identity $d_A F = 0$ automatically implies the Yang-Mills equation $d_A \star F = 0$. Thus our new action is closely related to Yang-Mills theory and self-duality. In fact, a vast amount of physics and mathematics arises as spinoffs of this fact! Self-dual solutions of the Yang-Mills equation are called **instantons**, and instantons are crucial in quantum chromodynamics, as well as the profound work of Donaldson and others on differential topology in 4 dimensions. For more on these topics we refer the reader to the Notes.

More generally, if our spacetime M is $2n$-dimensional, we could take as an action
$$S(A) = \int_M \operatorname{tr}(F^n),$$
where the nth **Chern form** $\operatorname{tr}(F^n)$ is the trace of the n-fold wedge product:
$$\operatorname{tr}(\underbrace{F \wedge \cdots \wedge F}_{n \text{ factors}}).$$
Unfortunately, this Lagrangian gives completely trivial equations — i.e., *every* vector potential is a critical point for this action! To see this, simply compute, copying what we did for the Yang-Mills case:

$$\begin{aligned} \delta S &= \delta \int_M \operatorname{tr}(F^n) \\ &= n \int_M \operatorname{tr}(\delta F \wedge F^{n-1}) \\ &= n \int_M \operatorname{tr}(d_D \delta A \wedge F^{n-1}) \\ &= n \int_M \operatorname{tr}(\delta A \wedge d_D F^{n-1}). \end{aligned}$$

Finally, use the Bianchi identity and the Leibniz law to note that
$$\begin{aligned} d_D F^{n-1} &= d_D F \wedge F^{n-2} + F \wedge d_D F \wedge F^{n-3} + \cdots \\ &= 0, \end{aligned}$$
so $\delta S = 0$ for all A.

We can turn this sad defeat into a victory of another sort, however, if we realize that $\delta S = 0$ is simply saying that $S(A)$ is independent of A, so it depends only on the bundle $\pi \colon E \to M$. In other words, there is an invariant of bundles E over an oriented manifold M given by

$$\int_M \operatorname{tr}(F^n)$$

where F is the curvature of *any* connection on E.

In fact, there is much more invariant information in the Chern forms than in the integral above. To see this, first of all note that the Chern

Chern Classes

forms are all closed:

$$\begin{aligned} d\operatorname{tr}(F^k) &= \operatorname{tr}(d_D F^k) \\ &= \operatorname{tr}(d_D F \wedge F^{k-1} + F \wedge d_D F \wedge F^{k-2} + \cdots) \\ &= 0, \end{aligned}$$

where we have used Exercise 118 and the Bianchi identity. This means that the kth Chern form defines a cohomology class in $H^{2k}(M)$ (see Chapter 6 of Part I). Now, the Chern form itself depends on the connection A, but its cohomology class does not — that is, if we change A, the Chern form changes by an exact form. To see this, note that

$$\begin{aligned} \delta \operatorname{tr}(F^k) &= \operatorname{tr}(\delta F^k) \\ &= \operatorname{tr}(\delta F \wedge F^{k-1} + F \wedge \delta F \wedge F^{k-2} + \cdots + F^{k-1} \wedge \delta F) \\ &= k \operatorname{tr}(\delta F \wedge F^{k-1}) \\ &= k \operatorname{tr}(d_D \delta A \wedge F^{k-1}) \\ &= k \operatorname{tr}(d_D(\delta A \wedge F^{k-1})) \\ &= k\, d \operatorname{tr}(\delta A \wedge F^{k-1}), \end{aligned}$$

using the graded cyclic property of the trace and the Bianchi identity, among other things. Now suppose A' is any other vector potential with curvature F'. Setting

$$\delta A = A' - A, \qquad A_s = A + s\delta A,$$

and letting F_s be the curvature of A_s, the difference of Chern forms

$$\begin{aligned} \operatorname{tr}(F'^k) - \operatorname{tr}(F^k) &= \int_0^1 \frac{d}{ds} \operatorname{tr}(F_s^k)\, ds \\ &= k \int_0^1 d\operatorname{tr}(\delta A \wedge F_s^{k-1})\, ds \\ &= k\, d \left(\int_0^1 \operatorname{tr}(\delta A \wedge F_s^{k-1})\, ds \right) \end{aligned}$$

is exact, as desired.

We thus can define the **kth Chern class** $c_k(E)$ of the vector bundle E over M to be the cohomology class of $\operatorname{tr}(F^k)$, where F is the curvature of any connection on E. These invariants are a very important tool for classifying vector bundles, and show up all throughout mathematics and physics. Using a more topological definition of the Chern classes, one can show that when properly normalized, their integrals over any compact oriented manifold mapped into M are *integers*. We thus say that they are **integral** cohomology classes. To be precise, if E is a complex vector bundle,

$$\frac{(i/2\pi)^k}{k!} \operatorname{tr}(F^k)$$

is an integral class. This means, for example, that when M is compact and oriented,

$$\frac{(i/2\pi)^n}{n!} \int_M \operatorname{tr}(F^n)$$

is an integer.

As we will see, the integrality of the Chern classes turns out to be very important in Chern-Simons theory. It also has a nice application to monopoles, as follows. We have seen that the magnetic field B is a 2-form on space. As we saw in Chapter 6 of Part I, there are solutions of the equations of vacuum magnetostatics on $\mathbb{R}^3 - \{0\}$,

$$dB = d \star B = 0,$$

having arbitrary magnetic flux through the unit sphere about the origin:

$$\int_{S^2} B = m.$$

We can think of this sort of solution as representing a monopole with magnetic charge m. However, if $B = dA$ for some 1-form A, the magnetic flux must vanish! To see this, divide S^2 into northern and southern hemispheres D_1 and D_2. Both of these have the equator as boundary, but they induce opposite orientations on the equator. Thus, if we use γ to denote a loop going around the equator counterclockwise when viewed from the north pole, Stokes' theorem gives

$$\int_{S^2} B = \int_{D^1} B + \int_{D^2} B = \int_\gamma A - \int_\gamma A = 0.$$

It might seem, therefore, that we cannot have magnetic monopoles of this sort when the magnetic field comes from a vector potential.

However, we *can* obtain monopoles if we generalize electromagnetism slightly. As we saw in the previous chapter, the Yang-Mills equations reduce to Maxwell's equations in the case of U(1)-connections on a U(1) bundle E with standard fiber given by the fundamental representation, as long as E is trivial. What if E is not trivial? Then the Yang-Mills equations for U(1)-connections on E look *locally* just like Maxwell's equations, since E is still locally trivial. Globally, however, interesting topological effects can occur. In particular, if A is the vector potential for a U(1)-connection on such a bundle over $\mathbb{R}^3 - \{0\}$, and F is the curvature of A, $\text{tr}(F)$ equals i times a real-valued 2-form on $\mathbb{R}^3 - \{0\}$. Call this 2-form B, since when E is trivial this 2-form is just the magnetic field. By the integrality of the first Chern class, we must have

$$\int_{S^2} B = 2\pi N$$

for some integer N. In other words, the magnetic charge is quantized! In fact, by choosing appropriate bundles we can obtain monopoles of any magnetic charge $m = 2\pi N$ this way.

How does this result compare to what we said in Chapter 5 of Part I about monopoles and charge quantization? To make a comparison, recall that we are now working in units where $q/\hbar = 1$, where q represents the fundamental unit of charge. Putting these constants back in explicitly, the result we have just obtained is that $qm/\hbar = 2\pi N$, or

$$qm = Nh,$$

which is just the same as what we obtained before! It may seem mysterious that we obtained the same result by two seemingly very different arguments. Actually, with some work and cleverness one can turn the argument for charge quantization in Chapter 6 of Part I into a proof of the integrality of the first Chern class for a U(1)-connection. We leave this as a challenge to the reader.

Exercise 122. *Show that if E is a U(1)-bundle over M with standard fiber given by the fundamental representation of U(1), the first Chern class of E is integral.*

The Chern-Simons Form

Let $\pi: E \to M$ be a trivial vector bundle over M, and let D be a connection on E. Write
$$D = D^0 + A$$
where D^0 is the standard flat connection and the vector potential A is an $\text{End}(E)$-valued 1-form. Then we can write
$$d_D \omega = d\omega + A \wedge \omega$$
for any E-valued form ω, where we use the abbreviation
$$d = d_{D^0},$$
and similarly
$$d_D \eta = d\eta + [A, \eta]$$
for any $\text{End}(E)$-valued form η.

We have already seen that the second Chern form,
$$\text{tr}(F \wedge F),$$
is closed. In the present context, it is actually exact! This is easy to see abstractly. First, if we take $A = 0$, then $F = 0$, so that $\text{tr}(F \wedge F) = 0$ in this case. Second, we have proven that changing A simply changes $\text{tr}(F \wedge F)$ by an exact form. So $\text{tr}(F \wedge F)$ must be exact for all A.

However, the proof that changing the vector potential changes the Chern form by an exact form actually gives an *explicit* 3-form whose exterior derivative is $\text{tr}(F \wedge F)$. Namely, let $A_s = sA$ and let
$$F_s = s\,dA + s^2 A \wedge A$$
be the curvature of A_s. Repeating the steps of the proof in this special case, we get
$$\begin{aligned}
\text{tr}(F \wedge F) &= \int_0^1 \frac{d}{ds} \text{tr}(F_s \wedge F_s) \, ds \\
&= 2 \int_0^1 \text{tr}(\frac{dF_s}{ds} \wedge F_s) \, ds
\end{aligned}$$

The Chern-Simons Form

$$\begin{aligned}
&= 2d \int_0^1 \mathrm{tr}(A \wedge F_s)\, ds \\
&= 2d \int_0^1 \mathrm{tr}(sA \wedge dA + s^2 A \wedge A \wedge A)\, ds \\
&= d\,\mathrm{tr}(A \wedge dA + \frac{2}{3} A \wedge A \wedge A),
\end{aligned}$$

where in the final step we did the integral over s.

We call the 3-form

$$\mathrm{tr}(A \wedge dA + \frac{2}{3} A \wedge A \wedge A)$$

the **Chern-Simons form**. It is nice to check directly that its exterior derivative is the second Chern form. Using Exercise 118 and the graded cyclic property of the trace we have

$$d\,\mathrm{tr}(A \wedge dA + \frac{2}{3} A \wedge A \wedge A) = \mathrm{tr}(dA \wedge dA + 2A \wedge A \wedge dA),$$

but

$$\mathrm{tr}(A \wedge A \wedge A \wedge A) = 0$$

by the graded cyclic property, so

$$\begin{aligned}
d\,\mathrm{tr}(A \wedge dA + \frac{2}{3} A \wedge A \wedge A) &= \mathrm{tr}(dA \wedge dA + 2A \wedge A \wedge dA) \\
&= \mathrm{tr}((dA + A \wedge A) \wedge (dA + A \wedge A)) \\
&= \mathrm{tr}(F \wedge F).
\end{aligned}$$

In fact, Chern-Simons forms exist much more generally. We are mainly interested in the 4-dimensional case because, as we shall see, it is related to knot theory and quantum gravity.

Exercise 123. *Let E be a trivial bundle over the manifold M and let*

$$D = D^0 + A,$$

*where D^0 is the standard flat connection and A is any vector potential. Generalize the above construction and obtain an explicit formula for the kth **Chern-Simons form**, a form whose exterior derivative is $\mathrm{tr}(F^k)$, where F is the curvature of D.*

Another way to think about the Chern-Simons form is as a 'boundary term' that shows up when we integrate the second Chern form over a 4-dimensional spacetime of the form

$$M = [0,1] \times S.$$

We can think of M as a piece of the spacetime $\mathbb{R} \times S$. Suppose that S (hence M) is oriented, so that we can do integrals over it, and let E be a trivial vector bundle over M. If F is the curvature of a connection on E, we can form the integral of the second Chern form over M. Since the second Chern form is the exterior derivative of the Chern-Simons form, by Stokes' theorem this integral is equal to the integral of the Chern-Simons form over ∂M, that is, the surfaces $t=0$ and $t=1$.

Actually, we can use this idea to compute the Chern-Simons form. This is easiest if we assume the vector potential A for D is in temporal gauge. Thus for each time t, A is really just an $\text{End}(E)$-valued form A_t on $\{t\} \times S$. Splitting d into space and time parts, we have

$$\begin{aligned} F &= dA + A \wedge A \\ &= d_S A_t + dt \wedge \partial_t A_t + A_t \wedge A_t \end{aligned}$$

It follows that

$$\int_M \text{tr}(F \wedge F) = 2 \int_M \text{tr}(dt \wedge \partial_t A_t \wedge d_S A_t + dt \wedge \partial_t A_t \wedge A_t \wedge A_t),$$

where we have left out terms involving $dt \wedge dt$ or a 4-form on S, since they vanish, and we have grouped together terms that are equal by the graded cyclic property of the trace. With a bit more work, this yields

$$\begin{aligned} \int_M \text{tr}(F \wedge F) &= \int_M \text{tr}\left(dt \wedge \partial_t(A_t \wedge d_S A_t + \frac{2}{3} A_t \wedge A_t \wedge A_t)\right) \\ &= \int_0^1 \partial_t \left(\int_S \text{tr}(A_t \wedge d_S A_t + \frac{2}{3} A_t \wedge A_t \wedge A_t)\right) dt. \end{aligned}$$

Doing the t integral, we find

$$\int_M \text{tr}(F \wedge F) = \int_S \text{tr}(A_t \wedge d_S A_t + \frac{2}{3} A_t \wedge A_t \wedge A_t)\Big|_0^1$$

or

$$\int_M \text{tr}(F \wedge F) = S_{CS}(A_1) - S_{CS}(A_0)$$

The Chern-Simons Form

where the **Chern-Simons action** $S_{CS}(A)$ of the vector potential A on the space S is given by

$$S_{CS}(A) = \int_S \text{tr}(A \wedge d_S A + \frac{2}{3} A \wedge A \wedge A).$$

We speak of the Chern-Simons 'action' because it can be used as the action for a field theory in 3 dimensions, called **Chern-Simons theory**. Classically, this theory is not very interesting, because the Euler-Lagrange equations

$$\delta S_{CS} = 0$$

say simply that A must be flat. To see this, we compute the variation of the action. In what follows we will simply write d for d_S.

We find

$$\begin{aligned} \delta S_{CS} &= \delta \int_S \text{tr}(A \wedge dA + \frac{2}{3} A \wedge A \wedge A) \\ &= 2 \int_S \text{tr}((dA + A \wedge A) \wedge \delta A) \end{aligned}$$

which only vanishes for all variations δA if

$$F = dA + A \wedge A = 0.$$

However, the quantum version of Chern-Simons theory is very interesting, as we shall see in the next chapter.

The special thing about Chern-Simons theory is that the Chern-Simons action is invariant under all orientation-preserving diffeomorphisms of S, and it is also *almost* gauge-invariant. The diffeomorphism-invariance is easy to see; since we have defined integration in a coordinate invariant manner we always have

$$\int \omega = \int \phi^* \omega$$

when ω is a differential form and ϕ is an orientation-preserving diffeomorphism, so in particular

$$S_{CS}(A) = S_{CS}(\phi^* A),$$

where we can pull back the $End(E)$-valued 1-form A in an obvious manner when E is trivial.

Now suppose that E is a trivial G-bundle, and restrict our attention to G-connections. The Chern-Simons action is not quite gauge-invariant, but it is invariant under gauge transformations g that are **connected to the identity**, meaning that there is a smooth 1-parameter family of gauge transformations g_s, $s \in [0,1]$ such that $g_0 = 1$ and $g_1 = g$. In physics these are often called **small** gauge transformations.

Consider such a 1-parameter family of gauge transformations g_s. Let A_s denote the gauge-transformed vector potential

$$A_s = g_s A g_s^{-1} + g_s d(g_s^{-1}).$$

Then we claim

$$\frac{d}{ds} S_{CS}(A_s) = 0.$$

With no loss of generality, it suffices to show this for $s = 0$. As a preliminary for this, let us write T for the section of $End(E)$ given by

$$T = \frac{d}{ds} g_s \Big|_{s=0}.$$

Noting that

$$0 = \frac{d}{ds} g_s g_s^{-1} = \left(\frac{d}{ds} g_s\right) g_s^{-1} + g_s \frac{d}{ds} g_s^{-1},$$

we also have

$$\frac{d}{ds} g_s^{-1} \Big|_{s=0} = -T.$$

Then note that

$$\begin{aligned}
\frac{d}{ds} A_s \Big|_{s=0} &= \frac{d}{ds} \left(g_s A g_s^{-1} + g_s d(g_s^{-1})\right)\Big|_{s=0} \\
&= \left(T A g_s^{-1} - g_s A T + T d(g_s^{-1}) - g_s dT\right)\Big|_{s=0} \\
&= [T, A] - dT.
\end{aligned}$$

This formula expresses how the vector potential transforms under 'infinitesimal gauge transformations'. Using it together with Stokes' theorem and the graded cyclic property of the trace, we obtain

$$\frac{d}{ds} S_{CS}(A_s)\Big|_{s=0} = 2 \int_S \operatorname{tr}([T, A] \wedge dA + A \wedge A \wedge ([T, A] - dT)).$$

The Chern-Simons Form

Exercise 124. *Check the above calculation.*

Next, note that
$$\int_S \mathrm{tr}(A \wedge A \wedge [T, A]) = 0$$
by the graded cyclic property of the trace, so

$$\begin{aligned}
\frac{d}{ds} S_{CS}(A_s)\Big|_{s=0} &= 2\int_S \mathrm{tr}([T, A] \wedge dA - A \wedge A \wedge dT) \\
&= 2\int_S \mathrm{tr}(T \wedge A \wedge dA - A \wedge T \wedge dA - A \wedge A \wedge dT) \\
&= 2\int_S \mathrm{tr}(d(A \wedge T \wedge A)) \\
&= 2\int_S d\,\mathrm{tr}(A \wedge T \wedge A) \\
&= 0
\end{aligned}$$

by Stokes' theorem.

The Chern-Simons action is not invariant under **large** gauge transformations, that is, those that are not connected to the identity. However, under large gauge transformations it always changes by an integer multiple of $8\pi^2$. This is a consequence of the integrality of the second Chern class. Let us just give a rough sketch of the argument. Suppose that A and A' are two connections on E that differ by a large gauge transformation g. Let A_s be a 1-parameter family of connections on E given by
$$A_s = A + s(A' - A),$$
so that $A_0 = A$ and $A_1 = A'$. We can also think of A_s as defining a connection $\widetilde{A}$ in temporal gauge on $[0, 1] \times S$. In fact, since A_1 differs from A_0 only by the gauge transformation g, we can glue the two ends of $[0, 1] \times S$ together and get a vector bundle $\widetilde{E}$ over $S^1 \times S$ with a connection on it, say $\widetilde{A}$. (Here we are thinking of S^1 as the interval $[0, 1]$ with its two ends identified.) On the one hand, we know by the integrality of the second Chern class that
$$\int_{S^1 \times S} \mathrm{tr}(\widetilde{F} \wedge \widetilde{F}) = 8\pi^2 N$$
for some integer N. On the other hand, we have
$$\int_{S^1 \times S} \mathrm{tr}(\widetilde{F} \wedge \widetilde{F}) = \int_{[0,1] \times S} \mathrm{tr}(\widetilde{F} \wedge \widetilde{F}) = S_{CS}(A_1) - S_{CS}(A_0)$$

by our earlier result describing the Chern-Simons action as a boundary term. It follows that $S_{CS}(A')$ differs from $S_{CS}(A)$ by an integer multiple of $8\pi^2$. As we shall see at the end of the next chapter, this turns out to be good enough to develop a gauge-invariant quantum version of Chern-Simons theory!

Chapter 5

Link Invariants from Gauge Theory

I was led to the consideration of the form of knots by Sir W. Thomson's Theory of Vortex Atoms, and consequently the point of view which, at least at first, I adopted was that of classifying knots by the number of their crossings.
— *P. G. Tait*

Knots and Links

Now we will turn to another theme of this book: knot theory. The goal of knot theory — simply put, to classify all knots — has its origins in the physics of the late 1800's. At this time, atoms were a mystery. Why should there be these apparently indestructible particles of so many different types, able to combine into molecules as they do, producing all the wonders of chemistry? At the time, the most beautiful equations in all physics were Maxwell's new equations for electromagnetism, so perhaps it was natural to attempt to explain atoms purely in terms of electromagnetism, even though we now know this was doomed to failure. In the late 1800's, electromagnetic waves were widely thought to be vibrations of a medium known as the 'luminiferous ether'. The rest frame of the ether was thought to provide a notion of absolute rest. Only later, due to the experiments of Michelson and Morley, did it become clear that motion with respect to this ether was undetectable.

In special relativity, the ether hypothesis was dropped along with the notion of absolute rest.

This was a difficult conceptual step: waves without something waving! Previously, there had always been a desire to understand electromagnetism using mechanical analogies. Maxwell, for example, had spent some time seeking to understand Faraday's electric and magnetic field lines in terms of "fine tubes of variable section carrying an incompressible fluid." One reason for this was that the equations of vacuum electrostatics

$$\nabla \cdot \vec{E} = 0, \qquad \nabla \times \vec{E} = 0,$$

are *also* the equations for the flow of an incompressible frictionless fluid with no viscosity and no 'vorticity', or curl:

$$\nabla \cdot \vec{v} = 0, \qquad \nabla \times \vec{v} = 0,$$

where $\vec{v}$ is the velocity field. More generally, in the situation where the vorticity $\nabla \times \vec{v}$ is not zero, Hermann von Helmholtz showed in 1858 that the vortex lines — that is, the lines of $\nabla \times \vec{v}$ — move in the direction of $\vec{v}$ as if they had an existence of their own. These vortex lines cannot have ends, but they can form loops. In 1867 the mathematician P. G. Tait (a follower of Hamilton and champion of quaternions) devised an ingenious way to demonstrate this fact by cutting a circular hole in a box, filling the box with smoke, and pressing the air out of the hole to form smoke rings. He showed these to his friend the physicist William Thomson (perhaps better known by his other name, Lord Kelvin). Kelvin soon noticed the analogy with electromagnetism and proposed a theory in which atoms were vortices in the ether! He also hypothesized that different kinds of atoms might correspond to differently knotted vortices! Tait began trying to list knots according to their numbers of crossings when drawn on the plane in the most efficient possible way. The very beginning of Tait's list is shown below. (In this table we do not distinguish between knots and their mirror images — more on that in a bit.)

Fig. 1. Knots with ≤ 5 crossings, as listed by Tait

The beauty of the vortex atom theory was that it seemed to relate the continuous world of fluid flow, or Maxwell's equations, to the discrete collection of different kinds of atoms. A main difficulty with the theory was the remarkable *stability* of atoms. By 1905 Kelvin admitted, "After many years of failure to prove that the motion in the ordinary Helmholtz circular ring is stable, I came to the conclusion that it is essentially unstable, and that its fate must be to become dissipated as now described." Indeed, the stability of atoms was one of the puzzles that gave birth to quantum mechanics. Once it was discovered that atoms consisted of electrons revolving about a central nucleus, the challenge was to explain why the electrons did not spiral into the nucleus, radiating energy in the form of light. Eventually Bohr had the courage to simply *postulate* that the electrons were confined to certain discrete energy levels or 'orbitals', which was later seen to be a consequence of Schrödinger's equation.

With the acceptance of special relativity, 'ether' became a synonym for a concept designed to explain something that needed no explanation, and with quantum mechanics, Kelvin's vortex theory of atoms was largely forgotten. Knot theory had taken on a life of its own though, in part because of some conjectures Tait had made but could not prove. These conjectures were only proven in the late 1980's, as a spinoff of an amazing set of new developments connecting knot theory and physics.

To proceed, we will need to define knots more carefully. To mathematicians, a knot is a particular way the circle, S^1, can sit inside ordinary space, $\mathbb{R}^3$. One way to make this precise is by defining a **knot** to be a submanifold of $\mathbb{R}^3$ that is diffeomorphic to S^1. One should imagine such a knot as a the result of taking a length of (very thin) rope, tying it up somehow, and then gluing the two ends together to prevent it from coming undone. The simplest example is the **unknot**:

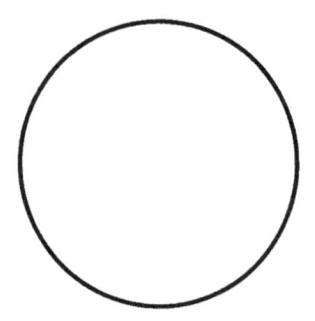

Fig. 2. The unknot

In other words, the unknot is just

$$\{(x,y,z) \in \mathbb{R}^3 \colon x^2 + y^2 = 1,\ z = 0\}.$$

This is the knot left out of Tait's list! It is one of those degenerate cases that only the most pedantic mathematician can explain without snickering. A more interesting knot is the **trefoil**.

Fig. 3. The trefoil knot

In fact, no less interesting than knots are certain collections of knots called 'links'. A **link** is a submanifold of $\mathbb{R}^3$ that is diffeomorphic to a disjoint union of circles. The circles themselves are called the

components of the link. A simple link with two components is the **Hopf link** shown below.

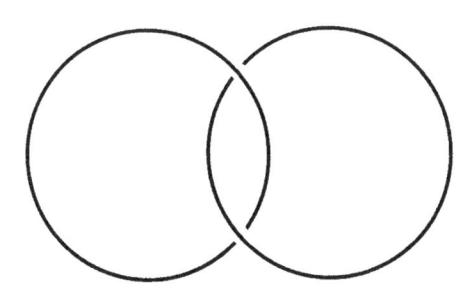

Fig. 4. The Hopf link

Links do not actually need to be 'linked'. For example, the 'distant union' of two unknots is a perfectly fine link!

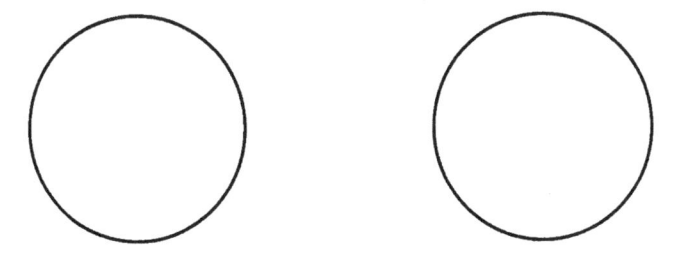

Fig. 5. The distant union of two unknots

Also, a knot is simply a link with exactly one component. Still more trivial is the **empty link**, shown in Figure 6, which has no components!

Fig. 6. The empty link

A more interesting link is the **Borromean rings,** with 3 components, no two of which would be linked to each other if the third were removed.

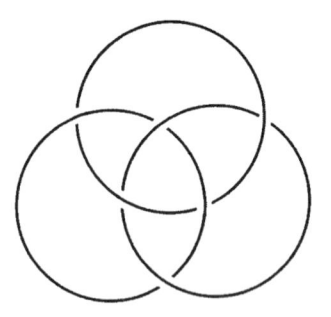

Fig. 7. The Borromean rings

Of course, knots and links have been used by fishermen, sailors, artists, and many other people for millenia before mathematicians became interested in them. For example, consider the **figure-eight knot** — shown, naturally, in Figure 8.

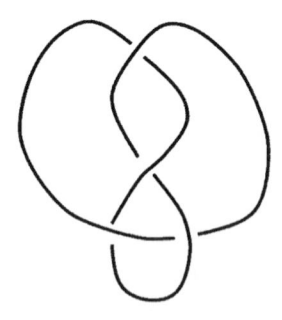

Fig. 8. Figure-eight knot

In practice, this is used in different forms, most of which have *loose ends*, and these forms have many different names: the figure eight bend, the slipped figure-eight knot, the figure eight lanyard knot, the figure eight loop, the figure eight mohair knot, and so on. However, mathematicians are primarily interested in the properties of a knot that persist no matter how one bends it, stretches it, and so on. This is why we work with knots that do not have loose ends; otherwise, all

Knots and Links

knots could be untied with sufficient persistence, so they would all be the same! Still, we need a precise definition of when one knot can be deformed to look like another. One way to proceed is as follows. We say that two links L and L' are **ambient isotopic** (or just 'isotopic') if there is a smooth map $\alpha: [0,1] \times \mathbb{R}^3 \to \mathbb{R}^3$ such that for each value of $t \in [0,1]$ the map $\alpha_t = \alpha(t,\cdot): \mathbb{R}^3 \to \mathbb{R}^3$ is a diffeomorphism, and α_0 is the identity map on $\mathbb{R}^3$, while α_1 maps L to L'. The map α is called an **ambient isotopy**. Note that it is crucial that α_t be a diffeomorphism for all t, to prevent the knot from 'passing through itself' as we deform it. If two links L and L' are isotopic, we write $L \sim L'$. We can draw an ambient isotopy as a kind of 'movie', with frames showing what happens as the 'time' parameter t passes.

Fig. 9. A knot that is ambient isotopic to the unknot

We should add that there is a different way of thinking about links, not as submanifolds but as maps. Since a link L is a submanifold of $\mathbb{R}^3$ diffeomorphic to $S^1 \cup \cdots \cup S^1$ (a bunch of circles), there must be a diffeomorphism
$$\gamma: S^1 \cup \cdots \cup S^1 \to L \subset \mathbb{R}^3.$$
Such a diffeomorphism is called an **embedding** of $S^1 \cup \cdots \cup S^1$ in $\mathbb{R}^3$. (This notion applies more generally whenever one has a submanifold of any manifold.) Sometimes it is nice to think of the link as being the embedding γ, rather than its range L. Of course, there will be many *different* embeddings with the same range, so this other way of treating links is somewhat different.

The easiest way to describe a knot or link is often by drawing it, but as drawings are 2-dimensional they must have 'crossings'. More precisely, given a link L, we obtain a **diagram** of L by taking a projection p of $\mathbb{R}^3$ onto some plane such that near any point of the plane, $p(L)$, as a subset of the plane, looks like one of the 3 scenes below. In each scene, the dashed circle represent the boundary of a small disk in the plane.

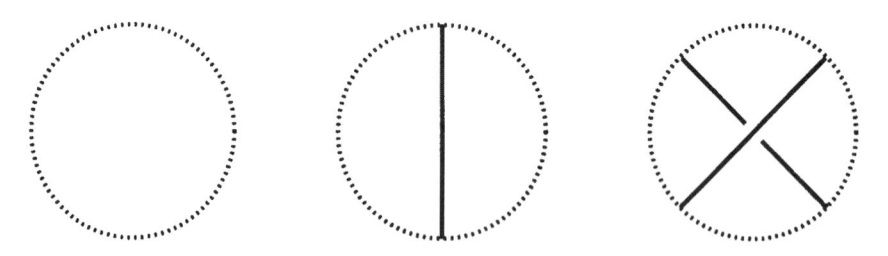

Fig. 10. Allowed local scenes in a diagram of a link

That is, each point in the plane has a chart containing it such that (up to diffeomorphism) $p(L)$ is either empty, a line, or two lines crossing at right angles. When drawing crossings we record which line lies on top, as in Figures 3–8, in order to keep track of the 3-dimensional information contained in the link. One can recover the link up to ambient isotopy from a diagram of this form.

It is a nontrivial exercise in differential topology — which we will skip — to show that we can find such a projection for any link.

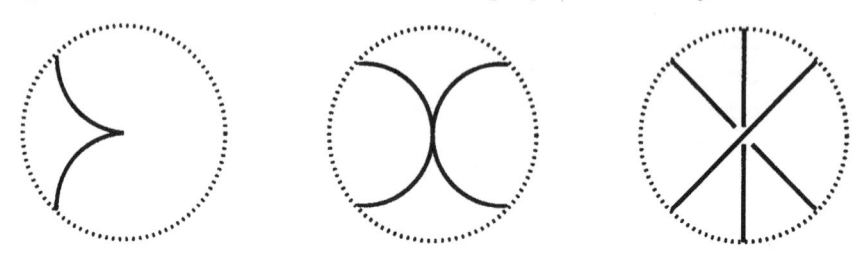

Fig. 11. Disallowed local scenes in a diagram of a link

In particular, as shown in Figure 11, we want to rule out projections that give rise to pictures of L having tangencies, cusps, triple crossings and the like. The point is that all these nastier sorts of 'singularities' are not generic — that is, they can be eliminated by changing the

Knots and Links

projection by an arbitrarily small amount. Generically, a projection gives rise to a diagram of L.

An equivalence class of links with ambient isotopy as the equivalence relation is called, naturally enough, an **isotopy class**. Of course there may be many different diagrams corresponding to the same isotopy class; for example, all the pictures in Figure 9 are diagrams of knots in the same isotopy class. By 1884, Tait had compiled a list of knots according to least number of crossings possible in the diagrams for them, starting with 3 crossings (since he forgot the unknot) and going up to 7 crossings. More recently, Thistlethwaite has been tabulating knots with the aid of a computer; at the time this book was written he had found all knots with ≤ 15 crossings. Tabulation efforts focus on 'prime' knots, where a knot is **prime** if it is not the **connected sum** of two other knots; a connected sum of the trefoil and the figure-eight being shown in Figure 12. Even with the help of a computer, listing knots quickly becomes difficult, since the number of knots with a given minimal number of crossings grows very rapidly. For example, there are 253,334 prime knots with minimal crossing number 15 (barring some error in the computation), not counting the difference between knots and their mirror images; of these, only one appears to be isotopic to its mirror image, or **amphicheiral**.

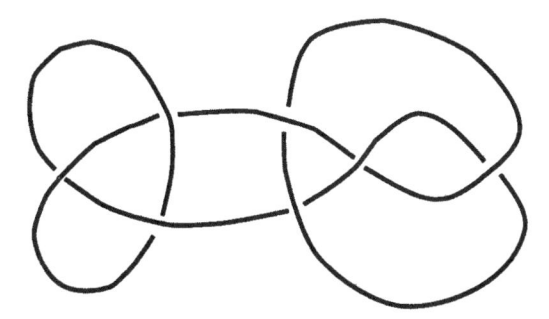

Fig. 12. A connected sum of the trefoil and the figure-eight

Clearly tabulation alone is of limited value; to gain real understanding of knots or links one must seek **isotopy invariants**, that is, quantities one can calculate from a link that do not change under ambient isotopy, or in other words, functions of the isotopy class of the link.

Indeed, in practice one needs to use such invariants when tabulating knots, to make sure one has not counted certain knots twice. A famous example, the **Perko pair,** was listed as two different knots by the knot tabulator Little, and was only revealed to be two diagrams of the *same* knot in the 1970s by Perko!

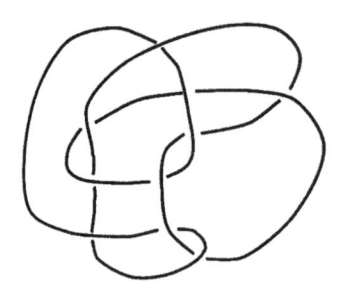

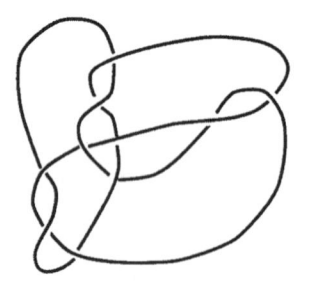

Fig. 13. The Perko pair

In listing knots Tait noticed some interesting patterns, which he formalized as conjectures. Let us call a diagram of a link **alternating** if, as we walk around any component, its crossings alternate between over and under. For example, the diagram of the trefoil in Figure 3 is alternating, as is that of the Hopf link in Figure 4. The simplest of Tait's conjectures was that if a link has any alternating diagram, the diagram of the link that has the fewest possible crossings is an alternating diagram. He also conjectured that any two alternating diagrams of the same link have the same number of crossings as long as neither diagram has a removable (or **nugatory**) crossing as shown in Figure 14. In this figure, the gray blobs denote an arbitrary mess of stuff.

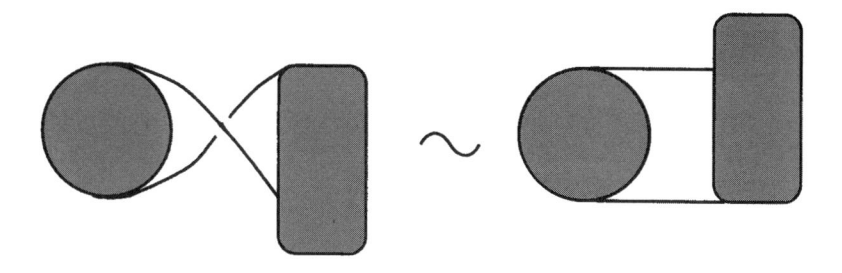

Fig. 14. Nugatory crossing in a link diagram

While Kelvin's theory of atoms as knotted electric field lines is only of historical interest now, there is an important kernel of insight in it. As we will see in the next section, electromagnetism *does* have an interesting connection to knot theory. As early as 1833, Gauss used electromagnetism to obtain a very important invariant of links called the 'linking number'. In fact, Gauss' result can be most elegantly derived in terms of U(1) Chern-Simons theory, rather than Maxwell's equations. A massive generalization of Gauss' result was found by Witten, who showed that Chern-Simons theory gives a a link invariant for each choice of compact gauge group and each choice of (finite-dimensional) representation. The simplest nonabelian example is the 'Jones polynomial', which comes from the spin-$\frac{1}{2}$ representation of SU(2). In a case of true poetic justice, this invariant turned out to play a crucial role in proving the Tait conjectures. It also appears naturally in the study of quantum gravity. In fact, the 'loop representation' of quantum gravity bears a mild resemblance to Kelvin's old theory of vortex atoms! We discuss the relation between knots and quantum gravity a bit more at the end of Part III.

For now let us start with a very basic question. We noted that there are many different diagrams that represent the same isotopy class of links. When do two different diagrams represent the same isotopy class? This question has a beautiful answer. Two diagrams represent the same isotopy class if and only if one can get from the first to the second by a sequence of maneuvers known as the **Reidemeister moves**. The zeroth Reidemeister move consists simply of an isotopy of the plane. That is, we say two diagrams are **isotopic** if there is a smooth map $\alpha \colon [0,1] \times \mathbb{R}^2 \to \mathbb{R}^2$ such that α_0 is the identity, α_1 takes the first diagram to the second (taking over-crossings to over-crossings and under-crossings to under-crossings), and α_t is a diffeomorphism for all t. An example of the zeroth Reidemeister move is shown below.

Fig. 15. Zeroth Reidemeister move

The other three Reidemeister moves, usually called I, II, and III, are less trivial. Each of these moves consists of modifying a small portion of a link diagram while keeping the rest fixed. We will call a small portion of a component of a link, as it appears in a diagram, a **strand**. Move I consists of modifying a diagram in a neighborhood containing only a single strand by putting a 'twist' in the strand.

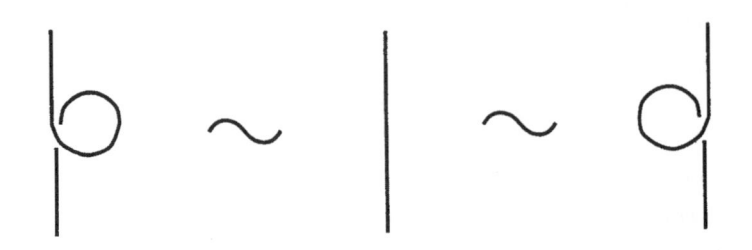

Fig. 16. First Reidemeister move

We emphasize that this move, as well as the second and third moves, is a 'local' move, in which we modify only the portion we have shown

Knots and Links

of an otherwise arbitrary diagram.

Move II consists of 'cancelling' a pair of crossings, as shown in Figure 17. Move III consists of sliding a strand under a crossing as shown in Figure 18. (Note that we can also interpret this move as sliding a strand *over* a crossing.)

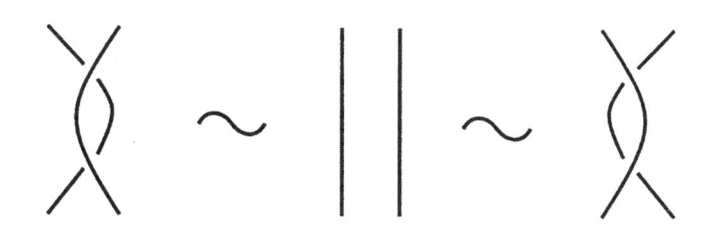

Fig. 17. Second Reidemeister move

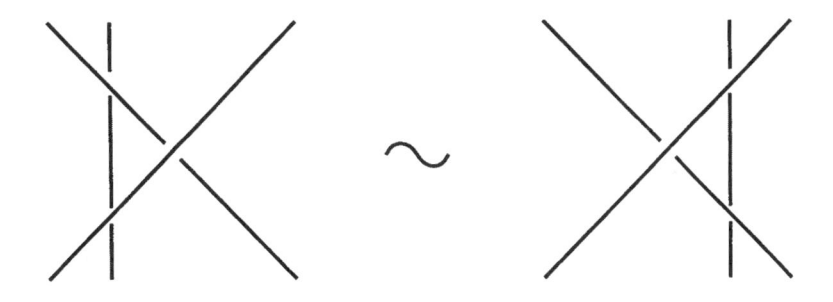

Fig. 18. Third Reidemeister move

We will not prove that these Reidemeister moves are sufficient to get between any two diagrams representing the same isotopy class of link (see the Notes), so the reader should try to gain confidence in them by using them. However, the *idea* of the proof is simple; in performing an isotopy of a link, its projection onto the plane will occasionally encounter certain 'catastrophes' where it does something nongeneric, as in Figure 11; if one looks at the link diagram before and after one of these catastrophes one will see that a Reidemeister move has occurred.

The work is to show that, by using a small perturbation if necessary, one can always make the isotopy only encounter the three *simplest* sorts of catastrophes shown in Figure 11, since more complicated catastrophes such as quadruple points are 'still less generic' and can be avoided.

It is interesting to play around with the Reidemeister moves and discover that some knots are isotopic to their mirror images, while others are not. The trefoil is *not* isotopic to its mirror image, so there are really two forms of trefoil, the right-handed and left-handed trefoil, shown in Figure 19. We should emphasize that Tait's knot table, part of which we reproduced in Figure 1, did not separately lists knots and their mirror images.

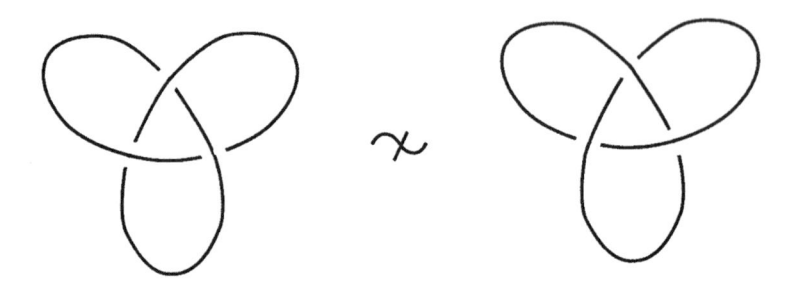

Fig. 19. The trefoil is not isotopic to its mirror image

On the other hand, the figure-eight knot *is* isotopic to its mirror image; such knots are said to be **amphicheiral**.

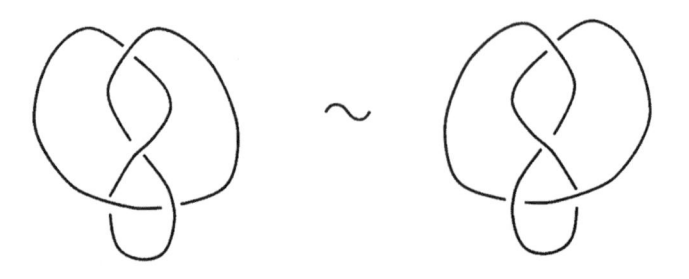

Fig. 20. The figure-eight knot is isotopic to its mirror image

Knots and Links   305

Exercise 125. *Find a sequence of Reidemeister moves taking the figure-eight knot to its mirror image — see Figure 20.*

Exercise 126. *Show using Reidemeister moves that the Perko pair consists of two isotopic knots. (Hint: it might help to make a model with string.)*

One nice way to get ahold of link invariants is to give a recipe for calculating a number from a diagram of a link, and then prove that this number is unchanged by the Reidemeister moves. The simplest example is the linking number. However, for this example, and others to come, it is necessary to consider links equipped with extra structures: so-called 'oriented' and 'framed' links.

Since a link is a manifold (a submanifold of $\mathbb{R}^3$) it makes sense to give it an orientation; a link with an orientation is called an **oriented** link. Two oriented links are said to be isotopic if there is an ambient isotopy taking one to the other and also taking the orientation of the first to the orientation of the second. We can represent the orientation of a link by a nowhere vanishing vector field tangent to the link. All that matters about this vector field is which way it points, not its magnitude, so there are 2^n ways to give a link with n components an orientation. In the figure below we give the Hopf link an orientation in two different ways:

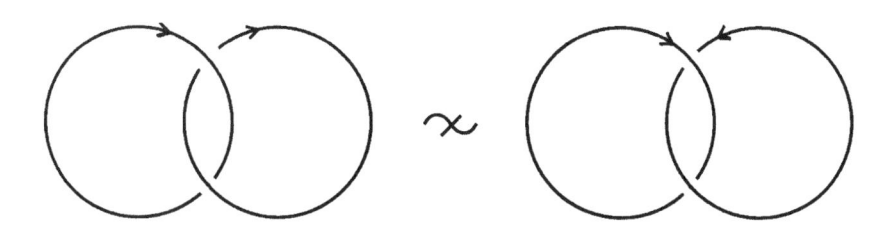

Fig. 21. Two orientations for the Hopf link

As we shall soon show, these two different oriented links are not isotopic. There are Reidemeister moves for oriented links, and they are just the same as the basic Reidemeister moves, except that we must keep track of orientations, and all possible orientations are allowed in the Reidemeister moves.

Exercise 127. *If one allows all possible orientations, there are many oriented versions of the first Reidemeister moves. Find a minimal set of oriented Reidemeister moves from which the rest can be derived.*

Just as an orientation may be represented by a nonvanishing vector field tangent to the link, a 'framing' of a link is a vector field that is *nowhere* tangent to the link. More precisely, note that the tangent space of the link L at any point p is a subspace of the tangent space of $\mathbb{R}^3$, which in turn may be identified with $\mathbb{R}^3$:

$$T_pL \subseteq T_p\mathbb{R}^3 \cong \mathbb{R}^3.$$

We will call a smooth function from L to $\mathbb{R}^3$ simply a vector field on L, being careful to realize that it is not necessarily a tangent vector field. Then a **framing** of a link L is a vector field v on L such that $v_p \notin T_pL$ for all $p \in L$. A link equipped with a framing is called a **framed link**. It is often handy to visualize framed links as ribbons, as in Figure 22 below, which shows a particular framing of the unknot. Note that the ribbon must make an integer number of 2π twists as we march around any component of the link — no Möbius strips allowed!

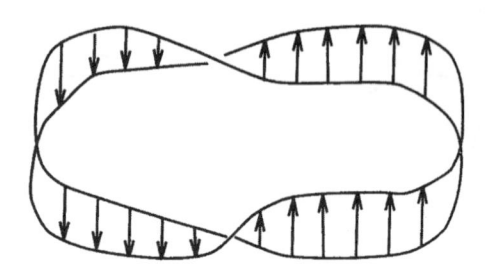

Fig. 22. A framing of the unknot

Given a diagram of a link, there is a standard framing, called the **blackboard framing**. This consists of the unit vector field that is everywhere orthogonal to the plane into which the link has been projected. More prosaically, we simply draw a diagram of the link on the

Knots and Links 307

blackboard, and imagine the framing to be pointing perpendicular to the blackboard, into the classroom.

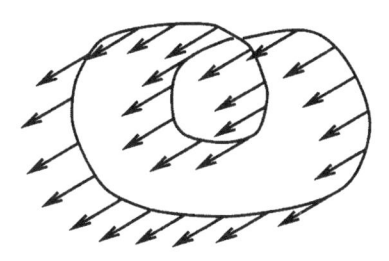

Fig. 23. The blackboard framing

We say two framed links are isotopic if there is an ambient isotopy taking the first link to the second and also taking the framing of the first to the framing of the second. Thus, for example, in Figure 24 below we show two ways of making the unknot into a framed link that are not isotopic as framed links. In fact, there are infinitely many ways, corresponding to the different numbers of twists in the framing.

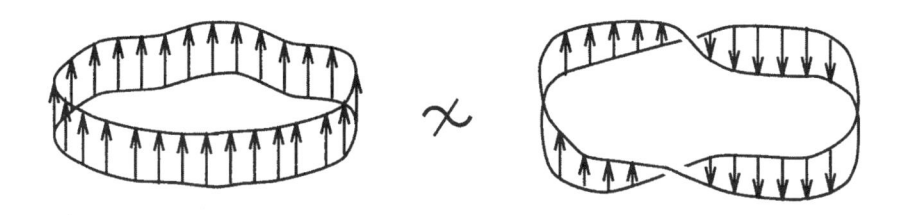

Fig. 24. Two framings of the unknot

Given two diagrams of framed links (with the blackboard framing), when are they isotopic? The key thing to note is that the first Reidemeister move no longer holds, since it introduces an extra twist in the framing.

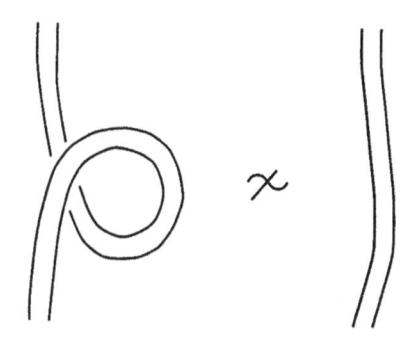

Fig. 25. The first Reidemeister moves fails for framed links

For framed links, we need to replace the first Reidemeister move by a modified version, shown in Figure 26, which we call I'.

Fig. 26. Modified first Reidemeister move for framed links

Exercise 128. *Check that the modified first Reidemeister move really gives an isotopy of framed links. (Hint: one can do so either using equations, or using a little piece of ribbon. The latter is definitely more enlightening!)*

In fact it is a theorem that two diagrams represent the same framed link (with blackboard framing) if one can get from one to the other using the **framed Reidemeister moves**, that is, the moves 0, I', II, and III. Part of proving this theorem is noticing the **Whitney trick** whereby one cancels two opposite twists in the framing using only moves 0, II, and III, as in Figure 27. This devious trick illustrates the power of the

Knots and Links

Reidemeister moves! By the way, if one can get from one link diagram to another using only moves 0, II, and III, one says they are **regular isotopic**. This is an equivalence relation on link diagrams, not on links — we mention it here only because it is frequently discussed.

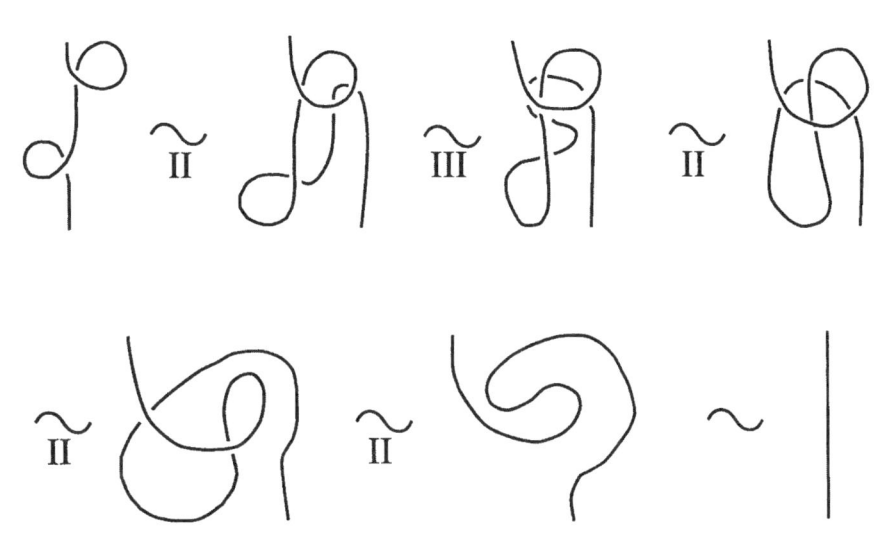

Fig. 27. The Whitney trick

Exercise 129. *Show using the framed Reidemeister moves that the figure-eight knot and its mirror image in Figure 20 are regular isotopic, hence isotopic as framed knots, giving both the blackboard framing. (Hint: this takes work, and it uses the Whitney trick.)*

The reader may find the idea of a framed link rather artificial. Actually, it corresponds quite nicely to what one can do with an actual piece of rope, since a rope, not being infinitesimally thin, remembers when it has been twisted. For very similar reasons, we will later see that framings arise naturally when considering 'regularized' Wilson loops in quantum field theory. Moreover, as we shall see when studying the Jones polynomial, one can sometimes get invariants of links without framing from invariants of framed links.

The Linking Number and Writhe

Thus, if two vortex-rings were once created in a perfect fluid, passing through each other like links of a chain, they never could come into collision, or break each other, they would form an indestructible atom; every variety of combinations might exist. — Lord Kelvin

The simplest link invariants — apart from the number of components — are the linking number and the writhe. As we shall see, these are closely related to electromagnetism, while certain subtler link invariants correspond to subtler gauge theories. We will define the linking number for oriented links with 2 components, as follows. Given a diagram of an oriented link, we can distinguish between **right-handed** and **left-handed** crossings as in Figure 28 below. Note that it takes an orientation to make this distinction invariant under the zeroth Reidemeister move! We define the **sign** of a crossing to be $+1$ if it is right-handed and -1 if it is left-handed.

Right-handed   Left-handed

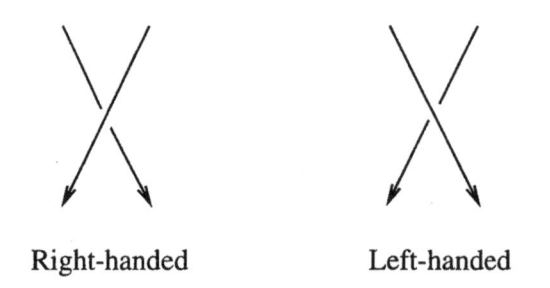

Fig. 28. Right-handed and left-handed crossings

We define the **linking number** as half the sum of the signs of all crossings where different components of the link cross each other. For example, the two differently oriented Hopf links in Figure 21 have linking numbers 1 and -1, respectively.

It is easy to show that the linking number is an isotopy invariant of oriented·links using the Reidemeister moves. Move 0 does not change right-handed crossings into left-handed crossings, or vice versa, so it preserves the linking number. Move I creates or destroys a single cross-

The Linking Number and Writhe

ing, but one in which a component crosses itself, so the linking number is unaffected. Move II creates or destroys a *pair* of crossings, but they have opposite handedness, so the linking number is unaffected. Finally, move III does not change the number of crossings or their handedness.

The linking number is a very important link invariant, but unfortunately not extremely powerful. We can use the linking number to prove that the two Hopf links in Figure 21 are not isotopic. But the **Whitehead link**, shown below, has linking number zero even though it is not isotopic to the distant union of two unknots.

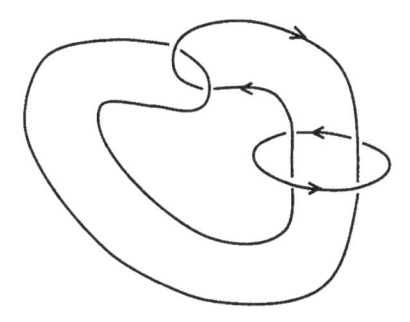

Fig. 29. The Whitehead link

There is related invariant of links equipped with both framing and orientation, called the **writhe** or **self-linking number**. To calculate the writhe $w(L)$ of a link L from a diagram, one simply sums the signs of all crossings! Two diagrams of framed oriented links represent isotopic links (where isotopy preserves both framing and orientation) if and only if one can get from one to the other by the framed Reidemeister moves, so to prove that the writhe is really invariant as claimed, we need only check the Reidemeister moves. A key point here is that while Reidemeister move I changes the writhe, the move I' does not.

Exercise 130. *Show that the writhe is invariant under Reidemeister moves 0, I', II, and III.*

Exercise 131. *Show that if L is a link with components K_i, then*

$$w(L) = \sum_{i \neq j} \mathcal{L}(K_i, K_j) + \sum_i w(K_i)$$

This is one reason why the writhe is also called the self-linking number.

$$W(L) = \#\left(\diagdown\!\!\!\!\diagup\right) - \#\left(\diagup\!\!\!\!\diagdown\right)$$

Fig. 30. Formula for the writhe

The writhe goes by the name it does because it counts how much the link 'writhes', or twists about. For example, below is a diagram of an unknot that, when given the blackboard framing, has writhe equal to 3.

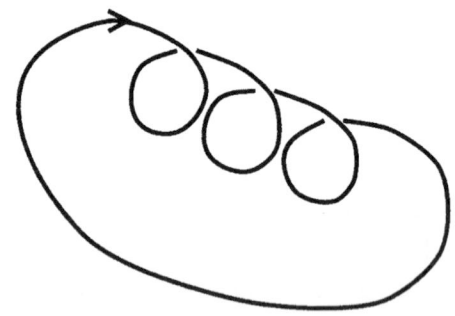

Fig. 31. A framed oriented unknot with writhe 3

There is a way to define the writhe in terms of what knot theorists call **skein relations**, that is, formulas saying how a link invariant is affected by modifying some small portion of the diagram of a link while keeping the rest of the diagram fixed. The skein relations for the writhe are so trivial as to seem a bit silly, but later we will meet link invariants with more tricky skein relations, so it is worthwhile starting with an easy case. Clearly, if we change any left-handed crossing to a right-handed crossing in the diagram of link, its writhe increases by 2. It is handy to blur the distinction between links and their writhe, allowing

The Linking Number and Writhe

us to write this skein relation simply as below.

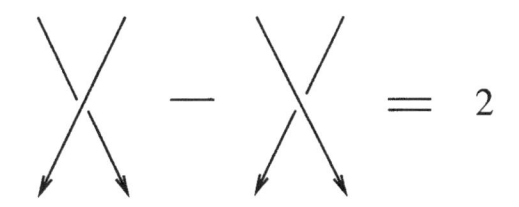

$$\asymp - \asymp \; = \; 2$$

Fig. 32. First skein relation for the writhe

Similarly, if we eliminate a right-handed twist as below, the writhe decreases by 1.

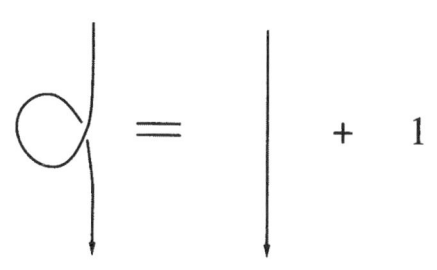

$$\text{(right twist)} = \; | \; + \; 1$$

Fig. 33. Second skein relation for the writhe

Finally, if our link diagram contains a circle embedded in the plane, we can get rid of it without changing the writhe. We can write this symbolically as follows:

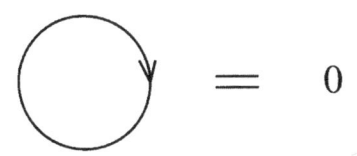

$$\bigcirc \; = \; 0$$

Fig. 34. Third skein relation for the writhe

We can use the skein relations to calculate the writhe of the trefoil knot as in Figure 35. We apply the first skein relation, then the second

Reidemeister move, then the second skein relation, and then the third skein relation to obtain the answer. The ideas is that skein relations allow one to calculate link invariants *recursively* by repeatedly simplifying the link diagram. They are clearly not always the most efficient way to calculate the invariants, but they can be very good for proving things about them.

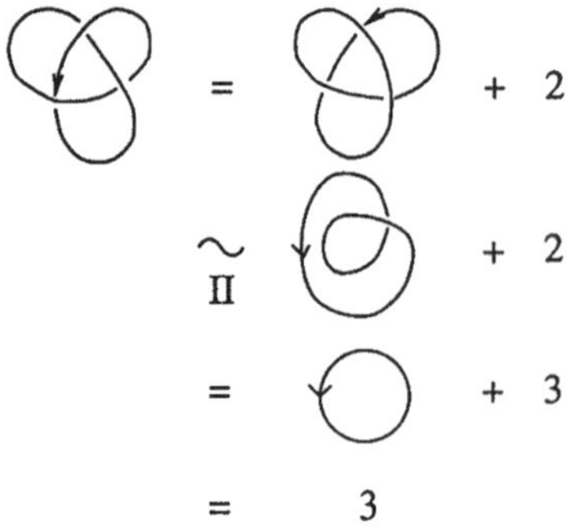

Fig. 35. Computing the writhe of the trefoil knot

The linking number satisfies similar skein relations to those for the writhe:

Exercise 132. *Deduce the skein relations for the linking number shown in Figure 36. Note that the first skein relation consists of two cases: the linking number increases by 1 if we change a left-handed crossing to a right-handed crossing when the two strands that cross belong to different components, but does not change when they belong to the same component.*

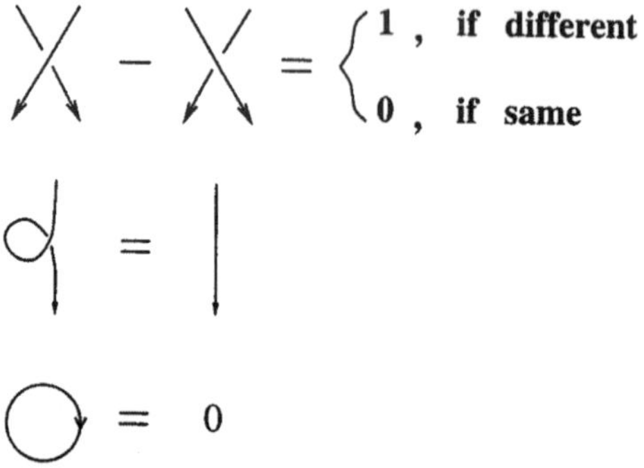

Fig. 36. Skein relations for the linking number

The Linking Number and Writhe

Now let us explain what the linking number has to do with electromagnetism. Actually there are a number of relationships; the first one was discovered by Gauss. Consider an oriented link with two components. Think of each component as a wire. Let an electrical current of unit strength flow around one of the wires, in the direction given by the orientation. The current will produce a magnetic field. Do a line integral of the magnetic field around the other wire. The resulting number is the linking number!

Let us make this a bit more precise using differential forms. Let us call the two components of our link K and K', since each one is a knot. Replace the knot K by a wire of finite thickness — that is, an embedded solid torus $S^1 \times D^2$ that does not bump into the knot K'. Let a unit current j flow around this solid torus in the direction given by the orientation of K.

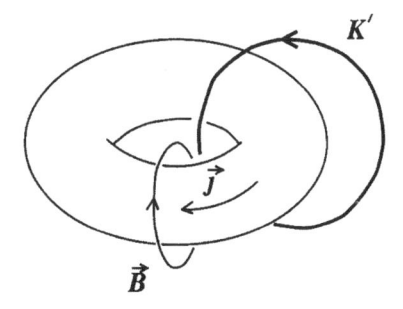

Fig. 37. The Gauss integral formula for the linking number

The magnetic field B must satisfy

$$d \star B = \star j, \qquad dB = 0,$$

since we are assuming there are no electric fields around. Note that the 'line integral of the magnetic field around K'' is too sloppy a way of talking when we are working with differential forms; since knots are 1-dimensional manifolds, we can integrate 1-forms over them, but B is a 2-form. Luckily we can use the Hodge star operator to turn B into a 1-form. We then claim that the **Gauss integral**

$$\int_{K'} \star B$$

equals $\mathcal{L}(K, K')$, the linking number of the link with components K and K'.

Something even better is true, too. The equation $dB = 0$ is actually utterly irrelevant to the problem at hand. Any 2-form B with $d{\star}B = {\star}j$ will give the same value for the Gauss integral. In other words, we can add to B any 1-form C with $d \star C = 0$ without changing the value of $\int_{K'} B$. Proving this fact uses a bit of deRham theory. Since $\mathbb{R}^3$ is simply connected, the fact that $d \star C = 0$ implies that $\star C = df$ for some function f. It then follows by Stokes' theorem that

$$\int_{K'} C = \int_{K'} df = \int_{\partial K'} f = 0$$

since K' has no boundary. Thus such a 1-form C contributes nothing to the Gauss integral.

At this point the reader should review some things we said in the section on the Bohm-Aharonov effect in Chapter 6 of Part I. There we saw how the equations $d \star B = \star j$ and $dA = B$ were related by the change of variables $\star B \to A$, $\star j \to B$. This gives us a second relationship between electromagnetism and the linking number. Namely, suppose we have a link with two components K and K'. Thicken up K to obtain an embedded solid torus T that does not bump into K', and let B be a **flux tube** running around T that contains unit magnetic flux. In other words, using the coordinates (t, r, θ) on $T \cong S^1 \times D^2$, where t goes around the S^1 direction, let

$$B = f(r, \theta)\, r\, dr \wedge d\theta$$

where $f = 0$ for $r \geq 1$, and

$$\int_0^1 \int_0^{2\pi} f(r, \theta)\, r\, dr \wedge d\theta = 1$$

Let A be a vector potential for B, that is, a 1-form with

$$dA = B.$$

Note that such an A always exists, because B is closed and $H^2(\mathbb{R}^3) = 0$. Then we claim that

$$\mathcal{L}(K, K') = \int_{K'} A.$$

The Linking Number and Writhe 317

Of course, we have not proved this result yet, but the reader should see from what we have said that this result is *equivalent* to Gauss' original formula.

The flux tube version of the formula for the linking number is mathematically more elegant than Gauss' version, since it does not involve the Hodge $\star$ operator, and thus requires no metric on space. It may seem more abstract, since wires are easier to visualize than flux tubes. However, flux tubes *do* exist in nature. Superconductive materials — which have essentially zero electrical resistance due to quantum effects — tend to exclude magnetic fields; if one puts a so-called type II superconductor in a magnetic field whose strength exceeds the 'lower critical field', it penetrates, but only through small tubes, also called 'vortex lines'. The magnetic flux flowing along each these vortex lines is precisely

$$\Phi = \frac{h}{2q}$$

where q is the electron charge, h is Planck's constant, and we work in units where the speed of light is 1, as always. The reason is that the electrons in superconductors form 'Cooper pairs' with charge $2q$, and for the wavefunction of these pairs to be preserved by parallel transport around the flux tube, we must have

$$e^{-\frac{2qi}{\hbar}\int_\gamma A} = e^{-\frac{2qi}{\hbar}\int_D B} = e^{-\frac{2qi}{\hbar}\Phi} = 1$$

where γ is a loop winding once around the flux tube, and D is a disk bounded by γ. Of course, this does not explain why the flux is exactly *one* of these basic flux units $h/2q$, which is a consequence of the detailed energetics of the superconductor.

We cannot resist adding that superfluids such as liquid helium, characterized by near-zero viscosity, are mathematically analogous to superconductors in many respects. Just as superconductors tend to have $\nabla \times \vec{A} = 0$, superfluids tend to have $\nabla \times \vec{v} = 0$, where $\vec{v}$ is the fluid flow velocity. However, under certain circumstances $\nabla \times \vec{v}$ is nonzero along vortex lines having quantized angular momentum! These may be thought of as the quantum analogs of the smoke rings produced by Tait. To see them outside the laboratory, one should look at the core of a neutron star, such the pulsar in the Crab Nebula. These cores are made of an extremely dense superfluid consisting mostly of neutrons. The angular momentum of the core is completely contained in vortex lines, which form a hexagonal array. In the Crab nebula they are probably spaced about 0.0025 centimeters apart. For more on superconductors and superfluids, the reader is urged to the references in the Notes!

Let us now sketch an argument for the claim made above that $\int_{K'} A$ is the linking number of K and K'. This argument will be neither very rigorous nor very efficient, but it illuminates a number of interesting points. (For a rigorous proof, see the references in the Notes.) Note, by the way, that $\int_{K'} A$ looks as though it depends on a choice of an embedded solid torus T having K as its core, as well as a choice of vector potential A for B, where B is defined by the formula above. But given that this integral equals the linking number, these choices must be immaterial!

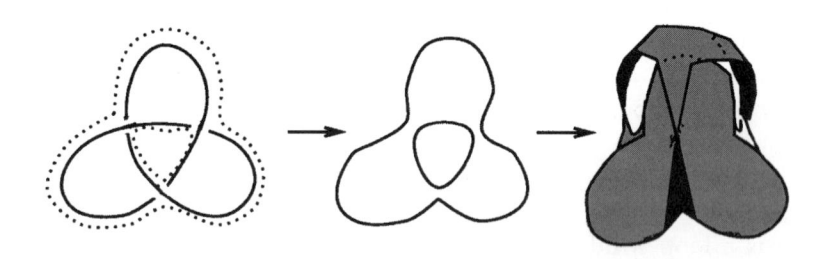

Fig. 38. The pancake proof

First, we need to use a remarkable fact: the knot K' is always the boundary of a surface $S \subset \mathbb{R}^3$, that is, a 2-dimensional submanifold with boundary. This can easily be seen by the 'pancake proof' shown in Figure 38. Draw a diagram of K' and think of it as a stack of pancakes. Connect the pancakes with twisting ramps and one has a surface S having the knot as boundary! In fact, this surface is orientable, and an orientation on K' determines an orientation on it.

Exercise 133. *By examining the pancake proof, show that S is orientable.*

Show that if K' is oriented there is a unique orientation on S compatible with the orientation on the K' in the sense explained in Chapter 6 of Part I — see also Figure 39.

The surface S is called a **Seifert surface** for K'.

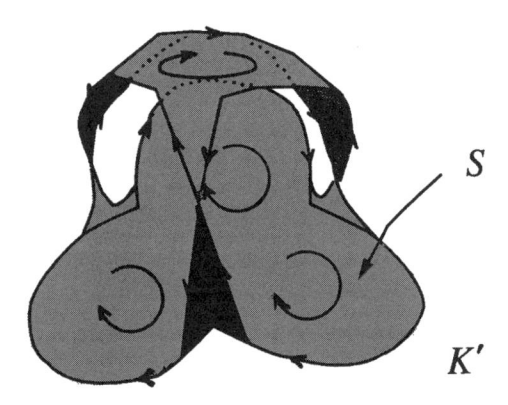

Fig. 39. Orientation on S compatible with orientation on $K' = \partial S$

Now, draw a diagram of the link and use the pancake proof to construct a Seifert surface S for K'. By deforming S slightly if necessary, we can assume that K intersects S **transversely** in finitely many points, that is, K is never tangent to S. We can give each point $p \in K \cap S$ a sign, say $\text{sign}(p) = \pm 1$, as follows. Pick a right-handed basis e_1, e_2 of $T_p S$. The coordinate vector field ∂_t on K gives $T_p K$ a basis. Then let $\text{sign}(p) = +1$ if the basis e_1, e_2, ∂_t is positively oriented relative to the standard orientation of $\mathbb{R}^3$ and let $\text{sign}(p) = -1$ if it is negatively oriented. Now by Stokes' theorem, the integral of A around K' is the magnetic flux through S:

$$\int_{K'} A = \int_S B.$$

Since the magnetic field B has a unit flux around the embedded torus T, the magnetic flux through S is a sum of contributions over the points p,

$$\int_S B = \sum_{p \in K \cap S} \text{sign}(p),$$

where the signs keep track of which way the magnetic field is flowing through S. The quantity on the right hand side, by the way, is called the **intersection number** of K and S.

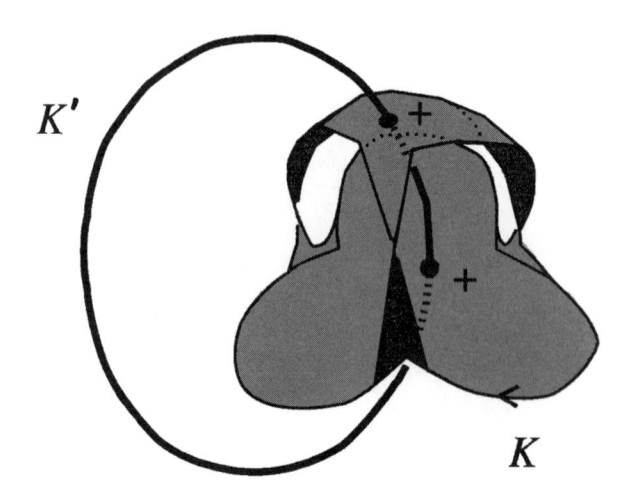

Fig. 40. The intersection number of K and S

To finish the argument, we need to show that

$$\mathcal{L}(K, K') = \sum_{p \in K \cap S} \text{sign}(p).$$

The easiest way to see this is to show that the intersection number satisfies the skein relations of the linking number, which we gave in Figure 36. The most interesting skein relation is the first one, in the case when the two strands belong to different components. We have shown how the intersection number changes by 1 in this case in Figure 41 below. There are two possibilities, since either line segment could be part of K; we have marked the intersection points $p \in K \cap S$ with their signs. We leave it to the reader to check the case where the two strands belong to the same component, as well as the other skein relations.

Exercise 134. *Check them.*

The Linking Number and Writhe   321

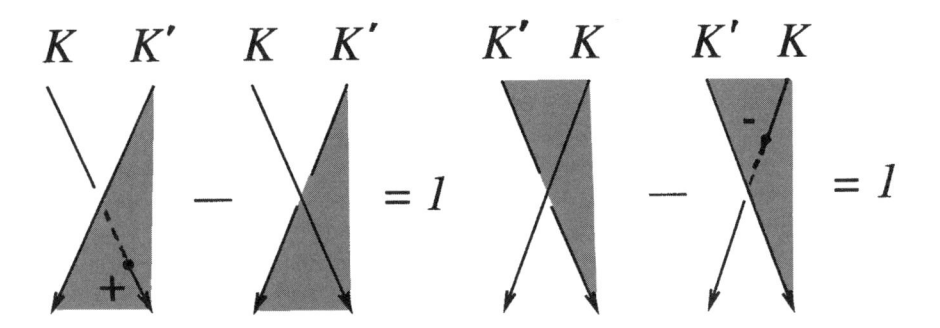

Fig. 41. Checking the first skein relation for the linking number

The Gauss formula for the linking number can be generalized to treat the writhe. Unlike the linking number, the writhe depends on a *framing*. The point is that while we can get an embedded solid torus by thickening up a knot, there are different ways to do so, and to choose a way so we need to choose a framing for the knot. The idea is that the framing tells us how many times the embedding of the solid torus in $\mathbb{R}^3$ 'twists' as one goes around the knot. More precisely, we can get an embedded solid torus from a framed oriented knot as follows. Think of the knot as an embedding

$$\gamma \colon S^1 \to \mathbb{R}^3,$$

and let v be the framing of K, that is, a vector field on K that is nowhere tangent to K. Give $S^1 \times D^2$ the coordinates (t, r, θ), where t goes around the S^1 direction and has $0 \le t \le 2\pi$. We can think of S^1 as a subset of $S^1 \times D^2$, namely the subset $\{r = 0\}$. We can also use Cartesian coordinates (x, y) on D^2, so that there is a vector field ∂_x on $S^1 \times D^2$. Then one can show that there is an embedding

$$\tilde{\gamma} \colon S^1 \times D^2 \to \mathbb{R}^3$$

of the solid torus in $\mathbb{R}^3$ with the properties listed below:

1) $\tilde{\gamma}$ restricted to $S^1 = \{r = 0\}$ is equal to γ.

2) For all $p \in S^1$, $d\gamma((\partial_x)_p) = v_{\gamma(p)}$.

If these hold, we say that the knot γ is the **core** of the embedded solid torus corresponding to $\tilde{\gamma}$. The embedding $\tilde{\gamma}$ is not at all unique. However, one can show that it is unique up to isotopy: that is, given any two embeddings satisfying 1) and 2) above, there is a ambient isotopy carrying one to the other. In fact, the ambient isotopy classes of framed oriented links are in one-to-one correspondence to the ambient isotopy classes of embedded collections of solid tori in $\mathbb{R}^3$, where the tori are required not to intersect. (For proofs of these facts, see the reference in the Notes.)

Now suppose L is a framed oriented link with components K_i. We can thicken all the components to nonintersecting embedded solid tori $T_i \subset \mathbb{R}^3$ as in the previous paragraph. It will be handy sometimes to write $\tilde{\gamma}_i$ for the actual embedding

$$\tilde{\gamma}_i \colon S^1 \times D^2 \to \mathbb{R}^3$$

corresponding to T_i. We can use $\tilde{\gamma}_i$ to transfer the coordinates (t, r, θ) from $S^1 \times D^2$ to T_i. For each solid torus T_i choose a magnetic flux tube B_i as we did in the proof of Gauss' formula for the linking number. Thus B_i is a 2-form vanishing outside T_i and given by

$$B_i = f(r, \theta) r\, dr \wedge d\theta$$

in T_i, where the function f has

$$\int_0^1 \int_0^{2\pi} f(r, \theta) r\, dr \wedge d\theta = 1.$$

Let the total magnetic field be the sum of these flux tubes:

$$B = \sum B_i.$$

If we choose a vector potential A_i for B_i, so that $dA_i = B_i$, the sum

$$A = \sum_i A_i$$

has $dA = B$. Then we claim that the writhe of L is given by

$$w(L) = \int_{\mathbb{R}^3} A \wedge B.$$

The Linking Number and Writhe

Before we sketch the proof of this fact, note the relationship to Chern-Simons theory! In general, the Chern-Simons action is given by

$$S_{CS}(A) = \int \text{tr}(A \wedge dA + \frac{2}{3} A \wedge A \wedge A),$$

but in the case of electromagnetism, taking the trace is unnecessary and $A \wedge A = 0$, so one has

$$S_{CS}(A) = \int A \wedge dA = \int A \wedge B.$$

Now let us outline how one can prove the integral formula for the writhe. First, note that

$$\int_{\mathbb{R}^3} A \wedge B = \sum_{i,j} \int_{\mathbb{R}^3} A_i \wedge B_j$$

where both i and j range over all the components of the link L. If $i \neq j$, we have

$$\int_{\mathbb{R}^3} A_i \wedge B_j = \int_{T_j} A_i \wedge f(r) r dr \wedge d\theta$$

in terms of the coordinates (t, r, θ) on T_j. If we write

$$A_i = (A_i)_t dt + (A_i)_r dr + (A_i)_\theta d\theta$$

on T_j, this reduces to

$$\int_{\mathbb{R}^3} A_i \wedge B_j = \int_{T_j} (A_i)_t dt \wedge f(r) r dr \wedge d\theta$$
$$= \int_0^{2\pi} \int_0^1 \int_0^{2\pi} (A_i)_t f(r) r \, dt dr d\theta.$$

Now, the integral

$$\int_0^{2\pi} (A_i)_t dt$$

is just the integral of A_i over the knot $\alpha: S^1 \to \mathbb{R}^3$ given by

$$\alpha(t) = \tilde{\gamma}_j(t, r, \theta).$$

By Gauss' formula this is just the linking number of K_i and the knot α, which equals the linking number of K_i and K_j. We thus have

$$\int_{\mathbb{R}^3} A_i \wedge B_j = \int_0^{2\pi} \int_0^1 \int_0^{2\pi} (A_i)_t f(r) r \, dt \, dr \, d\theta$$
$$= \int_0^{2\pi} \int_0^1 \mathcal{L}(K_i, K_j) f(r) r \, dr \, d\theta$$
$$= \mathcal{L}(K_i, K_j)$$

when $i \neq j$.

This argument breaks down when $i = j$, since Gauss' formula only applies when we have two *distinct* knots that are components of a *link*. The $i = j$ case is crucial; until we deal with it, all we can conclude is that

$$\int_{\mathbb{R}^3} A \wedge B = \sum_{i \neq j} \int_{\mathbb{R}^3} A_i \wedge B_j + \sum_i \int_{\mathbb{R}^3} A_i \wedge B_i$$
$$= \sum_{i \neq j} \mathcal{L}(K_i, K_j) + \sum_i \int_{\mathbb{R}^3} A_i \wedge B_i.$$

Luckily, from Exercise 131 we have

$$w(L) = \sum_{i \neq j} \mathcal{L}(K_i, K_j) + \sum_i w(K_i),$$

so to finish the proof that

$$w(L) = \int_{\mathbb{R}^3} A \wedge B,$$

all we need to show is that

$$w(K_i) = \int_{\mathbb{R}^3} A_i \wedge B_i.$$

This formula for the writhe of the knot K_i in terms of the magnetic flux tube B_i and its vector potential A_i is a kind of generalization of Gauss' formula to the case of *self-linking*.

In sketching the proof of this formula, we might as well drop the subscript i, since no knots except K_i are involved. Thus, we want to show that

$$w(K) = \int_{\mathbb{R}^3} A \wedge B$$

The Linking Number and Writhe 325

where K is any framed oriented knot, T is an embedded solid torus having K as its core, B is the magnetic flux tube running around T, and $dA = B$.

Exercise 135. *Check that this integral does not depend on which vector potential A we choose such that $dA = B$.*

We use the following trick: we cover D^2 with lots of embedded disks D_α, and use this to subdivide T into lots of thinner solid tori T_α as follows:

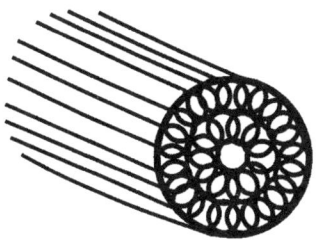

Fig. 42. Subdividing an embedded torus

We call these thinner tori **cables**. Let K_α denote the core of the cable T_α. If $\alpha \neq \beta$, we have

$$w(K) = \mathcal{L}(K_\alpha, K_\beta).$$

For example, if we took a framed oriented unknot K with writhe 1, thickened it up to a solid torus T, and looked at two cables in T, we would see something like this:

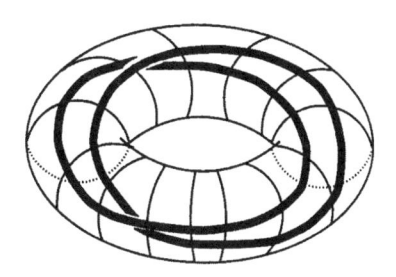

Fig. 43. Writhe as the linking of cables

The cores of these two cables have linking number 1.

Using a partition of unity we can write the function f on D^2 as a sum of functions f_α, where f_α is zero outside D_α. We can then write $B = \sum B_\alpha$, where
$$B_\alpha = f_\alpha \, r \, dr \wedge d\theta$$
is a flux tube running around the cable T_α. If we choose vector potentials A_α such that $B_\alpha = dA_\alpha$, then $A = \sum A_\alpha$ has $dA = B$. Then
$$\int_{\mathbb{R}^3} A \wedge B = \sum_{\alpha,\beta} \int_{\mathbb{R}^3} A_\alpha \wedge B_\beta.$$

When $\alpha \neq \beta$, our previous results imply
$$\int_{\mathbb{R}^3} A_\alpha \wedge B_\beta = \left(\int_{D^2} f_\alpha \, r \, dr \wedge d\theta\right)\left(\int_{D^2} f_\beta \, r \, dr \wedge d\theta\right) \mathcal{L}(K_\alpha, K_\beta)$$
$$= \left(\int_{D^2} f_\alpha \, r \, dr \wedge d\theta\right)\left(\int_{D^2} f_\beta \, r \, dr \wedge d\theta\right) w(K).$$

Exercise 136. *Check this computation.*

The terms with $\alpha = \beta$ are still a nuisance, but the point is that if we subdivide T into a large number of cables, say n, there will be n^2 terms in the sum over α and β, but only n terms for which $\alpha = \beta$. Using some analysis, one can arrange things so that as $n \to \infty$, the terms with $\alpha = \beta$ go to zero, so that
$$\lim_{n \to \infty} \sum_{\alpha \neq \beta} \int_{\mathbb{R}^3} A_\alpha \wedge B_\beta = \lim_{n \to \infty} \sum_{\alpha,\beta} \int_{\mathbb{R}^3} A_\alpha \wedge B_\beta$$
$$= \int_{\mathbb{R}^3} A \wedge B$$

and also
$$\lim_{n \to \infty} \sum_{\alpha \neq \beta} \left(\int_{D^2} f_\alpha \, r \, dr \wedge d\theta\right)\left(\int_{D^2} f_\beta \, r \, dr \wedge d\theta\right) =$$
$$\lim_{n \to \infty} \sum_{\alpha,\beta} \left(\int_{D^2} f_\alpha \, r \, dr \wedge d\theta\right)\left(\int_{D^2} f_\beta \, r \, dr \wedge d\theta\right) =$$
$$\left(\int_{D^2} f \, r \, dr \wedge d\theta\right)\left(\int_{D^2} f \, r \, dr \wedge d\theta\right) = 1.$$

As a result, one obtains

$$\int_{\mathbb{R}^3} A \wedge B = \lim_{n \to \infty} \sum_{\alpha \neq \beta} \int_{\mathbb{R}^3} A_\alpha \wedge B_\beta$$
$$= \lim_{n \to \infty} \sum_{\alpha \neq \beta} (\int_{D^2} f_\alpha \, r dr \wedge d\theta)(\int_{D^2} f_\beta \, r dr \wedge d\theta) \, w(K)$$
$$= w(K)$$

as desired! In short, the mysterious 'self-linking' terms with $\alpha = \beta$ go to zero in the limit as $n \to \infty$, and the writhe of K is completely accounted for by the terms

$$\int_{\mathbb{R}^3} A_\alpha \wedge B_\beta$$

with $\alpha \neq \beta$.

The Jones Polynomial

The linking number is, in fact, just the tip of a very interesting iceberg. First of all, there is a famous link invariant known as the **Alexander-Conway polynomial**. This is an isotopy invariant of oriented links; it is a polynomial in one variable, say z. It is customary to write the Alexander-Conway polynomial of a link L as $\nabla_L(z)$. The easiest way to define this polynomial is by its skein relations:

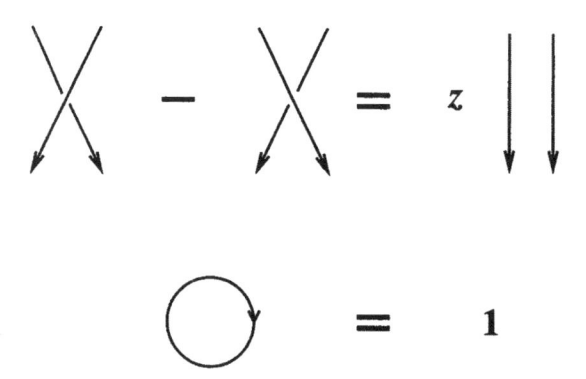

Fig. 44. Skein relations for the Alexander-Conway polynomial

The first relation relates the polynomial for three different link diagrams having either a right-handed crossing, a left-handed crossing, or *no* crossing in some region, but otherwise the same. The second relation says that the polynomial applied to the unknot equals 1. We leave it as a puzzle to use the first relation to show that if a link contains an unlinked unknot, its Alexander polynomial is zero. Given this, we can calculate the Alexander-Conway polynomial of the Hopf link as in Figure 45.

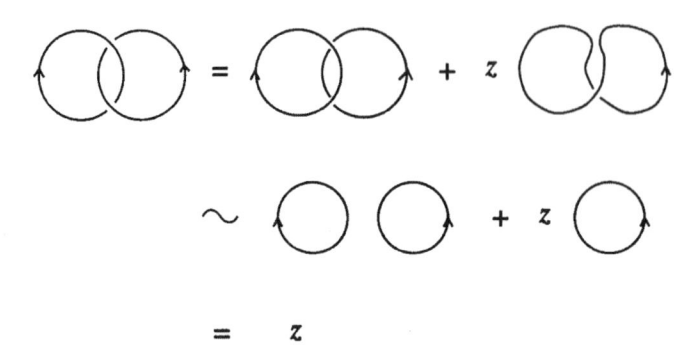

Fig. 45. The Alexander-Conway polynomial of the Hopf link

If the reader has not seen the Alexander-Conway polynomial before, it should be *utterly mysterious* why the skein relations uniquely define the polynomial of any link diagram. For all one knows, after all, different ways of applying the skein relations might give different answers. It should also be mysterious why the polynomial is invariant under the Reidemeister moves! We will not explain these mysteries; to do so would be too much of a digression. (See the reference in the Notes.) Instead, we simply want to note that the resemblance of these skein relations to those for the linking number are no coincidence. The reader can show the following using the skein relations:

Exercise 137. *Write*

$$\nabla_L(z) = \sum_{i=0}^{\infty} a_i z^i.$$

Show that a_0 is 1 if L has exactly one component, and 0 otherwise. Show that a_1 is the linking number of L if L has exactly two components, and 0 otherwise.

Exercise 138. *Given an oriented link L, let L^* denote its mirror image. Show that $\nabla_L(z) = \nabla_{L^*}(z)$. Thus the Alexander polynomial is unable to distinguish between links and their mirror images.*

We should also say that the Alexander-Conway polynomial is a 'classical' invariant of links, meaning that it was discovered by Alexander in 1923 when algebraic topology was first being developed, and that it is best understood in these terms, in particular, in terms of homology theory. Alexander only noted the skein relations in passing, and the polynomial was given its present normalization and simple-looking definition in terms of skein relations by Conway in 1970.

In a paper that appeared in 1985, Vaughan Jones described a remarkable new link invariant, now called the Jones polynomial. He was not even working on knot theory at the time! He was working on operator theory — closely related to some aspects of quantum field theory — and came across the application to knots more or less by accident. The remarkable thing was that this new invariant could also be defined using very simple skein relations; however, it could *not* be explained in terms of existing ideas in algebraic topology. One sign of this is that it very easily distinguishes between certain knots and their mirror images — something that was traditionally a bit tricky.

This development led to an amazing burst of activity. People invented new link invariants and tried to understand them. The new invariants were closely related to physics in many ways, only a few of which we will discuss. Work on these new invariants is not done yet! There is much we do understand about them, and much that we do not. One thing that is quite clear by now is that knot theory and gauge theory are very closely connected. Just as the linking number turns out to be related to a U(1) gauge theory (electromagnetism), it was shown by Witten that the new link invariants are related to gauge theories with other, nonabelian gauge groups. In fact, there is one new link invariant for each finite-dimensional representation of each semisimple Lie group. The Jones polynomial is the simplest of all these invariants, because it is associated to the spin-$\frac{1}{2}$ representation of SU(2).

To keep life simple, we will concentrate on the Jones polynomial and only make passing reference to some of the other new link invariants. Moreover, we will take advantage of Kauffman's work and give his very simple construction of the Jones polynomial. This proceeds as follows: first one defines an invariant of framed links called the 'Kauffman bracket', and then one uses this together with the writhe to define the Jones polynomial, which is an invariant of oriented links.

Let L be the diagram of an oriented link. The Kauffman bracket of L, written $\langle L \rangle$, will start out being function of three variables, A, B, and d. Later we will see that by making a special choice for B and d, namely

$$B = A^{-1}, \qquad d = -(A^2 + A^{-2}),$$

the Kauffman bracket will be invariant under the framed Reidemeister moves.

The invariants we discussed in the previous section were defined as sums over crossings of L. The Kauffman bracket is given by a subtler sum, called a 'state sum'. A **state** σ of L assigns to each crossing p of L a number σ_p that is either A or B. Thus a link with n vertices has 2^n possible states.

Next, given any state σ of L, go to each crossing p of L, and view it from some angle so that it looks like this:

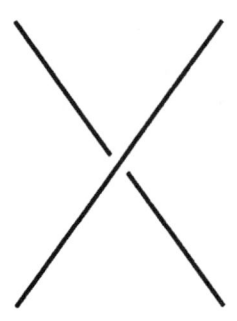

Fig. 46. Correct view for computing the Kauffman bracket

If $\sigma_p = A$, replace the crossing by two roughly vertical arcs that do not cross each other, while if $\sigma_p = B$, replace it by two roughly horizontal arcs that do not cross each other:

The Jones Polynomial

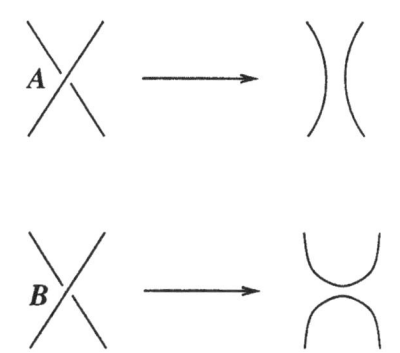

Fig. 47. Eliminating crossings according to whether σ_p is A or B

Exercise 139. *Check that this process is really well-defined, that is, there is no ambiguity about what to do!*

The result is a new link diagram having no crossings whatsoever; thus it consists of a finite set of circles (topologically speaking) embedded in the plane. The number of circles is called the **loop number** of σ and is written as $\|\sigma\|$.

The Kauffman bracket is then defined by a sum over all states, or **state sum**, as follows:

$$\langle L \rangle = \sum_{\text{states } \sigma} d^{\|\sigma\|} \prod_{\text{crossings } p} \sigma_p.$$

Let us illustrate this with the Hopf link. The Hopf link, as diagrammed in Figure 48, has 2 vertices, hence 4 possible states, which we have shown.

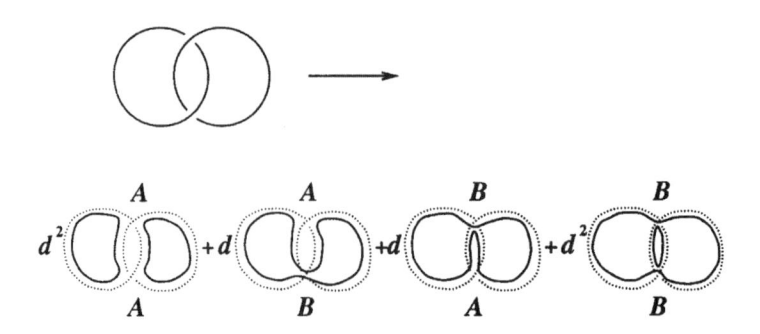

Fig. 48. Computing the Kauffman bracket of the Hopf link

For each state we have eliminated the crossings according to the rule in Figure 47 and worked out the loop number. We obtain the following sum over 4 states:

$$\langle L \rangle = d^2 A^2 + dAB + dAB + d^2 B^2.$$

Before we check that the Kauffman bracket really is an invariant of framed links — with the appropriate choices of B and d — let us comment on the formula for it, which may seem arbitrary and puzzling. Kauffman was led to it in part by analogies with statistical mechanics. Suppose we have a physical system that can be in finitely many states s, each state having some energy $E(s)$. Then in equilibrium, the probability that the system will be in any state s is proportional to

$$e^{-\beta E(s)},$$

where

$$\beta = \frac{1}{kT},$$

T being the temperature and k being a constant known as Boltzmann's constant. Thus, the probability of being in a given state diminishes exponentially with the energy of that state, with the exponential decreasing most rapidly at low temperatures. Of course, the probabilities should sum to 1, so the actual probability of being in a given state s is

$$\frac{1}{Z(\beta)} e^{-\beta E(s)},$$

The Jones Polynomial

where the **partition function** $Z(\beta)$ is given by

$$Z(\beta) = \sum_{\text{states } s} e^{-\beta E(s)}.$$

However, far from being merely a boring normalization factor, the partition function turns out to contain a large amount of information about the system. For example:

Exercise 140. *Show that*

$$-\frac{d}{d\beta} \ln Z(\beta) = \frac{1}{Z(\beta)} \sum_{\text{states } s} E(s) e^{-\beta E(s)}.$$

This is the expected value of the energy of the system at temperature T, usually written $\overline{E}$.

Exercise 141. *Show that*

$$\frac{d\overline{E}}{dT} = k\beta^2 \frac{d^2}{d\beta^2} \ln Z(\beta).$$

This quantity, which measures the change in expected energy with change in temperature, is called the **specific heat** *of the system at temperature T.*

In the statistical mechanics of crystals one obtains partition functions very similar to the state sum formula for the Kauffman bracket. Imagine, for example, a crystalline magnet, which is a lattice of atoms, each having an electron whose spin can point either up or down. Thus a state s of such a system assigns to each point in the lattice, or **site**, a spin s_p. The energy $E(s)$ of the state typically consists of two parts: a sum over sites p of some energy $E(s_p)$ depending only on the spin at p, and an 'interaction energy' $E_{int}(s)$ due to the interaction of spins at different sites. That is,

$$E(s) = E_{int}(s) + \sum_{\text{sites } p} E(s_p).$$

The partition function is thus

$$Z(\beta) = \sum_{\text{states } s} e^{-\beta E(s)}$$
$$= \sum_{\text{states } s} e^{-\beta E_{int}(s)} \prod_{\text{sites } p} e^{-\beta E(s_p)}.$$

Note how this resembles the formula for the Kauffman bracket,

$$\langle L \rangle = \sum_{\text{states } \sigma} d^{\|\sigma\|} \prod_{\text{crossings } p} \sigma_p.$$

One can think of the crossings of the knot as analogous to sites, and two allowed values A and B of σ_p as corresponding to the values of $\exp(-\beta E(s_p))$ when the spin s_p is up and down, respectively. The quantity $d^{\|\sigma\|}$, which is more global in nature (as it depends on the loop number), is analogous to the term depending on the interaction energy.

In fact, this analogy is a deep one. One can use ideas from knot theory to exactly solve certain problems in two-dimensional statistical mechanics; conversely, one can apply techniques from statistical mechanics to gain information about knot theory! This is a fascinating and active subject, but we will have to leave off here and direct the interested reader to the Notes.

Now we return to the problem of proving that under appropriate conditions the Kauffman bracket is an invariant of framed links. To do so, it is very useful to note that it satisfies some skein relations. These are shown below:

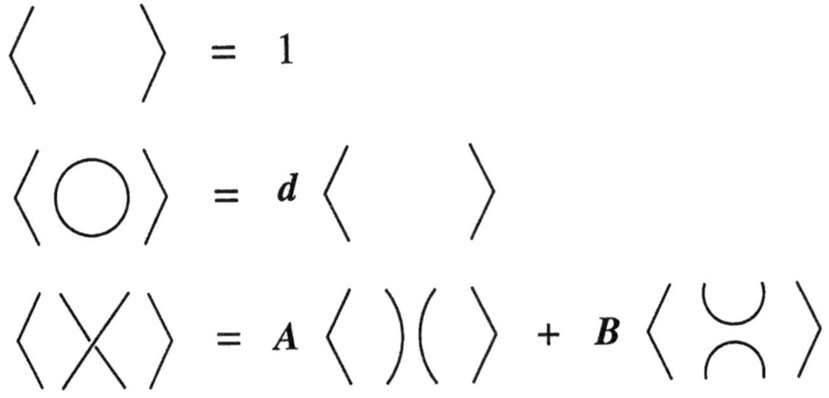

$$\left\langle \;\right\rangle = 1$$

$$\left\langle \bigcirc \right\rangle = d \left\langle \;\right\rangle$$

$$\left\langle \times \right\rangle = A \left\langle \;\right\rangle\left\langle \;\right\rangle + B \left\langle \asymp \right\rangle$$

Fig. 49. Skein relations for the Kauffman bracket

The first relation is due to the product over crossings in the formula

The Jones Polynomial

for the Kauffman bracket,

$$\langle L \rangle = \sum_{\text{states } \sigma} d^{\|\sigma\|} \prod_{\text{crossings } p} \sigma_p,$$

together with how one uses the two choices $\sigma_p = A, B$ to decide how to eliminate the intersection when calculating the loop number, as in Figure 47. The second relation means that the presence of an embedded circle in a link diagram has the effect of *multiplying* the value of the Kauffman bracket by d. The reason for this is that the embedded circle contributes nothing to the product over crossings (since it has no crossings), but it contributes 1 to the loop number $\|\sigma\|$ no matter what the state σ is. We call d the **loop value** of the Kauffman bracket. These skein relations should be supplemented by the rule that the Kauffman bracket of the empty link is 1 (by convention).

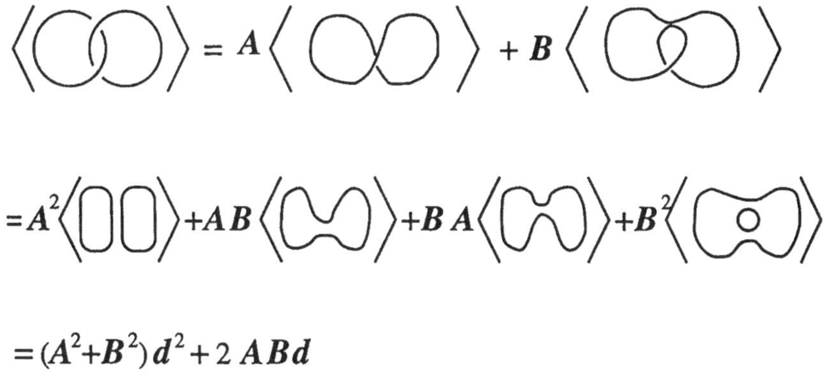

$$= (A^2 + B^2) d^2 + 2\, ABd$$

Fig. 50. The Kauffman bracket of the Hopf link using skein relations

In Figure 50 we calculate the Kauffman bracket of the Hopf link using the skein relations. Compare this calculation to that appearing in Figure 48 and the text that follows. It should become clear that the skein relations are simply a handy way of working out the state sum a bit at time.

We should emphasize that while the writhe is an **additive** link invariant, meaning that the writhe of a disjoint union $L \cup L'$ satisfies

$w(L \cup L') = w(L) + w(L')$, the Kauffman bracket is **multiplicative**, meaning that $\langle L \cup L' \rangle = \langle L \rangle \langle L' \rangle$. Thus in Figure 34 we indicated that an unlinked unknot *adds* zero to the writhe, while in Figure 49 we mean that an unlinked unknot *multiplies* the Kauffman bracket by d.

Now let us check invariance of the Kauffman bracket under the framed Reidemeister moves 0, I', II and III. Recall that move 0 simply consists of an isotopy of the plane. The formula for the Kauffman bracket makes it clear that $\langle L \rangle$ only depends on the link diagram L up to isotopies of the plane, not the actual geometry of the diagram, so we have invariance under move 0. We check invariance under move I' in the figure below.

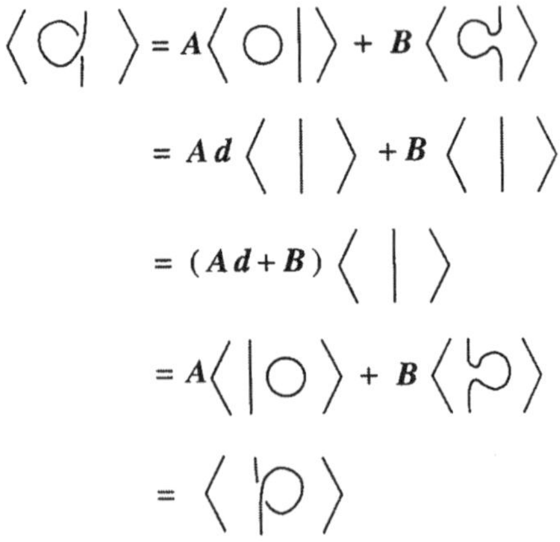

$$\langle \text{⌒} \rangle = A \langle \text{○|} \rangle + B \langle \text{⌒} \rangle$$
$$= Ad \langle \text{|} \rangle + B \langle \text{|} \rangle$$
$$= (Ad + B) \langle \text{|} \rangle$$
$$= A \langle \text{|○} \rangle + B \langle \text{⌒} \rangle$$
$$= \langle \text{⌒} \rangle$$

Fig. 51. Invariance of the Kauffman bracket under move I'

Here we get rid of the twist by first using the skein relation that eliminates crossings, and then the one that eliminates embedded circles. We see that a right-handed twist of either type has the effect of multiplying the Kauffman bracket by the same factor, $Ad + B$.

The Jones Polynomial

Likewise, we check invariance under move II in the following figure, using the rule for right-handed twists that we just derived.

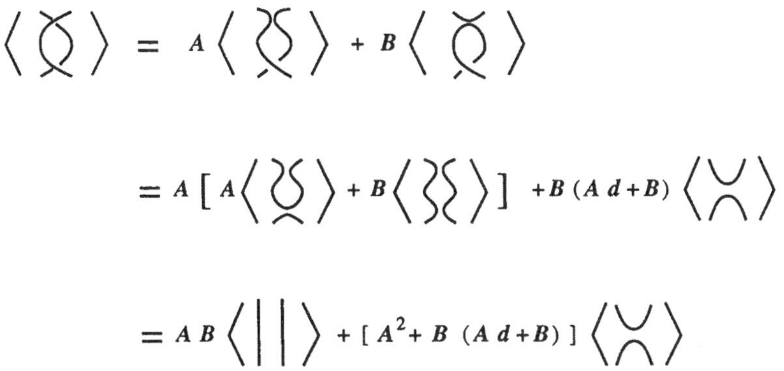

$$\left\langle \vcenter{\hbox{⧖}} \right\rangle = A \left\langle \vcenter{\hbox{⧖}} \right\rangle + B \left\langle \vcenter{\hbox{⧗}} \right\rangle$$

$$= A\left[A\left\langle \vcenter{\hbox{⧖}} \right\rangle + B\left\langle \vcenter{\hbox{)(}} \right\rangle \right] + B(A\,d+B)\left\langle \vcenter{\hbox{⌣⌢}} \right\rangle$$

$$= AB\left\langle \vcenter{\hbox{||}} \right\rangle + [A^2 + B(A\,d+B)]\left\langle \vcenter{\hbox{⌣⌢}} \right\rangle$$

Fig. 52. Invariance of the Kauffman bracket under move II

We see that an unwanted factor of AB appears, as does a bad term that vanishes only if

$$A^2 + B(Ad + B) = 0.$$

The factor of AB goes away if we set

$$B = A^{-1}.$$

This means that

$$A^2 + B(Ad + B) = A^2 + A^{-2} + d,$$

so that the bad term vanishes if we also set

$$d = -(A^2 + A^{-2}).$$

Henceforth we will choose B and d in this manner.

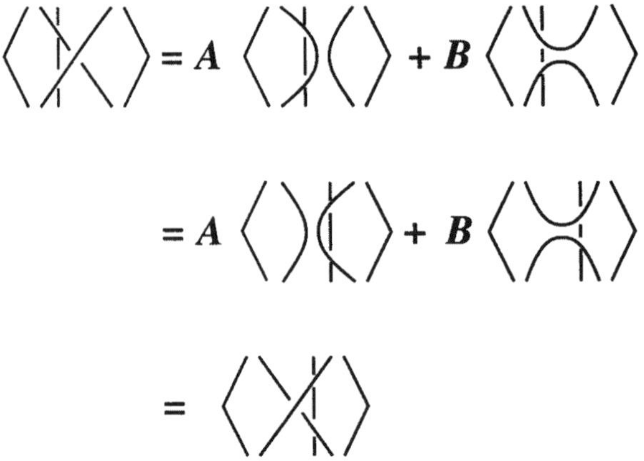

$$\left\langle \vcenter{\hbox{⧖}} \right\rangle = A \left\langle \vcenter{\hbox{)(}} \right\rangle + B \left\langle \vcenter{\hbox{⌣⌢}} \right\rangle$$

$$= A \left\langle \vcenter{\hbox{)(}} \right\rangle + B \left\langle \vcenter{\hbox{⌣⌢}} \right\rangle$$

$$= \left\langle \vcenter{\hbox{⧗}} \right\rangle$$

Fig. 53. Invariance of the Kauffman bracket under move III

Finally, let us check invariance under Reidemeister move III. We do this in Figure 53. First we use the skein relation to eliminate a crossing, then we use invariance under move II to slide the vertical strand to the right, and then we use the skein relation to reinstall the crossing! This beautiful argument (due to Kauffman) completes the proof that the Kauffman bracket is an invariant of framed links, with appropriate choices of B and d. In Figure 54 we list the skein relations for the bracket with the correct choices of B and d. We also include the skein relation that allows us to get rid of a right-hand twist in the framing while multiplying by

$$Ad + B = -A(A^2 + A^{-2}) + A^{-1} = -A^3.$$

Exercise 142. *Show that one can get rid of a left-handed twist in the framing while multiplying by $-A^{-3}$. (Hint: one can do this directly or by reducing it to the right-handed case via the Whitney trick.)*

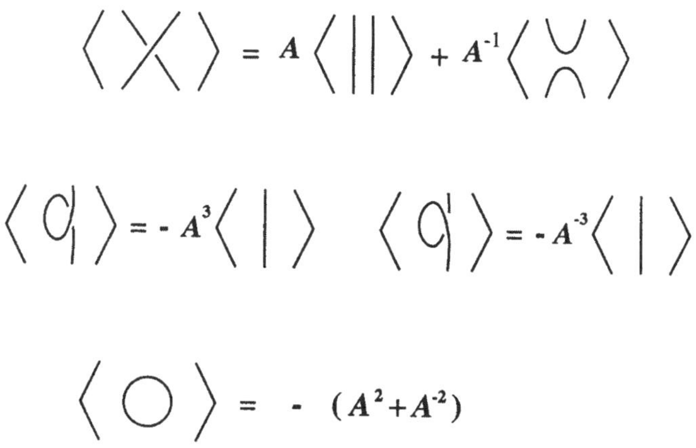

The Jones Polynomial

$$\left\langle \times \right\rangle = A \left\langle \;\vert\vert\; \right\rangle + A^{-1} \left\langle \asymp \right\rangle$$

$$\left\langle \overset{\frown}{\cap} \right\rangle = -A^3 \left\langle \;\vert\; \right\rangle \qquad \left\langle \overset{\frown}{\cap} \right\rangle = -A^{-3} \left\langle \;\vert\; \right\rangle$$

$$\left\langle \bigcirc \right\rangle = -(A^2 + A^{-2})$$

Fig. 54. Skein relations for the Kauffman bracket link invariant

For practice with the skein relations, try the following:

Exercise 143. *Show that the Kauffman bracket of the trefoil knot shown in Figure 3 equals $-(A^2 + A^{-2})(-A^5 - A^{-3} + A^{-7})$. Show that the Kauffman bracket of the unknot (with an arbitrary choice of framing) is $-(A^2 + A^{-2})(-A^3)^w$, w being its writhe. Conclude that the trefoil is not isotopic to the unknot. I.e., the trefoil knot is really knotted!*

Exercise 144. *Calculate the Kauffman bracket of the mirror image of the previous trefoil knot. Conclude that the trefoil is not isotopic to its mirror image.*

Exercise 145. *Show that for any framed link L, the mirror image L^* has*

$$\langle L^* \rangle(A) = \langle L \rangle(A^{-1}).$$

Exercise 146. *Calculate $\langle K \rangle$ for the figure-eight knot K shown in Figure 8. Check that $\langle K \rangle(A) = \langle K \rangle(A^{-1})$, which is consistent with the above exercise and the fact, shown in Exercise 129, that K is regular-isotopic to its mirror image.*

It should be clear from these exercises that there is a profound relation between the Kauffman bracket and chirality, or handedness. This is of course built into its definition, either in terms of the state sum or the skein relations. We will, however, see a deeper explanation of this fact in the next section: the Kauffman bracket is associated to Chern-Simons theory, which has intrinsic chirality-dependence.

Now let us turn to the Jones polynomial. Again, this is an invariant of oriented links. We will first define it for links equipped with *both* a framing *and* an orientation, and then prove that it is independent of the framing. The definition is simple: let L be a framed oriented link; then the **Jones polynomial** $V_L(A)$ is given by

$$V_L(A) = (-A^{-3})^{w(L)} \langle L \rangle (A)$$

where $w(L)$ is the writhe of L. This is clearly an invariant of framed oriented links, since both the writhe and Kauffman bracket are; all we need to show is that it is invariant under Reidemeister move I. This follows from the skein relations for the Kauffman bracket in Figure 54; the point is that the framing-dependence of the Kauffman bracket is precisely cancelled by the factor of $(-A^{-3})^{w(L)}$.

One can easily derive skein relations for the Jones polynomial from those for the writhe and the Kauffman bracket. These are written below in terms of the variable $q = A^4$ which is commonly used in this context.

Exercise 147. *Derive the skein relations for the Jones polynomial.*

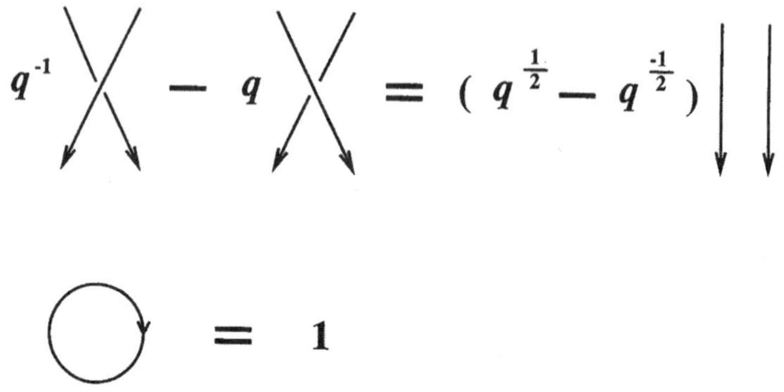

Fig. 55. Skein relations for the Jones polynomial

Chern-Simons Theory

Finally, let us sketch the relationship between some of the link invariants we have described and Chern-Simons theory. This relationship was first worked out by Witten in a paper that appeared in 1989, and subsequently there has been a flood of work relating quantum field theory and topology in 3 dimensions. As we shall show in Chapter III, these results are also relevant to quantum gravity in 4 dimensions!

To begin with, we must say a bit about the Lagrangian, or path-integral, approach to quantum field theory. We warn the reader that this is a large subject which is rather famous for its subtleties. We already mentioned the path-integral approach to quantum mechanics in Chapter 6 of Part I; the basic idea was that transition amplitudes could be computed as integrals over the space of paths using the 'measure'

$$e^{\frac{i}{\hbar}S(\gamma)}\mathcal{D}\gamma,$$

where S is the action and $\mathcal{D}\gamma$ is 'Lebesgue measure' on the space of paths. As we have noted, these 'measures' are difficult to define in a rigorous way, so the path integral is merely a heuristic device until one puts in a lot of work to figure out how to compute it approximately, or prove theorems about it! We will not go into these issues in any detail, instead referring the reader to the Notes for Chapter 3 of this part for more information on path integrals.

It is easiest to describe path integrals in field theory in the case where the spacetime M is a Riemannian manifold: one recovers results for the physically more realistic Lorentzian spacetimes by a process known as 'Wick rotation', which in simple cases essentially amounts to making a substitution $t \to it$. Consider Yang-Mills theory, for example. Let E be a G-bundle over the oriented Riemannian manifold M; if A denotes the vector potential for a connection on E, the Yang-Mills action is given by

$$S_{YM}(A) = \frac{1}{2}\int_M \operatorname{tr}(F \wedge \star F).$$

In *classical* field theory, we were interested in solutions of the Yang-Mills equation, and we have seen that they are given by critical points of the action. In *quantum* field theory one uses the action to compute the 'vacuum expectation values' of observables, that is, their average measured values in the vacuum state. (Various tricks allow one to express many physically interesting quantities as vacuum expectation values.) Mathematically, we may think of an observable in this formalism as a gauge-invariant function on the space $\mathcal{A}$ of all G-connections on E. The **vacuum expectation value** of the observable f, denoted by $\langle f \rangle$, is then given by

$$\langle f \rangle = \frac{1}{Z} \int_{\mathcal{A}} f(A) \, e^{\frac{i}{\hbar} S_{YM}(A)} \, \mathcal{D}A.$$

(A minus sign one often sees here has been absorbed into our definition of $S_{YM}(A)$.) Here $\mathcal{D}A$ denotes 'Lebesgue measure' on $\mathcal{A}$, and the normalization constant Z is given by

$$Z = \int_{\mathcal{A}} e^{\frac{i}{\hbar} S_{YM}(A)} \, \mathcal{D}A,$$

and is called the **partition function**, in analogy to statistical mechanics. In fact, the analogy between path integrals in quantum field theory and state sums in statistical mechanics has been very fruitful both for physics and pure mathematics.

We should note that since an observable in the path-integral approach to gauge theory f is a *gauge-invariant* function on $\mathcal{A}$, one often thinks of it as a function on the space $\mathcal{A}/\mathcal{G}$ of connections modulo gauge transformations. One then writes

$$\langle f \rangle = \frac{1}{Z} \int_{\mathcal{A}/\mathcal{G}} f(A) \, e^{\frac{i}{\hbar} S_{YM}(A)} \, \mathcal{D}A$$

where $\mathcal{D}A$ now denotes the 'measure' on $\mathcal{A}/\mathcal{G}$ obtained from pushing forward Lebesgue measure on $\mathcal{A}$ by the map $\mathcal{A} \to \mathcal{A}/\mathcal{G}$ that sends each connection to its gauge equivalence class, and A now denotes a gauge equivalence class of connections.

The simplest observables in gauge theory are the Wilson loops: given a loop γ in M, the Wilson loop $W(\gamma, A)$ is the trace of the holonomy around γ of the connection having vector potential A. We can think of $W(\gamma, A)$ as a function on $\mathcal{A}$, which we write as $W(\gamma)$ for short, and its vacuum expectation value is given by

$$\langle W(\gamma) \rangle = \frac{1}{Z} \int_{\mathcal{A}} W(\gamma, A) \, e^{\frac{i}{\hbar} S_{YM}(A)} \, \mathcal{D}A.$$

Alternatively, since $W(\gamma)$ is gauge-invariant, we can think of it as a function on $\mathcal{A}/\mathcal{G}$ and write

$$\langle W(\gamma) \rangle = \frac{1}{Z} \int_{\mathcal{A}/\mathcal{G}} W(\gamma, A) \, e^{\frac{i}{\hbar} S_{YM}(A)} \, \mathcal{D}A.$$

Since Yang-Mills theory involves the metric on M, the vacuum expectation value of a Wilson loop in Yang-Mills theory depends not just on the topology of the loop (e.g. its ambient isotopy class, if it is a knot), but also on its actual geometry.

Kenneth Wilson introduced Wilson loops in a paper published in 1974, in order to study the issue of 'confinement' in quantum chromodynamics. 'Confinement' is the name for the fact that one never sees free quarks or other objects of nonzero color except at extremely high temperatures. Instead, all one sees is hadrons, which are color-neutral collections of quarks: either **mesons**, which are quark-antiquark pairs, or **baryons**, which consist of three quarks bound together. For example, the pions are mesons, while the proton and neutron are baryons. Naively, one hopes to explain confinement for mesons by showing that there is force binding its two constituent quarks that does not diminish with distance. This would be equivalent to saying that the energy of the pair of quarks is proportional to their distance. Now, with some work one can transform this statement into a statement about vacuum expectations of Wilson loops! We will skip the heuristic derivation and just state the result: confinement is equivalent to the fact that for a rectangular loop γ, the expectation value $\langle W(\gamma) \rangle$ decreases approximately as an exponential function of the *area* of the rectangle. (We turn the reader to the Notes for more details.) A vast amount of work using analytical methods and computer calculations has gone into checking that $\langle W(\gamma) \rangle$ really does satisfy this area law in quantum chromodynamics. While the numerical evidence seems convincing, a rigorous proof of this fact — or even a rigorous formulation of the theory! — is still lacking.

The path-integral approach is, at least in principle, applicable to any field theory for which one has a Lagrangian. If one has a Lagrangian for a gauge theory, one can write down a formula analogous to those above for the vacuum expectation values of Wilson loops in the theory. In his 1989 paper, Witten attempted to understand the new link invariants in a systematic way by relating them to the vacuum expectation values of Wilson loops in Chern-Simons theory. Using the invariance of the Chern-Simons action under orientation-preserving diffeomorphisms, these vacuum expectation values should be isotopy invariants! For expository purposes, we will first sketch this idea in its beautiful simplicity as if the path integral presented no problems. Then we will mention some of the problems and briefly describe some ways people have dealt with them. For the details of this fascinating business, which is still not completely understood, we will have to refer the reader to the Notes.

One beautiful feature of the Chern-Simons construction of link invariants is that it works on manifolds other than $\mathbb{R}^3$. The basic definitions of knot theory are the same for links in arbitrary manifolds: a link in a manifold S is simply a submanifold of S diffeomorphic to a collection of circles, two links L, L' are said to be ambient isotopic if there is a smooth one-parameter family α_t of diffeomorphisms of S with α_0 equal to the identity and α_1 mapping L to L', and so on. For some purposes it is best to work on a *compact* oriented 3-dimensional manifold. Now $\mathbb{R}^3$ itself is not compact, but for the purposes of knot theory one can replace $\mathbb{R}^3$ by S^3 by adding a 'point at infinity'. Any link in $\mathbb{R}^3$ can then be regarded as a link in S^3, and two links in $\mathbb{R}^3$ are isotopic as links in S^3 if and only if they are isotopic as links in $\mathbb{R}^3$. Thus isotopy classes of links in S^3 are in one-to-one correspondence with those in $\mathbb{R}^3$. The same is true for links with orientation and/or framing.

So, let S be a compact oriented 3-dimensional manifold, and let E be a trivial G-bundle over S with standard fiber given by the vector space V on which G has a representation ρ. Let $\mathcal{A}$ denote the space of all G-connections on E. We can write any connection as a flat connection plus a vector potential since E is trivial, so we will simply think of the connection as *being* the vector potential. Now, as we have seen, the Chern-Simons action

$$S_{CS}(A) = \int_S \mathrm{tr}(A \wedge dA + \frac{2}{3} A \wedge A \wedge A)$$

Chern-Simons Theory

is invariant under orientation-preserving diffeomorphisms and also gauge invariant up to integer multiples of $8\pi^2$. It follows that the exponential

$$e^{\frac{ik}{4\pi}S_{CS}(A)},$$

where k is an integer called the **level**, is invariant under gauge transformations, so one can use this quantity in path integrals for Chern-Simons theory just as one would use

$$e^{\frac{i}{\hbar}S_{YM}(A)}$$

in Yang-Mills theory.

Suppose, for example, that L is an oriented link with components given by the loops $\gamma_1, \ldots, \gamma_n$. The vacuum expectation value

$$\langle W(\gamma_1) \cdots W(\gamma_n) \rangle$$

is given by

$$\frac{1}{Z(S)} \int_{\mathcal{A}} W(\gamma_1, A) \cdots W(\gamma_n, A)\, e^{\frac{ik}{4\pi}S_{CS}(A)}\, \mathcal{D}A$$

where the partition function $Z(S)$ is given by

$$Z(S) = \int_{\mathcal{A}} e^{\frac{ik}{4\pi}S_{CS}(A)}\, \mathcal{D}A.$$

Of course, one must use ingenuity to work with these formal expressions involving the 'Lebesgue measure' $\mathcal{D}A$. With some optimism, however — and mathematical physics is a discipline that requires endless optimism — one might expect that $\mathcal{D}A$ is diffeomorphism invariant, hence that the vacuum expectation value is invariant under all orientation-preserving diffeomorphisms of S. If this were the case, it would be an isotopy invariant of the link L.

The problem is thus to make enough sense of the path integral to be able to compute it, at least in some cases. This has been done rather thoroughly when G is compact, and to a lesser extent for certain non-compact semisimple groups. However, there are two significant caveats involved. First, it turns out that we can make sense of the path integral only if we equip L with a framing. This appears to be due to the need for 'regularization'. In quantum field theory, vacuum expectation values of products of fields are frequently rather singular. For this reason, the fields need to be 'regularized' or 'smeared', that is, multiplied by a function on spacetime and then integrated, before one multiplies them. Physically, this corresponds to the fact that we never measure the value of a field at a single point, but only integrals thereof, as our probes have finite size. Similarly, it appears that the path integral above is ill-defined unless we do something such as thickening each loop to a solid torus, as we did when discussing the writhe and electromagnetism, and using this to 'smear' the Wilson loop. Recall that a framing of a loop $\gamma\colon S^1 \to S$ gives us a way, unique up to ambient isotopy, to thicken it to an embedded solid torus $\tilde{\gamma}\colon S^1 \times D^2 \to S$. This embedded solid torus defines a family of loops γ^p in S, one for each point $p \in D^2$. In equations, we have

$$\gamma^p(t) = \tilde{\gamma}(t, p)$$

for all $t \in S^1$. Thus if we pick a function f on D^2 with $\int f = 1$, we can define the **smeared** Wilson loop

$$W(\tilde{\gamma}, A) = \int_{D^2} f(p) W(\gamma^p, A) \, \text{vol}$$

where vol is the usual volume form on the disk. It appears that only products of these smeared Wilson loops (or something along these lines) have truly well-defined vacuum expectation values in Chern-Simons theory.

The second caveat is that making sense of the 'measure'

$$e^{\frac{ik}{4\pi} S_{CS}(A)} \mathcal{D}A$$

requires a choice of **framing** of S, that is, an equivalence class of trivializations of the tangent bundle of S. If S is connected, any two framings differ by an integer number of 'twists' of a certain sort. If we change the framing by adding a twist, the definition of the 'measure' above can change by a phase factor. The reason for this is quite subtle and beyond the scope of this text! Let us simply note that when trying to make sense of path integrals, ambiguities that can only be resolved by additional information are rather frequent.

In any event, having chosen a framing for S, the idea is to obtain an invariant $\mathcal{L}$ of framed oriented links in S as follows: given a framed link L with components γ_i, thicken the γ_i to embedded tori using the framings, and set

$$\mathcal{L}(L) = \int_{\mathcal{A}} W(\tilde{\gamma}_1, A) \cdots W(\tilde{\gamma}_n, A) \, e^{\frac{ik}{4\pi} S_{CS}(A)} \, \mathcal{D}A.$$

(Here we have not bothered dividing by the partition function.) It turns out that one can make sense of this formula at least whenever G is the direct sum of a simply-connected compact Lie group and copies of $U(1)$ (and probably in other cases too). Thus one obtains a link invariant for every finite-dimensional representation of a group of this sort! Consider for example the case of S^3, which has a standard framing. When ρ is the fundamental representation of the group U(1), the invariant we obtain is simply a constant to the writhe of L:

$$\mathcal{L}(L) = e^{i\pi w(L)/k}.$$

When ρ is the fundamental, or spin-$\frac{1}{2}$, representation of SU(2), $\mathcal{L}(L)$ is, up to a sign, the Kauffman bracket evaluated at $A = q^{1/4}$, where

$$q = e^{\frac{2\pi i}{k+2}}.$$

From this point of view, the wonderful ability of the Kauffman bracket to detect the chirality of links is due to the fact that the Chern-Simons action is not preserved by diffeomorphisms that reverse orientation! For other representations we get other interesting link invariants. For example, from the fundamental representation of SU(n) one gets a generalization of the Kauffman bracket or Jones polynomial known as the HOMFLY polynomial, named after some of its discovers (Hoste, Ocneanu, Millet, Freyd, Lickorish and Yetter), while from the fundamental representation of SO(n) one gets another invariant known as the Kauffman polynomial. All these invariants satisfy simple skein relations. From more complicated representations one gets invariants that do not have such simple skein relations.

Now let us say a bit about how these results may be obtained. We will simply sketch a few approaches and provide more thorough references in the Notes. Perhaps the most obvious approach would be to figure out the nature of the 'Lebesgue measure' $\mathcal{D}A$ to the extent that one could calculate path integrals from first principles and compute the link invariants this way. Thanks to recent work on the loop representation of quantum gravity — an approach based on Wilson loops — a reasonable candidate for this 'Lebesgue measure' has emerged. This is actually a 'generalized measure' with respect to which products of Wilson loops have well-defined integrals even without regularization (see the Notes). However, the exponential of the Chern-Simons action is probably not integrable with respect to this generalized measure, except maybe in certain cases. Our evidence for this is primarily the funny framing-dependence of the path integrals. It may thus be necessary to interpret the whole expression

$$e^{\frac{ik}{4\pi}S_{CS}(A)}\mathcal{D}A$$

as a generalized measure of some sort, rather than trying to make sense of the parts separately; this is rather common in quantum field theory. In any event, while this sort of 'direct' approach to the path integral is worth pursuing, it has not yet been carried out.

An alternative approach is to calculate the path integral 'perturbatively' using Feynman diagrams. This is a very common technique in quantum field theory. The only path integrals that people are very good at doing are those in which the action is quadratic in the fields and their derivatives, because there are explicit formulas for the integrals involving Gaussians, and one can simply use those formulas without worrying about the nuances of infinite-dimensional integration. In Chern-Simons theory the action becomes quadratic whenever the gauge group is abelian, which makes the $A \wedge A \wedge A$ term vanish. For example, when $G = U(1)$ and ρ is the fundamental representation,

$$e^{\frac{ik}{4\pi}S_{CS}(A)} = e^{\frac{ik}{4\pi}\int_S A \wedge dA},$$

which is really the explanation for all the expressions involving $\int A \wedge B$ in the section before last. When the gauge group is not abelian, the Chern-Simons action is not quadratic, so a good idea is to write the action as a quadratic part plus a cubic part that is proportional to some sort of 'coupling constant', and then to work everything out as a Taylor

Chern-Simons Theory

series in this coupling constant. For example, if one rescales the vector potential A by writing $C = k^{1/2}A$, one has

$$\frac{ik}{4\pi}S_{CS}(A) = \frac{i}{4\pi}\int_S \text{tr}(C \wedge dC + \frac{2}{3}k^{-1/2}C \wedge C \wedge C)$$

which expresses the action as a part quadratic in C plus a part cubic in C that is proportional to $k^{-1/2}$. One can then attempt to compute the link invariant as a Taylor series in $k^{-1/2}$, whose terms are multiple integrals over S. Feynman developed a wonderful technique for keeping track of such integrals using diagrams, now called Feynman diagrams! See the cover of the book for an example — drawn in a rather more flamboyant style than is strictly necessary! In Chern-Simons theory, these diagrams consist of wiggly lines that connect various points of the link; these wiggly lines represent the field C, and vertices where three wiggly lines meet are allowed, corresponding to the $C \wedge C \wedge C$ 'interaction' term. Physically, these have a simple interpretation: we think of the link as a collection of loops traced out by *particles* moving around in the manifold S, and the wiggly lines represent the gauge bosons emitted and absorbed by these particles.

A difficulty with this perturbative approach is that, unless it is done carefully, it does not recover the curious shift $k \to k+2$ occuring in the definition of q above. Some perturbative calculations give the wrong answer, namely

$$q = e^{\frac{2\pi i}{k}} \quad \text{(wrong!)}$$

after which one must make the shift $k \to k+2$ on an *ad hoc* basis. We should note that it is typical in perturbative gauge theory to replace the integral over $\mathcal{A}$ with an integral over the space $\mathcal{A}/\mathcal{G}$ of connections modulo gauge transformations, using the gauge invariance of the integrand. Doing this in practice involves introducing additional 'ghost' fields into the Lagrangian, using a technique originally due to Faddeev and Popov but refined by Becchi, Rouet, Stora and Tyutin into what is now known as the BRST formalism.

A different sort of perturbative approach consists of trying to recover the skein relations by a perturbative calculation, rather than evaluating the invariant on a particular link. This approach depends on some beautiful formulas relating the change in the Wilson loop $W(\gamma, A)$ as one varies γ or A to the curvature of A, formulas which also play a key role in the loop representation of quantum gravity. However, this approach also has difficulty accounting for the $k \to k+2$ shift.

The original *nonperturbative* approach is that appearing in Witten's paper on quantum field theory and the Jones polynomial. This relies upon the relation between Chern-Simons theory and a quantum field theory in 2 dimensions known as the Wess-Zumino-Witten (or WZW) model. This is a conformal field theory, that is, a quantum field theory defined on a Riemann surface (an n-holed torus equipped with geometrical structure locally isomorphic to that of the complex plane). Conformal field theory is an extensive discipline in its own right, originating in string theory.

There are also a variety of nonperturbative approaches that make Witten's argument more rigorous and completely avoid path integrals. These tend to use mathematical machinery associated to conformal field theory. For example, the approach followed by Atiyah, Hitchin and others makes extensive use of Riemann surface theory to define the link invariants. An alternative style of approach due to Crane, Kohno, Reshetikhin, Turaev and others makes use of marvelous mathematical structures known as 'quantum groups'. These are not groups, but instead algebraic structures generalizing groups that are in one-to-one correspondence with the semisimple Lie algebras, but also depend on a parameter q. (In the case of $SU(2)$, this parameter is related to the level k by the formula above.) In the limit as $q \to 1$, quantum groups reduce to ordinary Lie groups, but for $q \ne 1$ they are closely related to knot theory, since their axioms are closely related to the Reidemeister moves. This approach lays bare the *algebraic* aspects of 3-dimensional topology in a most intriguing way. Generalizations of all of these approaches to cover gauge theories in higher dimensions — particularly the 4-dimensional spacetime we know and love — are being eagerly sought.

Clearly there is much more to say about these topics, but we wish instead to move on and discuss general relativity. The reader who wishes to learn more about the marvelous interplay between gauge fields and knots will have to turn to the references in the Notes. But we have not yet said all we mean to about Chern-Simons theory and the Kauffman bracket polynomial — as we shall see, they also have some relevance to treating gravity as a quantum field theory!

Notes to Part II

1. Symmetry

The quote from Yang is from his commentary to his paper 'Einstein's impact on theoretical physics', reprinted in his *Selected Papers, 1945–1980, with Commentary*, W. H. Freedman and Company, San Francisco, 1983. An excellent series of talks on the concept of symmetry appears in Hermann Weyl's classic *Symmetry*, Princeton U. Press, New Jersey, 1980. For a quick tour of Lie groups, Lie algebras, and their applications to physics, try Robert Hermann's *Lie Groups for Physicists*, Benjamin Cummings, New York, 1966. A more thorough explanation of Lie theory, which also covers many of the basics about Riemannian manifolds, is *Differential Geometry, Lie Groups, and Symmetric Spaces* by Sigurdur Helgason, Academic Press, New York, 1978. (Symmetric spaces are certain very nice manifolds that can be expressed as quotients G/H of Lie groups.) A nice text that puts more emphasis on representation theory is *Lie Groups, Lie Algebras, and Their Representations* by V. S. Varadarajan, Prentice-Hall, New Jersey, 1974. To start learning the deeper aspects of the representation theory of semisimple Lie groups, try Anthony Knapp's *Representation Theory of Semisimple Groups, an Overview Based on Examples*, Princeton University, New Jersey, 1986. To really appreciate the wide-ranging applications of groups and their fundamental role in quantum theory, read *Unitary Group Representations in Physics, Probability, and Number Theory* by George Mackey, Addison-Wesley, Massachusetts, 1989.

The theory of essential versus inessential cocycles is really part of a subject called group cohomology. This is a cohomology theory that applies to groups rather than spaces! For an introduction, try *An Introduction to Homological Algebra* by Joseph J. Rotman, Academic Press, New York, 1979. Homological algebra is abstract but it clarifies a diversity of phenomena. Cocycles also show up as 'anomalies' when one tries to quantize gauge theories. For more on anomalies, try *Current Algebra and Anomalies*, eds. Sam B. Treiman, Roman Jackiw, Bruno Zumino, and Edward Witten, Princeton University Press, New Jersey, 1985.

2. Bundles and Connections

The quote by Yang is from his review article 'Magnetic monopoles, fiber bundles, and gauge fields', reprinted in his *Selected Papers*, cited above. The quote by Yang and Mills is from their paper, 'Conservation of isotopic spin and isotopic gauge invariance', *Phys. Rev.* **96** (1954) 191-195, in which they presented the Yang-Mills equations, and which is reprinted in the same volume.

The standard model is a summary of a vast amount of research on the fundamental laws of physics, and some knowledge of the history of the quest for these laws is necessary to fully appreciate it. Also, a historical introduction is easier to start with than a quantum field theory text full of nasty equations! We recommend *Inward Bound: of Matter and Forces in the Physical World*, by Abraham Pais, Clarendon Press, New York, 1986, and *From X-Rays to Quarks: Modern Physicists and Their Discoveries* by Emilio Segre, W. H. Freeman, San Francisco, 1980, for a general overview of the physics of this century. The latter volume is a sequel to Segre's *From Falling Bodies to Radio Waves: Classical Physicists and Their Discoveries*, W. H. Freeman, New York, 1984, which is good for the reader wanting to start a bit earlier in the story. For the history of quarks in particular, a nice book is *Constructing Quarks: A Sociological History of Particle Physics* by Andrew Pickering, U. of Chicago Press, Chicago, 1984 — a book which is carefully neutral on the topic of whether quarks really exist, but which contains a lot of good physics!

For a more technical introduction to the standard model, try *Quarks, Leptons and Gauge Fields* by Kerson Huang, World Scientific, Singapore, 1982. This assumes some familiarity with quantum field theory (see the notes for the next chapter), but it is very lucid, particularly on the topological issues involved. Two excellent books that really get into the experimental results underlying the standard model are *Particle Physics and Introduction to Field Theory* by T. D. Lee, Harwood, New York, 1988, and *Leptons and Quarks*, by L. B. Okun, North-Holland, New York, 1982.

Vector bundles, connections and so on play a fundamental role in differential geometry and topology, so many of the references listed in the notes to Chapter 2 of Part I discuss these topics, and the reader should consult these. Also try *Connections, Curvature, and Cohomology* by Werner Greub, Stephen Halperin, and Ray Vanstone, published by Academic Press, New York, 1972-1976, which is a 3-volume series that explores the relations between connections, topology, and Lie group theory.

A very good advanced book on bundles is Dale Husemoller's *Fibre Bundles*, Springer-Verlag, New York, 1975. This contains a good introduction to K-theory, which is a cohomology theory constructed using bundles. The idea is that given a topological space X, we can form a group $K^0(X)$ by considering all formal differences $E - F$ of vector bundles over X, modulo the equivalence relation

$$E - F \sim E' - F' \iff E \oplus F' \oplus G \simeq E' \oplus F \oplus G$$

for some vector bundle G, and with the group operations being (in additive notation)

$$(E - F) + (E' - F') = (E \oplus F') - (E' \oplus F), \qquad -(E - F) = F - E.$$

(This is analogous to how one may construct the integers as differences of natural numbers.) With some more work one can construct a group called $K^1(X)$, and then one has for a compact manifold M

$$K^0(M) \otimes \mathbb{C} \cong \bigoplus_{p \geq 0} H^{2p}(M), \qquad K^1(M) \otimes \mathbb{C} \cong \bigoplus_{p \geq 0} H^{2p+1}(M),$$

where $H^p(M)$ is the deRham cohomology of M. A good book devoted to K-theory is Max Karoubi's *K-theory: An Introduction*, Springer-Verlag, New York, 1978.

Much of the most interesting work on Wilson loops appears in the physics literature. As early as 1962, S. Mandelstam used the concept to study electromagnetism, but the name 'Wilson loop' goes back to Kenneth G. Wilson's paper 'Confinement of quarks', *Phys. Rev.* **D10** (1974) 2445-2449. For a thorough review of the physics and mathematics of Wilson loops, try Renate Loll's paper 'Quantum chromodynamics and gravity as theories on loop space', to appear.

3. Curvature and the Yang-Mills Equation

The quote by Yang is from an interview of him by D. Z. Zhang, 'C. N. Yang and contemporary mathematics', *Mathematical Intelligencer*, 14 (1993) 13-21.

Curvature is discussed along with connections in the references for Chapter 2. A solid understanding of the *physics* of the Yang-Mills equations requires some preparation in quantum field theory, which in turn requires some quantum mechanics — see the notes to Chapter 6 of Part I for this.

A good book that bridges the gap between quantum mechanics and quantum field theory is J. J. Sakurai's *Advanced Quantum Mechanics*, Addison-Wesley, Redwood City, 1992. Another is *Relativistic Quantum Mechanics* by James D. Bjorken and Sidney D. Drell, McGraw-Hill, New York, 1964. To get going on quantum field theory proper, Bjorken and Drell's classic *Relativistic Quantum Fields*, McGraw-Hill, New York, 1965, is still worth reading (despite the the lack of any treatment of gauge theories other than quantum electrodynamics). A good introduction along more modern lines is *Quantum Field Theory* by Lewis H. Rider, Cambridge U. Press, Cambridge, 1985.

A good place to begin studying the physics of the Yang-Mills equations is Kerson Huang's book *Quarks, Leptons, and Gauge Fields*, cited in the notes for Chapter 2. A bit more advanced but still very readable is *Gauge Theory of Elementary Particle Physics* by Ta-Pei Cheng and Ling-Fong Li, Oxford University Press, Oxford, 1984. It is also worth looking at *Gauge Theories of Strong and Electroweak Interactions* by Peter Becher, Manfred Böhm and Hans Joos, Wiley, New York, 1984, or for more emphasis on the experimental data, *An Introduction to Gauge Theories and the 'New Physics'* by Elliot Leader and Enrico Predazzi, Cambridge U. Press, Cambridge, 1982. Still more advanced texts include J. Leite Lopes' *Gauge Field Theories, an Introduction*, Pergamon, New York, 1981. For more on the strong force, try *The QCD Vacuum, Hadrons and the Superdense Matter*, by E. V. Shuryak, World Scientific, Singapore, 1998, and F. J. Yndurâin's *Quantum Chromodynamics: An Introduction to the Theory of Quarks and Gluons*, Springer-Verlag, New York, 1983. The former has a lot on confinement and instantons, and also a well-organized bibliography of important papers on gauge theory. For a nice introduction to the electroweak force, try *Gauge Theories of Weak Interactions* by J. C. Taylor, Cambridge U. Press, Cambridge, 1976. For a survey of the role of 'anomalies' in gauge field theory, try the book edited by Treiman *et al*, as cited in the notes to Chapter 1.

None of the texts above is at all mathematically rigorous. The mathematically inclined reader is likely to wonder whether quantum field theory is really a theory or only a mirage, due to the many dubious manipulations used. In fact, as of the writing of this book there was still no rigorous construction of an interacting quantum field theory in 4-dimensional spacetime, despite decades of hard work on the subject! There is, however, a lot that has been rigorously shown concerning field theory. This work falls into two broad schools: 'axiomatic field theory' proves results about quantum fields starting from physically reasonable postulates, while 'constructive field theory' attempts to rigorously construct models meeting these axioms. For an introduction to quantum electrodynamics that pays attention to issues of rigor, try G. Scharf's *Finite Quantum Electrodynamics*, Springer-Verlag, New York, 1989. For some excellent tours of rigorous field theory, each emphasizing different aspects of the vast territory, try *PCT, Spin and Statistics, and All That* by R. F. Streater and A. S. Wightman, Addison-Wesley, Reading, 1989, *Introduction to Axiomatic Quantum Field Theory*, by N. N. Bogolubov, A. A. Logunov and I. T. Todorov, Benjamin, Reading, 1975, and *Local Quantum Physics: Fields, Particles, Algebras* by Rudolf Haag, Springer-Verlag, New York, 1992. For constructive quantum field theory from the Lagrangian viewpoint, try *Quantum Physics: A Functional Integral Point of View* by James Glimm and Arthur Jaffe, Springer-Verlag, New York, 1981; for the Hamiltonian viewpoint, try *Introduction to Algebraic and Constructive Quantum Field Theory*, by John C. Baez, Irving E. Segal, and Zhengfang Zhou, Princeton U. Press, New Jersey, 1992.

For the *mathematics* of curvature and the Yang-Mills equations, one should study the differential geometry books cited in the notes for the previous chapter and Chapter 2 of Part I, particularly those by Sternberg, Kobayashi and Nomizu, and Choquet-Bruhat *et al.* Or, for mathematics aimed at physicists, one can try *Topology and Geometry for Physicists* by Charles Nash and Siddhartha Sen, Academic Press, New York, 1983. A good treatment of the exterior covariant derivative, as well as a lot of other differential geometry, is *Differential Geometric Structures* by Walter Poor, McGraw-Hill, New York, 1981.

For a wide-ranging, sophisticated but readable tour of gauge fields, modern quantum field theory and topology, we strongly recommend Charles Nash's *Differential Topology and Quantum Field Theory*, Academic Press, New York, 1991.

4. Chern-Simons Theory

The quote by Chern is from Volume II of his *Selected Papers*, Springer-Verlag, New York, 1989.

As we have indicated, the calculus of variations springs out of classical mechanics, and a good understanding of the physics is necessary to appreciate the mathematics. If ones classical mechanics is rusty, one might start with a physics treatment such as Herbert Goldstein's *Classical Mechanics*, Addison-Wesley, Reading, 1980. It is also worth learning some more mathematically sophisticated approaches; here it is good to begin with *Mathematical Methods of Classical Mechanics* by V. I. Arnol'd, Springer-Verlag, New York, 1989, and then study *Foundations of Mechanics* by Ralph Abraham and Jerrold E. Marsden, Benjamin/Cummings, Reading, 1978.

For a discussion of 'fixed background structures' in physics, see for example the essay 'Space and time in the quantum universe' by Lee Smolin, pp. 228-291 of the volume *Conceptual Problems of Quantum Gravity* cited in the notes to Chapter 6 of Part III. This essay also has many other useful references.

For proof of the integrality of the Chern classes, and an excellent introduction to characteristic classes from the topological viewpoint, try *Characteristic Classes* by John W. Milnor and James D. Stasheff, Princeton U. Press. For a quick introduction to Chern-Simons classes, try the second volume of Choquet-Bruhat *et al* (see the notes for Chapter 2 of Part I). For insight into how Chern-Simons classes show up in physics, see the book edited by Treiman *et al*, as cited above in the notes to Chapter 1. More references on Chern-Simons theory appear in the notes for the next chapter.

For an introduction to instantons, try *Solitons and Instantons* by R. Rajaraman, North-Holland, New York, 1987. For their uses in the standard model, see the texts by Kerson Huang and E. V. Shuryak cited in the notes for Chapters 2 and 3. For an introduction to their applications to topology, try the book by Charles Nash cited in the notes for Chapter 3. A more thorough introduction to the relation between topology and the Yang-Mills equations in 4 dimensions, with an emphasis on the self-dual solutions, is *Instantons and Four-Manifolds*, by Daniel S. Freed and Karen K. Uhlenbeck, Springer-Verlag, 1984. A still more detailed and demanding text along these lines is *The Geometry of Four-Manifolds* by Simon K. Donaldson and P. B. Kronheimer, Oxford U. Press, Oxford, 1990.

5. Link Invariants from Gauge Theory

The quote by Tait is taken from *Life and Scientific Work of Peter Guthrie Tait*, by Cargill G. Knott (we kid you not!), Cambridge U. Press, Cambridge, 1911. Tait's *On Knots* was reprinted by Amphion Press, Washington D.C., in 1993. The quote by Maxwell is from his 1855 paper 'On Faraday's Lines of Force', which appears in *The Scientific Papers of James Clerk Maxwell*, ed. W. D. Niven, Dover, New York, 1966. The paper by Helmholtz on vortex lines is 'Über Integrale, der hydrodynamischen Gleichungen, welche den Wirbelbewegungen entsprechen', *Jour. Reine Angewandte Mathematik* **55** (1858), 25-55. The quote by Kelvin at the beginning of the chapter appears as a footnote in his paper 'Deep-Water Ship-Waves' in *Proc. Roy. Soc. Edinburgh*, **25** (1905), 562-587, while the quote appearing in the section on the linking number and writhe is from a letter of his to Helmholtz in 1867, quoted in 'Thomson, Maxwell and the universal ether in Victorian physics' by Daniel M. Siegel, in *Conceptions of Ether*, ed. G. N. Cantor and M. J. S. Hodge, Cambridge U. Press, Cambridge, 1981. For more on the history of ether, see the book by E. T. Whittaker cited in the notes for Chapter 1 of Part I. A nice introduction to the theory of vortex lines in fluid flow appears in the second volume of Feynman's lectures on physics, referred to in the notes for Chapter 1 of Part I.

For compendia of knots, try George Russell Shaw's *Knots — Useful and Ornamental*, Houghton Mifflin, Boston, 1924, or Clifford W. Ashley's *The Ashley Book of Knots*, Doubleday, Garden City, 1944. If you have always been unable to *tie* knots, try John Cassidy's *The Klutz Book of Knots*, Klutz Press, Palo Alto, 1985.

Good introductions to knot theory include Dale Rolfsen's *Knots and Links*, Publish or Perish, Berkeley, 1976, Gerhard Burde and Heiner Zieschang's *Knots*, W. De Gruyeter, New York, 1985, Siegfried Moran's *The Mathematical Theory of Knots and Braids: an Introduction*, North-Holland, New York, 1983, and Louis H. Kauffman's *On Knots*, Princeton U. Press, Princeton, 1987. For braids, which are intimately related to knots and links, try *Braids, Links, and Mapping Class Groups* by Joan S. Birman, Princeton U. Press, Princeton, 1974. Kauffman's book contains a proof that the Alexander polynomial is a knot invariant which uses only the skein relations, as well as the standard proof using homology theory, which is conceptually somewhat simpler. It also discusses Seifert surfaces.

Two books stand out when it comes to the relationship between knot theory and physics. Louis H. Kauffman's *Knots and Physics*, World Scientific, Singapore, 1991, is delightfully easy to read, loaded with pictures, and concentrates on the use of knot diagrams, state sums, and representations of the category of tangles to define and study the new knot polynomials, in particular the Kauffman bracket. Michael Atiyah's *The Geometry and Physics of Knots*, Cambridge U. Press, Cambridge, 1990, assumes a fair amount of mathematical sophistication and concentrates on Chern-Simons theory.

The recent burst of activity on knot theory and physics was triggered by Vaughan F. R. Jones' discovery of what is now called the Jones polynomial, published in 'A polynomial invariant of knots and links', *Bull. Amer. Math. Soc.* **12** (1985), 103-111. His work deals primarily with certain operator algebras known as factors, and a good introduction to it can be found in his slim book, *Subfactors and Knots*, American Mathematical Society, Providence, 1991. The connection between knot theory and Chern-Simons theory was made in Edward Witten's paper 'Quantum field theory and the Jones polynomial', *Comm. Math. Phys.* **121** (1989), 351-399.

The papers by Jones and Witten cited above both appear in the reprint collection *New Developments in the Theory of Knots*, ed. Toshitake Kohno, World Scientific, Singapore, 1990. This is an excellent source of information about the knot revolution of the late 1980s. It also contains Kunio Murasugi's proof of the Tait conjectures, which first appeared in 'Jones polynomials and classical conjectures in knot theory', *Topology* **26** (1987) 187-194. Another reprint volume, which concentrates on the statistical mechanics aspects, is *Braid Group, Knot Theory, and Statistical Mechanics*, eds. C. N. Yang and M. L. Ge, World Scientific, Singapore, 1989. The conference proceedings *Knots, Topology and Quantum Field Theories*, ed. L. Lusanna, World Scientific, Singapore, 1989, contains some nice papers on knot theory and quantum gravity. The volume *Proceedings of the Conference on Quantum Topology*, ed. David Yetter, World Scientific, Singapore, 1994, goes into many more recent developments in knot theory, Chern-Simons theory, and the like.

For a readable introduction to superconductors and the quantization of magnetic flux, try *Fundamentals of Superconductivity* by Vladimir Z. Kresin and Stuart A. Wolf, Plenum Press, New York, 1990; for more detail, especially on vortices, try *Foundations of Applied Superconductivity* by Terry P. Orlando and Kevin A. Delin, Addison-Wesley, Reading, 1991. For a much more mathematical approach to vortices and the Landau-Ginzburg equation (a phenomenological U(1) gauge theory describing superconductivity), try *Vortices and Monopoles: Structure of Static Gauge Theories* by Arthur Jaffe and Clifford Taubes, Birkhauser, Boston, 1980.

For a nice proof of Gauss' formula for the linking number using differential forms, see *Differential Forms in Algebraic Topology*, cited in the notes to Chapter 6 of Part I. For a proof that ambient isotopy classes of embedded solid tori are in one-to-one correspondence to ambient isotopy classes of framed oriented links, and applications of this fact to gauge theory, see 'Link invariants, holonomy algebras and functional integration', by J. Baez, to appear in *Jour. Funct. Anal.*.

For quantum groups try *A Guide to Quantum Groups*, by Vyjanathi Chari and Andrew Pressley, Cambridge U. Press, Cambridge, 1994. For various approaches to making Witten's work on Chern-Simons theory rigorous, see for example Louis Crane's paper '2-d physics and 3-d topology', *Comm. Math. Phys.* **135** (1991), 615-640, Nikolai Reshetikhin and Vladimir Turaev's paper 'Invariants of 3-manifolds via link-polynomials and quantum groups', *Invent. Math.* **103** (1991), 547-597, Toshitake Kohno's paper 'Topological invariants for 3-manifolds using representations of mapping class groups I', *Topology* **31** (1992) 203-230, Dror Bar-Natan's paper 'On the Vassiliev knot invariants', to appear, and Scott Axelrod and Isadore Singer's paper 'Chern-Simons perturbation theory', to appear. (Also see the books by Atiyah and Kauffman cited above, which contain further references.)

The relationship between Chern-Simons theory and quantum gravity in 3 dimensions was noted by Edward Witten in '2+1 dimensional gravity as an exactly soluble system', *Nucl. Phys.* **B311** (1988) 46-78, and a rigorous construction of this theory using quantum groups was provided by Vladimir Turaev and Oleg Viro in 'State sum invariants of 3-manifolds and quantum $6j$-symbols', *Topology* **31** (1992), 865-902. For an elementary approach to the Turaev-Viro theory based on the Kauffman bracket invariant, try the book by Louis Kauffman and Sóstenes Lins, *Temperley-Lieb Recoupling Theory and Invariants of 3-manifolds*, Princeton U. Press, New Jersey, 1994. For further discussion of the relationship to the loop representation of quantum gravity, see 'The basis of the Ponzano-Regge-Turaev-Viro-Ooguri quantum gravity model is the loop representation basis' by Carlo Rovelli, to appear.

Part III

Gravity

Chapter 1
Semi-Riemannian Geometry

Riemann has shown that as there are different kinds of lines and surfaces, so there are different kinds of space of three dimensions; and that we can only find out by experience to which of these kinds the space in which we live belongs. I hold in fact

(1) That small portions of space are in fact of a nature analogous to little hills on a surface which is on the average flat; namely, that the ordinary laws of geometry are not valid for them.

(2) That this property of being curved or distorted is continually being passed on from one portion of space to another after the manner of a wave.

(3) That this variation of the curvature of space is what really happens in that phenomenon which we call the motion of matter, *whether ponderable or etherial.*

(4) That in the physical world nothing else takes place but this variation, subject (possibly) to the law of continuity. — William Clifford, 1876.

Tensors

Part of the beauty of Einstein's theory of gravity is that it lives up to Clifford's dream, and explains gravity in purely geometrical terms as the curvature of spacetime. As we have tried to explain, Yang-Mills fields are geometrical in essence, since a connection is really just a rule for parallel translation, and the Yang-Mills field F, being just the curvature of a connection, measures the dependence of parallel translation on the path taken between two points. General relativity is even *more* geometrical, since it concerns, not just any old bundle, but the tangent bundle! The basic ingredient of general relativity is the metric on spacetime, but this metric defines a connection on the tangent bundle, and thus a curvature, and this curvature — the Riemann tensor — is the real star of the show. To understand the Riemann tensor we need to talk about 'tensor bundles' a bit.

Let M be an n-dimensional manifold. Recall that starting with the tangent bundle TM we can take its dual and get the cotangent bundle T^*M. Sections of TM are vector fields; sections of T^*M are 1-forms. If we are working with local coordinates x^μ on an open set $U \subseteq M$, we have a basis of vector fields ∂_μ and a dual basis of 1-forms dx^μ. More generally, any basis of vector fields e_μ on an open set of U gives a dual basis e^μ of 1-forms.

We define the bundle of (r,s) **tensors** to be the tensor product of r copies of the tangent bundle and s copies of the cotangent bundle:

$$\underbrace{TM \otimes \cdots \otimes TM}_{r} \otimes \underbrace{T^*M \otimes \cdots \otimes T^*M}_{s}$$

We call a section X of this bundle an (r,s) **tensor field**, or just an (r,s)-tensor for short. We define $(0,0)$ tensors to be simply functions on M; physicists also call these **scalar fields**.

By Exercise 72 in Part II, any (r,s) tensor field is a linear combination of ones that look like

$$v_1 \otimes \cdots \otimes v_r \otimes \omega_1 \otimes \cdots \otimes \omega_s$$

where $v_1, \ldots, v_r$ are vector fields and $\omega_1, \ldots, \omega_s$ are 1-forms. (Note: the subscripts here do *not* denote components; we are using them only to list the vector fields v_i and 1-forms ω_j.) Thus for defining and proving things it is often enough to consider tensor fields of this type. Alternatively, in local coordinates, we have a basis of (r,s) tensor fields given by

$$\partial_{\alpha_1} \otimes \cdots \otimes \partial_{\alpha_r} \otimes dx^{\beta_1} \otimes \cdots \otimes dx^{\beta_s}$$

where the indices α_i, β_j run independently over $1, \ldots, n$. Thus if we have any (r,s) tensor field X, we can write

$$X = X^{\alpha_1 \cdots \alpha_r}_{\beta_1 \cdots \beta_s} \, \partial_{\alpha_1} \otimes \cdots \otimes \partial_{\alpha_r} \otimes dx^{\beta_1} \otimes \cdots \otimes dx^{\beta_s}$$

in local coordinates, where the functions $X^{\alpha_1\cdots\alpha_r}_{\beta_1\cdots\beta_s}$ are the **components** of the tensor. We can also do the same thing using any basis e_μ of vector fields and dual basis e^μ of 1-forms.

Another way to think of an (r,s) tensor field X is a kind of machine that accepts r 1-forms $\omega_1, \ldots, \omega_r$ and s vector fields $v_1, \ldots, v_s$ as input, and outputs a function on M, in a manner that is $C^\infty(M)$-linear in each input. This works as follows in local coordinates:

$$X(\omega_1, \ldots, \omega_r, v_1, \ldots, v_s) = X^{\alpha_1\cdots\alpha_r}_{\beta_1\cdots\beta_s} \omega_{1\alpha_1} \cdots \omega_{i\alpha_r} v_1^{\beta_1} \cdots v_j^{\beta_s}.$$

For example, we can think of a semi-Riemannian metric g as a $(0,2)$ tensor, because we feed it two vector fields v, w and get a function $g(v,w)$, and it is $C^\infty(M)$-linear:

$$\begin{aligned} g(v+v', w) &= g(v,w) + g(v',w) \\ g(v, w+w') &= g(v,w) + g(v,w') \\ g(fv, w) &= g(v, fw) = fg(v,w) \end{aligned}$$

where $v, v', w, w' \in \text{Vect}(M)$ and $f \in C^\infty(M)$. In local coordinates, we have

$$g = g_{\alpha\beta} dx^\alpha \otimes dx^\beta$$

where

$$g_{\alpha\beta} = g(\partial_\alpha, \partial_\beta).$$

It is common, however, to be sloppy and leave out the $\otimes$ symbol here. Thus, in spherical coordinates (ϕ, θ) on S^2, the standard metric has

$$g(\partial_\phi, \partial_\phi) = 1, \quad g(\partial_\theta, \partial_\theta) = \sin^2\phi, \quad g(\partial_\theta, \partial_\phi) = 0,$$

so we should write it as

$$g = d\phi \otimes d\phi + \sin^2\phi \, d\theta \otimes d\theta$$

but instead we just write

$$g = d\phi^2 + \sin^2\phi \, d\theta^2.$$

Another important example of a tensor is a differential form! A p-form is really just another way of thinking about a completely antisymmetric $(0,p)$ tensor. This is because we can identify the p-form

$$\omega = \frac{1}{p!} \omega_{\alpha_1\cdots\alpha_p} dx^{\alpha_i} \wedge \cdots \wedge dx^{\alpha_p},$$

where we assume the coefficients $\omega_{\alpha_1\cdots\alpha_p}$ switch sign when we interchange any two indices, with the $(0,p)$ tensor

$$\frac{1}{p!}\omega_{\alpha_1\cdots\alpha_p}dx^{\alpha_i}\otimes\cdots\otimes dx^{\alpha_p}.$$

In other words, for any vector space we can think of $\Lambda^p V$ as a subspace of the p-fold tensor product $V\otimes\cdots\otimes V$, so we can think of the p-form bundle $\Lambda^p(T^*M)$ as a sub-bundle of the bundle of $(0,p)$-tensors.

There are a number of basic operations on tensors that are frequently used in general relativity. First, since we can take the tensor product of a section of a vector bundle E and a section of a vector bundle E' to get a section of $E\otimes E'$, we can take tensor products of tensor fields. In particular, the tensor product of an (r,s) tensor field and an (r',s') tensor field Y is a $(r+r',s+s')$ tensor field $X\otimes Y$. The components of $X\otimes Y$ are given by

$$(X\otimes Y)^{\alpha_1\cdots\alpha_{r+r'}}_{\beta_1\cdots\beta_{s+s'}} = X^{\alpha_1\cdots\alpha_r}_{\beta_1\cdots\beta_s}\,Y^{\alpha_{r+1}\cdots\alpha_{r+r'}}_{\beta_{s+1}\cdots\beta_{s+s'}}.$$

Second, since we can pair a vector field and a 1-form to get a function, if we start with an (r,s) tensor field X we can get an $(r-1,s-1)$ tensor field Y in many ways. For example, if we have the (r,s) tensor field

$$X = v_1\otimes\cdots\otimes v_r\otimes\omega_1\otimes\cdots\otimes\omega_s,$$

we can pair, or **contract**, the ith vector field with the jth 1-form to get the $(r-1,s-1)$ tensor field

$$Y = \omega_j(v_i)\,v_1\otimes\cdots\hat{v}_i\cdots\otimes v_r\otimes\omega_1\otimes\cdots\hat{\omega}_j\cdots\otimes\omega_s,$$

where the $\hat{\ }$ on top of v_i and ω_j means that those factors are left out. In terms of components, we can describe this process as follows: we obtain Y from X by by contracting one of the upper indices with one of the lower ones. Namely, if the same index μ shows up as both the ith superscript and the jth subscript of the components of the tensor X, the components of Y are given by

$$Y^{\alpha_1\cdots\hat{\alpha}_i\cdots\alpha_r}_{\beta_1\cdots\hat{\beta}_j\cdots\beta_s} = X^{\alpha_1\cdots\mu\cdots\alpha_r}_{\beta_1\cdots\mu\cdots\beta_s},$$

where we use the Einstein summation convention to sum over μ from 1 to n.

Tensors

Exercise 1. *Show that this follows.*

There are a few things we should say at this point about working with tensors. Contracted (repeated) indices are also called **dummy indices** and can be labeled with any letters we want, as long as we use the same one for both indices. For example, in a bit we will be talking a lot about the Riemann curvature tensor, which is a $(1,3)$-tensor with components $R^\alpha{}_{\beta\gamma\delta}$. (We write the components this way to emphasize the fact that the index α comes 'first'.) We can contract this to get something called the Ricci tensor, a $(0,2)$-tensor given by

$$R_{\beta\gamma} = R^\alpha{}_{\beta\alpha\gamma}.$$

But we could equally well write

$$R_{\beta\gamma} = R^\sigma{}_{\beta\sigma\gamma}$$

since α was a dummy index.

Another powerful technique we can use when we work with tensors is **raising** and **lowering** of indices using a semi-Riemannian metric. We have already described this in Chapter 5 of Part I, where we introduced not only the metric, with components $g_{\alpha\beta}$, but also the inner product on 1-forms, with components $g^{\alpha\beta}$, which satisfies

$$g^{\alpha\beta} g_{\beta\gamma} = \delta^\alpha_\gamma,$$

where δ is the Kronecker delta. The inner product on 1-forms is a $(2,0)$-tensor since it accepts two 1-forms as input and gives us a function, their inner product. For example, starting with the Ricci tensor above we can get a 'different version' of it, a $(1,1)$-tensor, by raising an index:

$$R^\alpha_\beta = g^{\alpha\gamma} R_{\beta\gamma}.$$

We can then contract this to get a $(0,0)$-tensor or function, the Ricci scalar:

$$R = R^\alpha_\alpha.$$

The trick to keep from screwing up while doing index gymnastics is to look at both sides of the equation and check that we have the same indices appearing as superscripts and subscripts on both sides, not counting dummy indices. In the equation

$$R^\alpha_\beta = g^{\alpha\gamma} R_{\beta\gamma},$$

we have on the left hand side α as a superscript and β as a subscript. On the right hand side we have the contracted (dummy) indices γ, which we can cancel in our minds, as well as an α on top and a β on bottom. This is thus a sensible equation, while something like $R^\alpha_\gamma = g^{\alpha\gamma} R_{\beta\gamma}$ would mean one has made a mistake somewhere.

Just to whet the reader's appetite, and to provide a nice example of tensor equation, we will write down Einstein's equation for general relativity:

$$R_{\mu\nu} - \frac{1}{2} g_{\mu\nu} R = 8\pi \kappa T_{\mu\nu}.$$

Here $T_{\mu\nu}$ is called the stress-energy tensor, which describes the flow of energy and momentum through a given point in spacetime, and κ is Newton's **gravitational constant**, about $6.67 \cdot 10^{-11}$ newton-meter^2 per kilogram^2. This is the constant appearing in Newton's inverse square force law for gravity.

Some readers may be mildly appalled by the 'debauch of indices', to use Cartan's famous phrase, in the equations above. Things will get worse! It is important to understand a few things about all these indices. First and most importantly, there are so many operations on tensors that *any* notation able to describe all these operations is bound to be a bit messy, and the index notation is in fact the simplest one known that handles all these operations. Secondly, while we appear to be introducing coordinate-dependence when writing the components of a tensor, such as $g_{\alpha\beta}$, rather than the tensor itself, such as g, there is a way of thinking that avoids this conclusion. Namely, we can think of the superscripts and subscripts as **abstract indices** whose sole purpose is to tell us what kind of tensor we are dealing with and to efficiently describe the process of contraction. Most people in general relativity have adopted this view, and part of why we will not flinch from using index-ridden notation is to get the reader used to how relativists think. For more about this, see the Notes.

The Levi-Civita Connection

In general relativity the basic 'field' is the metric, which lets us measure distances and angles. However, general relativity is also a bit like Yang-Mills theory and other gauge theories, since the metric gives rise to a connection on the tangent bundle, the Levi-Civita connection. In other words, a metric on spacetime allows us to parallel translate tangent vectors in a unique 'best' way. What do we mean by 'best'? It turns out that there are two conditions that make the Levi-Civita condition best: it is metric preserving and torsion free. The condition that it be metric preserving is simple: it just says that a tangent vector does not change length when we parallel translate it.

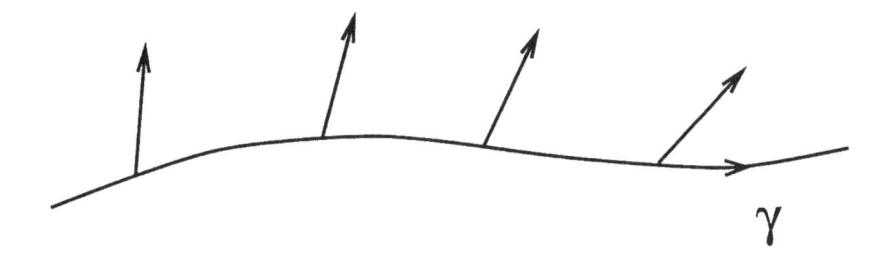

Fig. 1. A metric-preserving connection

The condition that the Levi-Civita connection is torsion free is subtler. Basically, it means that a tangent vector does not rotate when we parallel translate it. It takes some work to make this condition precise, though! After all, how can we tell if the vector rotates except by comparing it to the original vector by parallel translation? The trick is as follows. Work in local coordinates so that we can freely identify vectors with points, as in $\mathbb{R}^n$. Consider a small square of size ϵ in the x^μ-x^ν plane. We can take the vector $\epsilon\partial_\mu$ and parallel translate it in the ∂_ν direction to the tip of $\epsilon\partial_\nu$, obtaining a vector v shown in Figure 2 below. Alternatively, we can parallel translate $\epsilon\partial_\nu$ in the ∂_μ direction to the tip of $\epsilon\partial_\mu$, obtaining a vector u. The condition that the connection be torsion free says that the tips of v and u touch — up to terms of order ϵ^3. The idea is that neither u nor v has rotated. Interestingly, this condition will then hold in *any* coordinate system.

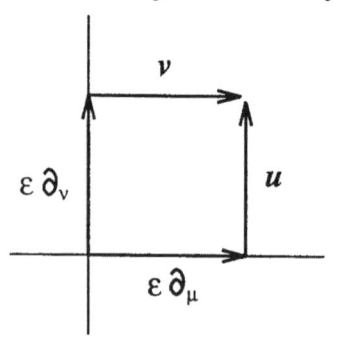

Fig. 2. A torsion free connection

Unfortunately, the standard definitions of 'metric preserving' and 'torsion free' hide these simple ideas a bit, because they speak in terms of the connection itself, rather than in terms of parallel translation. They go as follows. Suppose M is a manifold with semi-Riemannian metric g, and let D be a connection on the tangent bundle TM. The connection D allows us to take the derivative of a vector field v on M in the direction of a vector field u, obtaining a new vector field $D_u v$, in such a way that the usual rules for a connection hold. We say that D is **metric preserving** if for all $u, v, w \in \text{Vect}(M)$

$$ug(v,w) = g(D_u v, w) + g(v, D_u w).$$

If D were not metric preserving we would have a third term involving the derivative of g. We will show later that D being metric preserving in this sense implies that parallel translation preserves the lengths of vectors. We say D is **torsion free** if for all $v, w \in \text{Vect}(M)$

$$[v,w] = D_v w - D_w v.$$

Note that since the Lie bracket $[v,w]$ and the expression $D_v w - D_w v$ are both antisymmetric and involve derivatives of v and w, it is not unreasonable to hope that they are equal.

Now let us prove a wonderful result: for any metric g, there is precisely one connection on TM that is metric-preserving and torsion-free. This is the so-called **Levi-Civita** connection, and it is denoted by ∇.

The Levi-Civita Connection

Exercise 2. *The tangent bundle of $\mathbb{R}^n$ is trivial, with a basis of sections being given by the coordinate vector fields ∂_α; thus it has a standard flat connection D^0 as described in Chapter 2 of Part II. Show that this is the Levi-Civita connection for the standard metric of signature (p,q) on $\mathbb{R}^n$,*

$$g = dx_1^2 + \cdots + dx_p^2 - dx_{p+1}^2 - \cdots - dx_{p+q}^2.$$

In particular, this applies to Euclidean $\mathbb{R}^n$ or Minkowski spacetime. More generally, show it is true for any metric on $\mathbb{R}^n$ such that the components $g_{\alpha\beta}$ with respect to the coordinate vector fields are constant.

Suppose ∇ is a metric-preserving, torsion-free connection on TM. Let us work in local coordinates, and write

$$\nabla_\alpha = \nabla_{\partial_\alpha}.$$

Then since ∇ is metric preserving,

$$\partial_\alpha g_{\beta\gamma} = \partial_\alpha g(\partial_\beta, \partial_\gamma) = g(\nabla_\alpha \partial_\beta, \partial_\gamma) + g(\partial_\beta, \nabla_\alpha \partial_\gamma).$$

Similarly, by permuting indices as shown below we obtain:

$$\begin{aligned} \text{I}: \quad & \partial_\alpha g_{\beta\gamma} = \underbrace{g(\nabla_\alpha \partial_\beta, \partial_\gamma)}_{a} + \underbrace{g(\partial_\beta, \nabla_\alpha \partial_\gamma)}_{b} \\ \text{II}: \quad & \partial_\beta g_{\gamma\alpha} = \underbrace{g(\nabla_\beta \partial_\gamma, \partial_\alpha)}_{c} + \underbrace{g(\partial_\gamma, \nabla_\beta \partial_\alpha)}_{d} \\ \text{III}: \quad & \partial_\gamma g_{\alpha\beta} = \underbrace{g(\nabla_\gamma \partial_\alpha, \partial_\beta)}_{e} + \underbrace{g(\partial_\alpha, \nabla_\gamma \partial_\beta)}_{f} \end{aligned}$$

Now, since the connection is torsion free and the coordinate vector fields commute, we have

$$\nabla_\alpha \partial_\beta - \nabla_\beta \partial_\alpha = [\partial_\beta, \partial_\alpha] = 0,$$

so $\nabla_\alpha \partial_\beta = \nabla_\beta \partial_\alpha$. Together with the symmetry property of the metric, this implies that the terms above labeled b and e are equal, and similarly that c and f are equal. If we add and subtract the equations above to get I+II−III, the b and e terms cancel, as do c and f, and we get

$$\partial_\alpha g_{\beta\gamma} + \partial_\beta g_{\gamma\alpha} - \partial_\gamma g_{\alpha\beta} = g(\nabla_\alpha \partial_\beta, \partial_\gamma) + g(\partial_\gamma, \nabla_\beta \partial_\alpha) = 2g(\nabla_\alpha \partial_\beta, \partial_\gamma).$$

Since two vector fields can only have the same inner products with all the coordinate vector fields ∂_γ if they are equal, this formula determines the covariant derivatives $\nabla_\alpha \partial_\beta$, which means that ∇ is unique.

To show that ∇ exists, we should use the formula above to get an explicit formula for the Levi-Civita connection. Recall that when we had a connection D on a vector bundle E and worked in local coordinates with a local basis of sections e_i of E, we could define the components of the vector potential A as follows:

$$D_\alpha e_j = A^i_{\alpha j} e_i.$$

As a special case, we can define **Christoffel symbols** for the Levi-Civita connection by

$$\nabla_\alpha \partial_\beta = \Gamma^\gamma_{\alpha\beta} \partial_\gamma,$$

and then the covariant derivatives of any vector field w in the direction v are given by

$$\nabla_v w = v^\alpha (\partial_\alpha w^\beta + \Gamma^\beta_{\alpha\gamma} w^\gamma) \partial_\beta.$$

Rewriting the formula in the previous paragraph as

$$\partial_\alpha g_{\beta\gamma} + \partial_\beta g_{\gamma\alpha} - \partial_\gamma g_{\alpha\beta} = 2g(\Gamma^\delta_{\alpha\beta}\partial_\delta, \partial_\gamma) = 2\Gamma^\delta_{\alpha\beta}\, g(\partial_\delta, \partial_\gamma) = 2g_{\delta\gamma}\Gamma^\delta_{\alpha\beta}$$

and raising the index γ, we get a formula for the Christoffel symbols:

Exercise 3. *Show that for the Levi-Civita connection ∇, the Christoffel symbols are given by*

$$\Gamma^\gamma_{\alpha\beta} = \frac{1}{2} g^{\gamma\delta}(\partial_\alpha g_{\beta\delta} + \partial_\beta g_{\delta\alpha} - \partial_\delta g_{\alpha\beta}).$$

Exercise 4. *More generally, suppose we are working with an arbitrary basis of vector fields e_α, satisfying*

$$[e_\alpha, e_\beta] = c^\gamma_{\alpha\beta} e_\gamma.$$

Defining

$$\nabla_\alpha = \nabla_{e_\alpha}, \qquad \nabla_\alpha e_\beta = \Gamma^\gamma_{\alpha\beta} e_\gamma,$$

and

$$\Gamma_{\gamma\alpha\beta} = g_{\gamma\delta}\Gamma^\delta_{\alpha\beta}, \qquad c_{\gamma\alpha\beta} = g_{\gamma\delta} c^\delta_{\alpha\beta},$$

show that

$$\Gamma_{\gamma\alpha\beta} = \frac{1}{2}(\partial_\alpha g_{\beta\gamma} + \partial_\beta g_{\gamma\alpha} - \partial_\gamma g_{\alpha\beta} + c_{\gamma\alpha\beta} + c_{\gamma\beta\alpha} - c_{\alpha\beta\gamma}).$$

The Levi-Civita Connection

It is often handy to use a comma for partial derivatives, for example:

$$\partial_\gamma X^{\alpha_1\cdots\alpha_r}_{\beta_1\cdots\beta_s} = X^{\alpha_1\cdots\alpha_r}_{\beta_1\cdots\beta_s,\gamma}.$$

Using this notation, the above formula becomes a bit easier to remember if we juggle indices a bit and write it as:

$$\Gamma_{\alpha\beta\gamma} = \frac{1}{2}(g_{\alpha\beta,\gamma} + g_{\gamma\alpha,\beta} - g_{\beta\gamma,\alpha} + c_{\alpha\beta\gamma} + c_{\gamma\alpha\beta} - c_{\beta\gamma\alpha}).$$

Exercise 5. *Show that in a basis of coordinate vector fields we have*

$$\Gamma^\alpha_{\beta\gamma} = \Gamma^\alpha_{\gamma\beta}$$

while in an orthonormal basis, e.g. one in which $g(e_\alpha, e_\beta)$ is zero if $\alpha \neq \beta$ and ± 1 if $\alpha = \beta$, we have

$$\Gamma_{\alpha\beta\gamma} = -\Gamma_{\gamma\beta\alpha}.$$

Exercise 6. *Compute the Christoffel symbols on S^2 in spherical coordinates, with the standard metric*

$$d\phi^2 + \sin^2\phi\, d\theta^2.$$

Do the same for the spacetime $\mathbb{R}^4$ using spherical coordinates on space, with the metric

$$g = -f(r)^2 dt^2 + f(r)^{-2} dr^2 + r^2(d\phi^2 + \sin^2\phi\, d\theta^2)$$

(Up to a change of coordinates, this is basically a spacetime version of the wormhole metric considered in Chapter 6 of Part I.)

Now that we have the Levi-Civita connection on the tangent bundle TM, we automatically get connections on all the (r, s) tensor bundles using the recipes of Chapter 2, Part II. We call all these connections ∇ too. If we have an (r, s) tensor field X, we can use the aforementioned recipes to work out the components of $\nabla_\mu X$, that is,

$$(\nabla_\mu X)^{\alpha_1\cdots\alpha_r}_{\beta_1\cdots\beta_s},$$

in terms of partial derivatives of the components of X and Christoffel symbols. One gets, for example,

$$(\nabla_\mu X)^{\alpha\beta}_{\gamma\delta} = \partial_\mu X^{\alpha\beta}_{\gamma\delta} + \overbrace{\Gamma^\alpha_{\mu\lambda} X^{\lambda\beta}_{\gamma\delta} + \Gamma^\beta_{\mu\lambda} X^{\alpha\lambda}_{\gamma\delta}}^{(+)\text{ for vector indices}} - \underbrace{\Gamma^\lambda_{\mu\gamma} X^{\alpha\beta}_{\lambda\delta} - \Gamma^\lambda_{\mu\delta} X^{\alpha\beta}_{\gamma\lambda}}_{(-)\text{ for covector indices}}$$

in the case of a (2,2) tensor field.

Exercise 7. *Prove that this sort of formula holds for arbitrary (r,s)-tensors.*

We should note that

$$(\nabla_\mu X)^{\alpha_1\cdots\alpha_r}_{\beta_1\cdots\beta_s},$$

is usually sloppily written as

$$\nabla_\mu X^{\alpha_1\cdots\alpha_r}_{\beta_1\cdots\beta_s},$$

or sometimes as

$$X^{\alpha_1\cdots\alpha_r}_{\beta_1\cdots\beta_s;\mu}$$

where the semicolon denotes covariant differentiation just as a comma denotes a partial derivative (see Exercise 4).

Recall that an (r,s) tensor field X can be thought of as a machine that eats r 1-forms and s vector fields and spits out a function in a $C^\infty(M)$-linear manner. Since $\nabla_v X$ depends $C^\infty(M)$-linearly on the vector field v as well, we can define an $(r, s+1)$ tensor field ∇X, the **covariant derivative** of X, by

$$\nabla X(\omega_1,\ldots,\omega_r,v,v_1,\ldots,v_s) = (\nabla_v X)(\omega_1,\ldots,\omega_r,v_1,\ldots,v_s).$$

In local coordinates we have

$$\nabla X = dx^\mu \otimes \nabla_\mu X.$$

One can show that this covariant derivative has a bunch of nice properties, and in fact is uniquely determined by these properties:

Exercise 8. *Show that the covariant derivative ∇ satisfies linearity*

$$\nabla(cX) = c\nabla X, \quad \nabla(X + X') = \nabla X + \nabla X'$$

The Levi-Civita Connection

(where c is a scalar), the generalized Leibniz law

$$\nabla_\mu(X \otimes X') = \nabla_\mu X \otimes X' + X \otimes \nabla_\mu X',$$

and compatibility with contraction: if Y is obtained from X by contracting indices as follows,

$$Y^{\alpha_1 \cdots \hat{\alpha}_i \cdots \alpha_r}_{\beta_1 \cdots \hat{\beta}_j \cdots \beta_s} = X^{\alpha_1 \cdots \mu \cdots \alpha_r}_{\beta_1 \cdots \mu \cdots \beta_s},$$

then

$$\nabla_\rho Y^{\alpha_1 \cdots \hat{\alpha}_i \cdots \alpha_r}_{\beta_1 \cdots \hat{\beta}_j \cdots \beta_s} = \nabla_\rho X^{\alpha_1 \cdots \mu \cdots \alpha_r}_{\beta_1 \cdots \mu \cdots \beta_s}.$$

Also, we define the covariant derivative of a (0,0)-tensor to be its differential,

$$\nabla f = df,$$

and define it to agree with the Levi-Civita connection on (1,0)-tensors. Show that ∇ is uniquely determined by the above properties.

As an example of a little calculation involving the Levi-Civita connection, let us show that $\nabla_\alpha g_{\beta\gamma} = 0$. Since the Levi-Civita connection is metric preserving,

$$\begin{aligned}
\partial_\alpha g_{\beta\gamma} &= \partial_\alpha g(\partial_\beta, \partial_\gamma) \\
&= g(\nabla_\alpha \partial_\beta, \partial_\gamma) + g(\partial_\beta, \nabla_\alpha \partial_\gamma) \\
&= g(\Gamma^\mu_{\alpha\beta}\partial_\mu, \partial_\gamma) + g(\partial_\beta, \Gamma^\mu_{\alpha\gamma}\partial_\mu) \\
&= \Gamma^\mu_{\alpha\beta} g_{\mu\gamma} + \Gamma^\mu_{\alpha\gamma} g_{\beta\mu}
\end{aligned}$$

Thus

$$\nabla_\alpha g_{\beta\gamma} = \partial_\alpha g_{\beta\gamma} - \Gamma^\mu_{\alpha\beta} g_{\mu\gamma} - \Gamma^\mu_{\alpha\gamma} g_{\beta\mu} = 0.$$

Geodesics

The theory described here originates from the conviction that the proportionality between the inertial and gravitational mass of a body is an exact law of nature that must be expressed as a foundation principle of theoretical physics.
— *Albert Einstein and Marcel Grossman*

Einstein's explanation of the force of gravity was inspired by the observation that it is impossible to tell the difference between being in a gravitational field and being in an accelerated reference frame using only a local experiment. If, for example, one were in an elevator in a space station, as the elevator accelerated upwards one would be pushed down to the floor just as if one were standing in an elevator at rest in the earth's gravitational field. Another way of putting this is that while in Newtonian gravity mass plays two logically distinct roles — inertial mass measures the force required to accelerate a body a certain amount, while gravitational mass measures the force a gravitational field creates on a body — the ratio of inertial to gravitational mass appears to be the same for all objects (and is set to 1 in standard units).

Einstein's explanation of gravity says, first of all, that gravity is not really a force! In other words, the natural state of motion is free fall, and it takes a force to create a *deviation* from free fall. To put it dramatically but quite correctly, there is not really a force of gravity pulling us down as we stand on the floor; instead, the electromagnetic repulsion of the molecules on the floor are pushing our feet *up*! (It is amusing to get into the habit of feeling things this way.) Geometrically speaking, free fall traces out a path in spacetime that is as straight as possible at every step of the way, a so-called 'geodesic'. It is only the curvature of spacetime that makes these geodesics look different from straight lines in Minkowski spacetime. For this reason, the very notion of a 'gravitational field', as some sort of force field á la Newton, is misleading and not really appropriate to general relativity. There is only the metric, which, as we now show, determines the notion of a geodesic via the Levi-Civita connection.

We already know about parallel transport from our study of gauge theories. Suppose we have a path γ in M and at each point $\gamma(t)$ a

Geodesics

tangent vector $v(t)$. If we translate the definition from gauge theory:

$$D_{\gamma'(t)}u(t) = \frac{d}{dt}u(t) + A(\gamma'(t))u(t)$$

(see Chapter 2 of Part II) into the language of general relativity, we get the following formula in coordinates for the **covariant derivative** of $v(t)$ along γ:

$$D_{\gamma'(t)}v^\mu(t) = \frac{d}{dt}v^\mu(t) + \Gamma^\mu_{\nu\lambda}\frac{d\gamma^\lambda}{dt}v^\nu(t).$$

In particular, $v(t)$ is **parallel transported** along γ if

$$D_{\gamma'(t)}v(t) = 0,$$

and the path γ is a **geodesic** if the tangent vector $\gamma'(t)$ is parallel translated along γ, that is,

$$D_{\gamma'(t)}\gamma'(t) = 0.$$

In coordinates, this reads

$$\frac{d^2\gamma^\mu}{dt^2} + \Gamma^\mu_{\nu\lambda}\frac{d\gamma^\nu}{dt}\frac{d\gamma^\lambda}{dt} = 0.$$

The path of a particle in free fall satisfies this equation.

Exercise 9. *Show that the great circles on the sphere S^2 are geodesics with respect to its standard metric.*

Geodesics are the closest thing there are to straight lines on a semi-Riemannian manifold. If M is Riemannian, we can define the **length** of a path $\gamma\colon [0,T] \to M$ to be the integral

$$\int_0^T \sqrt{g(\gamma'(t),\gamma'(t))}\, dt.$$

We can then define the **distance** between two points to be the minimum (or more precisely, the greatest lower bound) of the lengths of all paths from one to the other. One can show that any two sufficiently nearby points have a unique path between them minimizing the distance, and that this is a geodesic. In the semi-Riemannian case we need to consider timelike geodesics separately since then the quantity in the square root is negative.

We will not go further in the study of parallel transport and geodesics except to show that the metric-preserving property of the Levi-Civita connection implies that the length of a vector does not change when one parallel transports it. More generally, if γ is a path and $v(t), w(t)$ are two vectors parallel transported along γ, we claim

$$\frac{d}{dt}g(v(t), w(t)) = 0.$$

To see this, note that from

$$ug(v, w) = g(\nabla_u v, w) + g(u, \nabla_u w)$$

we expect

$$\frac{d}{dt}g(v(t), w(t)) = g(\nabla_{\gamma'(t)}v(t), w(t)) + g(v(t), \nabla_{\gamma'(t)}w(t)) = 0.$$

One can confirm this with an explicit computation.

Exercise 10. *Do the computation.*

The Riemann Tensor

General relativity is all about curved spacetime. We now have all the tools in place to explain exactly what it means for spacetime to be curved. In general relativity, spacetime is a Lorentzian manifold. More generally, any semi-Riemannian manifold M has a 'best' connection on its tangent bundle — the Levi-Civita connection, ∇ — and we define the **Riemann curvature tensor** to be the curvature of this connection. Just as it is traditional to write the 'vector potential' for gravity as $\Gamma^\alpha_{\beta\gamma}$ rather than A, it is traditional to write the Riemann curvature as $R^\alpha{}_{\beta\gamma\delta}$ rather than F. Thus, given vector fields u, v, w on M, we have

$$R(u, v)w = (\nabla_u \nabla_v - \nabla_v \nabla_u - \nabla_{[u,v]})w,$$

The Riemann Tensor

or if we pick a local basis of vector fields e_α, we have the components $R^\alpha{}_{\beta\gamma\delta}$ given as follows:

$$R(e_\beta, e_\gamma)e_\delta = R^\alpha{}_{\beta\gamma\delta}\, e_\alpha$$

We can also think of the Riemann curvature as a machine that accepts a 1-form μ and three vector fields u, v, w on M and returns the function

$$\mu(R(u,v)w).$$

By the general theory of curvature, $\mu(R(u,v)w)$ depends $C^\infty(M)$-linearly on u, v, w and μ, so the Riemann curvature is a $(1,3)$ tensor field.

The physical, or geometrical, significance of the Riemann curvature is simple: working in coordinates, if we take the tangent vector ∂_δ and parallel translate it around a square of size ϵ in the x^β-x^γ plane, it comes back slightly 'rotated' if spacetime is curved. The change is given, up to second order in ϵ, by

$$-\epsilon^2 R^\alpha{}_{\beta\gamma\delta}\partial_\alpha.$$

This follows from a result of Chapter 3 in Part II.

As a sort of stretching exercise for the serious index gymnastics to follow, let us calculate the components of the Riemann curvature in terms of the Christoffel symbols. Together with the formulas in Exercises 3 and 4, this lets one calculate the Riemann curvature of any metric. Part of the trick is knowing when a basis of coordinate vector fields makes the calculations easier, and when an orthonormal basis is better. We will only work out $R^\alpha{}_{\beta\gamma\delta}$ in the case of a coordinate basis, leaving the general case to the reader. We begin as follows:

$$R(\partial_\beta, \partial_\gamma)\partial_\delta = (\nabla_\beta \nabla_\gamma - \nabla_\gamma \nabla_\beta)\partial_\delta,$$

since $[\partial_\beta, \partial_\gamma] = 0$. Then we use the definition $\nabla_\alpha \partial_\beta = \Gamma^\gamma_{\alpha\beta}\partial_\gamma$ repeatedly:

$$\begin{aligned}
R(\partial_\beta, \partial_\gamma)\partial_\delta &= \nabla_\beta(\Gamma^\sigma_{\gamma\delta}\partial_\sigma) - \nabla_\gamma(\Gamma^\sigma_{\beta\delta}\partial_\sigma) \\
&= (\partial_\beta \Gamma^\sigma_{\gamma\delta})\partial_\sigma + \Gamma^\sigma_{\gamma\delta}\nabla_\beta\partial_\sigma - (\partial_\gamma \Gamma^\sigma_{\beta\delta})\partial_\sigma - \Gamma^\sigma_{\beta\delta}\nabla_\gamma\partial_\sigma \\
&= (\partial_\beta \Gamma^\sigma_{\gamma\delta})\partial_\sigma + \Gamma^\sigma_{\gamma\delta}\Gamma^\tau_{\beta\sigma}\partial_\tau - (\partial_\gamma \Gamma^\sigma_{\beta\delta})\partial_\sigma - \Gamma^\sigma_{\beta\delta}\Gamma^\tau_{\gamma\sigma}\partial_\tau.
\end{aligned}$$

With a little reshuffling and relabeling of dummy indices, we get

$$R^\alpha{}_{\beta\gamma\delta} = \partial_\beta \Gamma^\alpha_{\gamma\delta} - \partial_\gamma \Gamma^\alpha_{\beta\delta} + \Gamma^\sigma_{\gamma\delta}\Gamma^\alpha_{\beta\sigma} - \Gamma^\sigma_{\beta\delta}\Gamma^\alpha_{\gamma\sigma}.$$

It is important to realize that we have not done anything new here: this is simply a horribly ugly way of writing our earlier result, good for the curvature of any connection, that

$$F_{\gamma\delta} = \partial_\gamma A_\delta - \partial_\delta A_\gamma + [A_\gamma, A_\delta],$$

or, even more elegantly,

$$F = dA + A \wedge A.$$

All the formalisms we developed in Part II still apply, and are very useful.

Something special we can do with the Riemann tensor is to contract it and get other tensors. By this process we get the **Ricci tensor**,

$$R_{\alpha\beta} = R^\gamma{}_{\alpha\gamma\beta}$$

and the **Ricci scalar** or **scalar curvature**,

$$R = R^\alpha_\alpha.$$

As we shall see, if we multiply this by the volume form associated to the metric, we get the Lagrangian for general relativity. Another very important tensor is the **Einstein tensor**

$$G_{\alpha\beta} = R_{\alpha\beta} - \frac{1}{2} R g_{\alpha\beta}.$$

This is the tensor that appears in one side of Einstein's equation for general relativity.

Exercise 11. *Calculate the Riemann tensor, the Ricci tensor, the Ricci scalar and Einstein tensor for the standard metric on S^2, starting with the results of Exercise 6. Do the same for the spacetime metric*

$$g = -f(r)^2 dt^2 + f(r)^{-2} dr^2 + r^2(d\phi^2 + \sin^2\phi\, d\theta^2).$$

This takes some work, but in the next chapter you can use these computations to work out the metric describing a black hole!

Symmetries of the Riemann Tensor

There are some important identities satisfied by the Riemann tensor which we need to understand to appreciate Einstein's equation. These identities also imply identities for the Ricci tensor and scalar.

Some of the symmetries of the Riemann tensor are easiest to state if we lower one index as follows:

$$R_{\alpha\beta\gamma\delta} = g_{\alpha\lambda} R^{\lambda}{}_{\beta\gamma\delta}.$$

It is important to note that upon lowering the index we put it *first*. We warn the reader that different people have different conventions for this and all other matters concerning the Riemann tensor! The virtue of the above convention is that

$$R_{\alpha\beta\gamma\delta} = g(e_\alpha, R(e_\beta, e_\gamma)e_\delta)$$

where e_α is the basis of vector fields used to define the components.

Exercise 12. *Show that $R_{\alpha\beta\gamma\delta} = g(e_\alpha, R(e_\beta, e_\gamma)e_\delta)$.*

The three basic symmetries of the Riemann tensor are:

$$\begin{aligned} 1: \quad & R^{\lambda}{}_{\beta\gamma\delta} = -R^{\lambda}{}_{\gamma\beta\delta} \\ 2: \quad & R_{\alpha\beta\gamma\delta} = -R_{\delta\beta\gamma\alpha} \\ 3: \quad & R^{\lambda}{}_{[\beta\gamma\delta]} = 0 \end{aligned}$$

The square brackets in the third relation mean that we are to **antisymmetrize**, that is sum over all the possible permutations of the indices enclosed by the brackets, multiplying each term in the sum by the sign of the permutation (+1 for an even permutation and −1 for an odd one), and then divide by $n!$ if n indices are enclosed by the brackets. This is standard general relativity notation, as are round brackets for **symmetrization**, where we leave out the signs.

Exercise 13. *Write the relation $R^{\lambda}{}_{[\beta\gamma\delta]} = 0$ in an explicit form, and simplify it using other symmetries.*

Let us start by proving relation 1. Note that this is equivalent to

$$R(e_\beta, e_\gamma)e_\delta = -R(e_\gamma, e_\beta)e_\delta,$$

where e_α is the basis of vector fields we are using to define components of tensors. This is a consequence of the fact that

$$R(v,w) = -R(w,v)$$

for any vector fields v, w; we saw in Chapter 3 of Part II that the curvature of any connection has this property.

Similarly, relation 2 is equivalent to

$$g(e_\alpha, R(e_\beta, e_\gamma)e_\delta) = -g(e_\delta, R(e_\beta, e_\gamma)e_\alpha),$$

and can be expressed in a coordinate-free manner as

$$g(u, R(v,w)z) = -g(z, R(v,w)u)$$

for any vector fields u, v, w, z. By linearity it suffices to show this for coordinate vector fields, in which case $[v, w] = 0$. To do this, first note that since the connection is metric preserving,

$$\begin{aligned} vw(g(u,z)) &= v(g(\nabla_w u, z) + g(u, \nabla_w z)) \\ &= g(\nabla_v \nabla_w u, z) + g(\nabla_w u, \nabla_v z) + \\ &\quad g(\nabla_v u, \nabla_w z) + g(u, \nabla_v \nabla_w z) \end{aligned}$$

and similarly

$$wv(g(u,z)) = g(\nabla_w \nabla_v u, z) + g(\nabla_w u, \nabla_v z) + g(\nabla_v u, \nabla_w z) + g(u, \nabla_w \nabla_v z).$$

Subtracting these and using the symmetry of the metric, we get

$$(vw - wv)(g(u,z)) = g([\nabla_v, \nabla_w]u, z) + g(u, [\nabla_v, \nabla_w]z),$$

or, using the definition of the Riemann tensor and the fact that $[v, w]$ vanishes,

$$0 = g(R(v,w)u, z) + g(u, R(v,w)z)$$

which gives the desired result.

Relation 3 is, by Exercise 13, equivalent to

$$R^\lambda{}_{\beta\gamma\delta} + R^\lambda{}_{\gamma\delta\beta} + R^\lambda{}_{\delta\beta\gamma} = 0,$$

or
$$R(e_\beta, e_\gamma)e_\delta + R(e_\gamma, e_\delta)e_\beta + R(e_\delta, e_\beta)e_\gamma = 0.$$
In coordinate-free form, it simply says that
$$R(u,v)w + R(v,w)u + R(w,u)v = 0$$
for all vector fields u, v, w. By linearity it suffices to show this when u, v, w are coordinate vector fields, in which case all their Lie brackets vanish and
$$R(u,v)w + R(v,w)u + R(w,u)v =$$
$$\nabla_u \nabla_v w - \nabla_v \nabla_u w + \nabla_v \nabla_w u - \nabla_w \nabla_v u + \nabla_w \nabla_u v - \nabla_u \nabla_w v.$$
Using the fact that the connection is torsion free, the above equals
$$\nabla_u [v, w] + \nabla_v [w, u] + \nabla_w [u, v],$$
which is zero since the Lie brackets vanish.

One can show that relations 1-3 are essentially all the symmetries possessed by the Riemann tensor. There are, however, nice symmetries that are simply algebraic consequences of these basic ones. As they are useful but not fundamental, we leave them as exercises.

Exercise 14. *Show that relations 1-3 imply $R_{\alpha\beta\gamma\delta} = R_{\gamma\delta\alpha\beta}$ and $R_{[\alpha\beta\gamma\delta]} = 0$.*

Now let us consider the Ricci tensor $R_{\alpha\beta}$. Using Exercise 14, we see that it is symmetric:
$$R_{\beta\delta} = R^\gamma{}_{\beta\gamma\delta} = g^{\gamma\alpha} R_{\alpha\beta\gamma\delta} = g^{\gamma\alpha} R_{\gamma\delta\alpha\beta} = R^\alpha{}_{\delta\alpha\beta} = R_{\delta\beta}.$$
This also implies the symmetry of the Einstein tensor:
$$G_{\alpha\beta} = G_{\beta\alpha},$$
where recall that $G_{\alpha\beta} = R_{\alpha\beta} - \frac{1}{2} R g_{\alpha\beta}$. Another handy identity allows us to express the Ricci tensor in terms of the Einstein tensor in all dimensions other than 2. Raising an index and contracting,
$$G^\mu_\mu = R - \frac{1}{2} R \delta^\mu_\mu = (1 - \frac{n}{2}) R$$
where n is the dimension of spacetime, so
$$R_{\alpha\beta} = G_{\alpha\beta} + \frac{1}{2} R g_{\alpha\beta} = G_{\alpha\beta} + \frac{1}{2-n} G^\mu_\mu g_{\alpha\beta}.$$

Exercise 15. *Show that all the $(0,2)$ tensors that can be constructed from the Riemann tensor by raising indices and contraction are proportional to the Ricci tensor.*

Exercise 16. *Show that in 2 dimensions*

$$R_{\alpha\beta} = \frac{1}{2} R g_{\alpha\beta}$$

so that $G_{\alpha\beta} = 0$. Show that in 3 dimensions

$$R_{\alpha\beta\gamma\delta} = g_{\alpha\gamma} R_{\beta\delta} + g_{\beta\delta} R_{\alpha\gamma} - g_{\beta\gamma} R_{\alpha\delta} - g_{\alpha\delta} R_{\beta\gamma} - \frac{1}{2}(g_{\alpha\gamma} g_{\beta\delta} - g_{\alpha\delta} g_{\beta\gamma}) R.$$

Chapter 2

Einstein's Equation

The sought for generalization will surely be of the form $\Gamma_{\mu\nu} = \kappa T_{\mu\nu}$, where κ is a constant and $\Gamma_{\mu\nu}$ is a contravariant tensor of the second rank that arises out of the fundamental tensor $g_{\mu\nu}$ through differential operations. — Albert Einstein and Marcel Grossman

The Stress-Energy Tensor

Einstein's equation says how spacetime is curved by the presence of matter, or more generally, anything possessing energy or momentum. The wonderful thing about this equation is that it is the simplest one relating curvature and energy-momentum for which the law of local conservation of energy-momentum is an automatic consequence! This is analogous to Maxwell's equation

$$\star d \star F = J$$

which automatically implies local conservation of electric charge:

$$d \star J = 0.$$

In the case of Maxwell's equation, the key to local conservation of electric charge is the identity $d^2 = 0$. In the case of Einstein's equation, the key is its natural generalization, the Bianchi identity.

In relativistic physics, the flow of energy and momentum through a given point of spacetime is summarized by the **stress-energy** tensor, which is a (0,2) tensor with components written $T_{\mu\nu}$. Suppose that spacetime is split into time and space by $\mathbb{R} \times S$, and pick local coordinates $x^0, x^1, \ldots, x^{n-1}$ where $x^0 = t$ is the time coordinate and x^i, $i = 1, \ldots, n-1$ are the space coordinates. Then T^{00} represents the **energy** density, the components T^{i0} represent the flow of energy in the ∂_i direction, T^{0j} represents the density of the jth component of **momentum**, and T^{ij} represents the flow of the jth component of momentum in the ∂_i direction. This tensor (at least in the theories we are interested in) is symmetric:

$$T_{\mu\nu} = T_{\nu\mu}.$$

On curved spacetime, the law of local conservation of energy-momentum is written

$$\nabla^\mu T_{\mu\nu} = 0,$$

where the left hand side is short for $g^{\mu\lambda}\nabla_\lambda T_{\mu\nu}$. We say that $T_{\mu\nu}$ is **divergence-free**. To understand the physical meaning of this condition, we should compare the situation with electromagnetism. As we mentioned, the current 1-form J satisfies $d \star J = 0$, but one can show that for any 1-form J,

$$\star d \star J = -\nabla^\mu J_\mu,$$

so we can also express this as $\nabla^\mu J_\mu = 0$.

Exercise 17. *Show that for any 1-form J on a Lorentzian manifold, $\star d \star J = -\nabla^\mu J_\mu$.*

In Minkowski spacetime, where the Levi-Civita connection is just the standard flat connection, this reduces to

$$\partial^\mu J_\mu = 0,$$

or the 'continuity equation'

$$\frac{\partial \rho}{\partial t} + \nabla \cdot \vec{j} = 0,$$

which says that any divergence in the current density must give a corresponding change in the charge density. Similarly, in Minkowski spacetime the equation $\nabla^\mu T_{\mu\nu} = 0$ gives four equations: for $\nu = 0$, we get conservation of energy

$$\frac{\partial T^{00}}{\partial t} + \frac{\partial T^{i0}}{\partial x^i} = 0,$$

while for $j = 1, 2, 3$ we obtain **conservation of momentum**

$$\frac{\partial T^{0j}}{\partial t} + \frac{\partial T^{ij}}{\partial x^i} = 0.$$

Different kinds of fields have different stress-energy tensors. This subject has some tricky aspects, and we refer the reader to the Notes for references. As an example, though, the Yang-Mills field has

$$T_{\mu\nu} = -\text{tr}(F_{\mu\lambda}F_{\nu}{}^{\lambda} - \frac{1}{4}g_{\mu\nu}F_{\alpha\beta}F^{\alpha\beta}).$$

We leave it to the reader to check that this is divergence-free, and move on to the other side of Einstein's equation.

Exercise 18. *Show that the Yang-Mills equations imply $\nabla^\mu T_{\mu\nu} = 0$ with $T_{\mu\nu}$ defined as above. Work out the components of $T_{\mu\nu}$ in terms of the Yang-Mills electric and magnetic fields, and compare T_{00} to the quantity discussed in Exercise 58 of Part I, keeping track of the fact that the vector potential of a $U(1)$-connection is an imaginary-valued 1-form.*

The Bianchi Identity

It is to be emphasized that the (generally covariant) laws of conservation are deduced from the field equations of gravitation, in combination with the postulate of general covariance (relativity) alone, without using the field equations for material phenomena. — Albert Einstein

To get equations for gravity that are consistent with conservation of energy it is natural to try something of the form

$$C_{\mu\nu} = T_{\mu\nu}$$

where $C_{\mu\nu}$ is a symmetric divergence-free tensor that depends only on the curvature of spacetime, that is, the Riemann tensor. As we shall show, the simplest choice is the Einstein tensor $G_{\mu\nu}$ (up to a constant factor). By analogy with Maxwell's equations and the Yang-Mills equations, we expect to get something divergence-free using the Bianchi identity, so we should look at various versions of the Bianchi identity for the Riemann curvature. Most of these are just special cases of the versions discussed in Chapter 3 of Part II, to which we refer the reader.

The simplest form to remember is that for any vector fields u, v, w on spacetime,

$$[\nabla_u, [\nabla_v, \nabla_w]] + [\nabla_v, [\nabla_w, \nabla_u]] + [\nabla_w, [\nabla_u, \nabla_v]] = 0,$$

which is just the Jacobi identity. From this it follows that

$$[\nabla_\alpha, R(\partial_\beta, \partial_\gamma)] + [\nabla_\beta, R(\partial_\gamma, \partial_\alpha)] + [\nabla_\gamma, R(\partial_\alpha, \partial_b)] = 0.$$

It is also illuminating to regard the curvature as an $\text{End}(TM)$-valued 2-form, say $\mathcal{R}$. This is just another way of talking about a $(1,3)$-tensor $R^\alpha{}_{\beta\gamma\delta}$ with $R^\alpha{}_{\beta\gamma\delta} = -R^\alpha{}_{\gamma\beta\delta}$. In these terms, the above version of the Bianchi identity becomes simply

$$d_\nabla \mathcal{R} = 0$$

where d_∇ is the exterior covariant derivative coming from the Levi-Civita connection.

There is yet another version of the Bianchi identity that is very useful in general relativity. This relies upon the fact that the exterior derivative of a p-form,

$$d(\omega_I dx^I) = (\partial_\mu \omega_I) dx^\mu \wedge dx^I,$$

is also equal to

$$(\nabla_\mu \omega_I) dx^\mu \wedge dx^I$$

where ∇ is the Levi-Civita connection. If we use ∇_μ this way to define the exterior derivative of ordinary differential forms, and thence to define d_∇ on $\text{End}(TM)$-valued differential forms, working out the Bianchi identity $d_\nabla \mathcal{R} = 0$ in local coordinates gives

$$\nabla_{[\alpha} R^\lambda{}_{\beta\gamma]\delta} = 0.$$

The Bianchi Identity

Exercise 19. *Check these claims.*

Now let us get 'boiled-down' versions of the Bianchi identity by contracting indices. If we contract once we get

$$\nabla_{[\alpha} R^{\alpha}{}_{\beta\gamma]\delta} = 0.$$

If we explicitly write out the antisymmetrization, we get

$$\nabla_{\alpha} R^{\alpha}{}_{\beta\gamma\delta} + \nabla_{\beta} R^{\alpha}{}_{\gamma\alpha\delta} + \nabla_{\gamma} R^{\alpha}{}_{\alpha\beta\delta} = 0$$

or

$$\nabla_{\alpha} R^{\alpha}{}_{\beta\gamma\delta} + \nabla_{\beta} R^{\alpha}{}_{\gamma\alpha\delta} - \nabla_{\gamma} R^{\alpha}{}_{\beta\alpha\delta} = 0,$$

which by the definition of the Ricci tensor and a little index juggling is just

$$\nabla^{\alpha} R_{\alpha\beta\gamma\delta} + \nabla_{\beta} R_{\gamma\delta} - \nabla_{\gamma} R_{\beta\delta} = 0.$$

We can boil down the Bianchi identity still further. First, using the symmetries of the Riemann tensor, we have

$$\nabla^{\alpha} R_{\delta\gamma\beta\alpha} + \nabla_{\beta} R_{\gamma\delta} - \nabla_{\gamma} R_{\beta\delta} = 0.$$

Then, raising the index β and contracting it with δ, we get

$$\nabla^{\alpha} R_{\gamma\alpha} + \nabla^{\beta} R_{\gamma\beta} - \nabla_{\gamma} R = 0$$

or relabelling dummy indices and dividing by 2,

$$\nabla^{\alpha} R_{\gamma\alpha} - \frac{1}{2} \nabla_{\gamma} R = 0$$

which we can also write as

$$\nabla^{\alpha} (R_{\gamma\alpha} - \frac{1}{2} g_{\gamma\alpha} R) = 0$$

since $\nabla^{\alpha} g_{\gamma\alpha} = 0$, as shown earlier. The tensor in parentheses is just the Einstein tensor. In other words, we have shown that

$$\nabla^{\mu} G_{\mu\nu} = 0.$$

The Einstein tensor is thus the simplest divergence-free symmetric (0,2)-tensor that depends only on the curvature of spacetime, so **Einstein's equation** for general relativity,

$$G_{\mu\nu} = 8\pi\kappa T_{\mu\nu},$$

automatically implies local conservation of energy and momentum! The **vacuum** Einstein equation is the special case when $T_{\mu\nu} = 0$. This says simply that the Einstein tensor vanishes, but except in dimension 2 this is equivalent to the vanishing of the Ricci tensor, as we saw in the previous chapter. So except in 2-dimensional spacetime, the vacuum Einstein equation simply says that the metric is **Ricci flat**, that is,

$$R_{\mu\nu} = 0.$$

At this point the reader should get a good book on general relativity and look at a bunch of solutions of Einstein's equation to see what it says about our world! The first solution to be studied is a spherically symmetric static vacuum solution, the **Schwarzschild solution**, which represents the gravitational field of a point mass.

Exercise 20. *Starting with the metric*

$$g = -f(r)^2 dt^2 + f(r)^{-2} dr^2 + r^2(d\phi^2 + \sin^2\phi\, d\theta^2),$$

use the results of Exercise 11 to show that Einstein's equation implies the differential equation for f,

$$\frac{d}{dr} r f(r)^2 = 1.$$

This has the solution

$$f(r)^2 = 1 - \frac{2M}{r},$$

which describes (in units where $\kappa = 1$) the metric produced by a point particle of mass M.

In fact, this solution is singular and describes a black hole! Still, it is a good enough approximation to the gravitational field of the sun for Einstein to have correctly used it to explain a mysterious small amount of precession in the orbit of Mercury, and to predict the bending of starlight as it passed by the sun. (Both of these are simply matters of computing *geodesics* in the Schwarzschild metric.)

However, there is another sort of divergence-free symmetric (0,2)-tensor that we can cook up starting from the metric, namely, the metric itself! We have already seen that

$$\nabla^\mu g_{\mu\nu} = 0$$

so we could, if we wanted, modify Einstein's equation by adding a term proportional to the metric, as follows:

$$G_{\mu\nu} + \Lambda g_{\mu\nu} = 8\pi\kappa T_{\mu\nu}.$$

The parameter Λ is called the **cosmological constant**, because Einstein introduced it when he noticed that his original equation predicted an expanding universe — obvious nonsense! Then Hubble's study of galactic redshifts led people to conclude that the universe *is indeed* expanding, and Einstein rejected the cosmological term, calling it the "biggest blunder of my life." Pushing it over to the right hand side of the equation, the cosmological term can be interpreted as the stress-energy tensor of the vacuum. There has continued to be interest in the cosmological term, for a variety of reasons; we will eventually see that Chern-Simons theory gives a solution to the *quantized* version of Einstein's equations with cosmological constant. (From this solution one can also construct solutions with zero cosmological constant.)

There is, of course, much more to say about all these matters, but let us wrap up this chapter by saying a bit about the dimension of spacetime. Nobody knows why spacetime is 4-dimensional, mainly because nobody knows what an answer to this question would amount to. Quite possibly it is a meaningless question. Still, it is interesting to ponder the things that are special about 4 dimensions. Earlier we have seen that Maxwell's equations and the Yang-Mills equation are special in 4 dimensions, because in that dimension F and $\star F$ are both 2-forms, so that duality becomes a powerful tool. Later we will see that self-duality is important in Einstein's equations as well, and possibly the key to developing a quantum theory of gravity. On the other hand, general relativity is not a very interesting theory in dimensions less than 4. To see this, we need only count the independent components of the Riemann and Ricci tensors, as follows.

One can show that the symmetries of the Riemann tensor $R^\alpha{}_{\beta\gamma\delta}$ reduces number of independent components from n^4, where n is the dimension of spacetime, down to

$$\frac{n^2(n^2-1)}{12}.$$

Exercise 21. *Show this.*

In particular, in dimension 1 the Riemann tensor has *no* independent components, because every connection on a 1-dimensional manifold is flat. In dimension 2 the Riemann tensor has just one independent component, which is proportional to the Ricci scalar R. Thus the curvature of a 2-manifold is completely described by a function on the manifold that at each point says how much it is 'positively curved', like a sphere, or 'negatively curved', like a saddle. In dimension 3 there are 6 independent components, and in dimension 4 there are 20.

As for the Ricci tensor $R_{\mu\nu}$, since it is symmetric one would expect that it has

$$\frac{n(n+1)}{2}$$

independent components. This is true except in dimension 1, where of course it must vanish, and in dimension 2, where it has only one independent component, since by Exercise 15 we have $R_{\alpha\beta} = \frac{1}{2}Rg_{\alpha\beta}$ in dimension 2. In dimension 3, the Ricci tensor has 6 independent components, which is just as many as the Riemann tensor. In dimension 4 it has 10 independent components.

It follows that in dimensions 1 and 2 the vacuum Einstein equation is *automatically* true for any metric, while in dimension 3 the vacuum Einstein equation implies that the Riemann tensor vanishes — i.e., spacetime is flat! Only in dimension 4 do things get interesting: there are, in fact, lots of non-flat solutions of the vacuum Einstein equations, some of which represent gravitational waves. A simple example is as follows:

Exercise 22. *Suppose the metric on* $\mathbb{R}^4$ *has the form*

$$g = L(u)^2(e^{2\beta(u)}dx^2 + e^{-2\beta(u)}dy^2) - du\,dv$$

where $u = t - z$, $v = t + z$. Show that the vacuum Einstein equations hold when
$$\frac{d^2 L(u)}{du^2} + \left(\frac{d\beta(u)}{du}\right)^2 L(u) = 0.$$
Study linear approximations to this equation when L is near 1 and β is small; note that $L = 1$, $\beta = 0$ gives the Minkowski metric. Show that solutions of the linearized equations represent propagating ripples in the metric.

Chapter 3

Lagrangians for General Relativity

Whenever any action occurs in nature, the quantity of action employed by this change is the least possible. — Pierre de Maupertuis, 1746

The Einstein-Hilbert Action

In this chapter we will derive Einstein's equations from an action principle, just as we derived the Yang-Mills equations from an action principle in Chapter 3 of Part II. In the case of the Yang-Mills equations we treated the spacetime metric as a fixed background structure, the presence of which allowed us to define the Hodge star operator and then the Yang-Mills Lagrangian, which is quadratic in the curvature F. The Lagrangian for Einstein's equations is somewhat different. First, in this case the only basic field is the metric, which we are now treating as a dynamical variable. Second, the Lagrangian we will use is *linear* in the curvature: this simpler possibility is available because gravity is 'all about the tangent bundle', while Yang-Mills theory concerns an arbitrary vector bundle. We will discuss two ways of formulating an action principle for gravity in this chapter, the Einstein-Hilbert action and the Palatini action. The first approach emphasizes the importance of the *metric*, while the second emphasizes the *connection*.

Suppose that M is an oriented manifold, 'spacetime', with a semi-

Riemannian metric g on it. The Lagrangian for general relativity, discovered by Hilbert five days before Einstein independently found it, is simply
$$R \, \text{vol}$$
where R is the Ricci scalar curvature of g and vol is the volume form associated to g. The **Einstein-Hilbert action** is thus
$$S(g) = \int_M R \, \text{vol}.$$
If M is not compact this integral may not converge, but we will see that as in the Yang-Mills case, the variation of S will still make sense if the variation of the metric vanishes outside of a compact set.

To emphasize the dependence of the volume form on the metric one can also write the Einstein-Hilbert action in local coordinates as
$$S(g) = \int_M R \sqrt{|\det g|} \, d^n x.$$
For a Lorentzian metric we have $|\det g| = -\det g$. We warn the reader that people often do not bother writing the 'det' in this formula.

Let us compute the variation of the Einstein-Hilbert action in the Lorentzian case (other signatures are just as easy, but it is simpler to be specific). Suppose g is a Lorentzian metric and δg is any symmetric $(0, 2)$-tensor that vanishes outside a compact set. If $s \in \mathbb{R}$ is small, $g + s\delta g$ will still be a Lorentzian metric, so as s varies, $g + s\delta g$ is a path of metrics through g. This allows us to define the variation of the action as
$$\delta S(g) = \frac{d}{ds} S(g + s\delta g)\Big|_{s=0},$$
and we can define the variation of any quantity depending on g in a similar manner. We have
$$\delta S = \int_M \delta(R \, \text{vol}) = \int_M (\delta R) \, \text{vol} + R \, \delta\text{vol}.$$
To continue this calculation, we must compute the variations of R and the volume form. Let us do the volume form first, as it is a bit easier. We need a little fact about linear algebra:

The Einstein-Hilbert Action

Exercise 23. *Show that for any matrix A, $\det(1 + sA)$ is equal to $1 + s\operatorname{tr}(A)$ up to terms of order s^2. (Hint: first consider the case where A is diagonalizable, and then use the fact that such matrices are dense in the space of all matrices.)*

Using the above exercise, it follows that for any two matrices A and B,

$$\frac{d}{ds}\det(A+sB)\Big|_{s=0} = \frac{d}{ds}\det(A)\det(1+sA^{-1}B)\Big|_{s=0}$$
$$= \det(A)\operatorname{tr}(A^{-1}B),$$

so
$$\delta(\det g) = (\det g)\operatorname{tr}(g^{-1}\delta g) = (\det g)g^{\alpha\beta}\,\delta g_{\alpha\beta}.$$

Alternatively, since $g^{\alpha\beta}g_{\alpha\beta}$ equals n in n dimensions, its variation is zero, so
$$g^{\alpha\beta}\,\delta g_{\alpha\beta} = -g_{\alpha\beta}\,\delta g^{\alpha\beta}.$$

Thus we can write
$$\delta(\det g) = -(\det g)g_{\alpha\beta}\,\delta g^{\alpha\beta}.$$

This implies that
$$\delta\sqrt{-\det g} = -\frac{1}{2}\sqrt{-\det g}\,g_{\alpha\beta}\,\delta g^{\alpha\beta},$$

hence
$$\delta\mathrm{vol} = -\frac{1}{2}g_{\alpha\beta}(\delta g^{\alpha\beta})\mathrm{vol},$$

which is the formula we will need.

Now, to compute δR we need to go back and compute the variation of the Christoffel symbols, the Riemann tensor, and the Ricci tensor! This takes some work, and we will only outline the computations, leaving the reader to fill in the details. First, recall how we defined the Christoffel symbols:

$$\Gamma^{\alpha}_{\beta\gamma} = \frac{1}{2}g^{\alpha\eta}(\partial_{\beta}g_{\gamma\eta} + \partial_{\gamma}g_{\beta\eta} - \partial_{\eta}g_{\beta\gamma})$$

From this, one can compute the variation

$$\delta\Gamma^{\alpha}_{\beta\gamma} = \frac{1}{2}g^{\alpha\eta}(\nabla_{\beta}\,\delta g_{\gamma\eta} + \nabla_{\gamma}\,\delta g_{\beta\eta} - \nabla_{\eta}\,\delta g_{\beta\gamma}).$$

Exercise 24. *Compute the variation of the Christoffel symbols.*

Similarly, starting from the definition of the Riemann tensor

$$R^\alpha{}_{\beta\gamma\eta} = \partial_\beta \Gamma^\alpha_{\gamma\eta} - \partial_\gamma \Gamma^\alpha_{\beta\eta} + \Gamma^\sigma_{\gamma\eta}\Gamma^\alpha_{\beta\sigma} - \Gamma^\sigma_{\beta\eta}\Gamma^\alpha_{\gamma\sigma}.$$

one can show that

$$\delta R^\alpha{}_{\beta\gamma\eta} = \nabla_\beta \, \delta\Gamma^\alpha_{\gamma\eta} - \nabla_\gamma \, \delta\Gamma^\alpha_{\beta\eta}.$$

Exercise 25. *Compute the variation of the Riemann tensor. (Hint: one can do this from scratch or by showing that it is a special case of the formula $\delta F = d_D \delta A$ given in Chapter 4 of Part II.)*

We thus have

$$\delta R_{\alpha\beta} = \delta R^\gamma{}_{\alpha\gamma\beta} = \nabla_\alpha \, \delta\Gamma^\gamma_{\gamma\beta} - \nabla_\gamma \, \delta\Gamma^\gamma_{\alpha\beta},$$

and if we use the formula for the variation of the Christoffel symbols, this gives

$$\delta R_{\alpha\beta} = \frac{1}{2}\Big(g^{\gamma\eta}\nabla_\alpha\nabla_\beta\delta g_{\gamma\eta} + g^{\gamma\eta}\nabla_\gamma\nabla_\eta\delta g_{\alpha\beta} - g^{\gamma\eta}\nabla_\gamma(\nabla_\beta \, \delta g_{\alpha\eta} + \nabla_\alpha \, \delta g_{\beta\eta})\Big).$$

Exercise 26. *Check this formula for the variation of the Ricci tensor.*

Finally, for the variation of the Ricci scalar, we obtain

$$\begin{aligned}\delta R &= \delta(g^{\alpha\beta}R_{\alpha\beta}) \\ &= (\delta g^{\alpha\beta})R_{\alpha\beta} + g^{\alpha\beta}\,\delta R_{\alpha\beta} \\ &= R_{\alpha\beta}\,\delta g^{\alpha\beta} + \nabla^\gamma\nabla_\gamma(g^{\alpha\beta}\,\delta g_{\alpha\beta}) - \nabla^\alpha\nabla^\beta\delta g_{\alpha\beta},\end{aligned}$$

or simply

$$\delta R = R_{\alpha\beta}\delta g^{\alpha\beta} + \nabla^\alpha \omega_\alpha$$

where the 1-form ω is given by

$$\omega_\alpha = g^{\gamma\eta}\nabla_\alpha \, \delta g_{\gamma\eta} - \nabla^\beta \delta g_{\alpha\beta}.$$

Exercise 27. *Check this computation of the variation of the Ricci scalar.*

The Einstein-Hilbert Action

Now we are ready to derive Einstein's equation from the Einstein-Hilbert action. We have:

$$\delta S = \int_M (\delta R)\,\text{vol} + R\,\delta\text{vol}$$
$$= \int_M (R_{\alpha\beta}\delta g^{\alpha\beta} + \nabla^\alpha \omega_\alpha - \frac{1}{2} R g_{\alpha\beta}\,\delta g^{\alpha\beta})\,\text{vol}$$

Ignoring the second term for a moment, this is just

$$\int_M (R_{\alpha\beta} - \frac{1}{2} R g_{\alpha\beta})(\delta g^{\alpha\beta})\,\text{vol},$$

which is zero for all variations $\delta g^{\alpha\beta}$ vanishing outside a compact set precisely when Einstein's equation holds:

$$R_{\alpha\beta} - \frac{1}{2} R g_{\alpha\beta} = 0.$$

What about that second term, however? Well, in Exercise 17 the diligent reader showed that for any 1-form ω we have

$$\nabla^\alpha \omega_\alpha = - \star d \star \omega,$$

so the term involving ω is just

$$-\int_M \text{vol} \wedge \star d \star \omega = \pm \int_M d \star \omega \wedge \star\text{vol} = \pm \int_M d \star \omega = 0$$

by Stokes' theorem, where the $\pm$ sign comes from the fact that $\star^2 = \pm 1$. In physics terminology, we say that an expression like $\nabla^\alpha \omega_\alpha$ is a **total divergence**. It is worth noting that when M has a boundary, this term does *not* equal zero; in fact it leads to interesting physics. For example, in the study of 'asymptotically Minkowskian' solutions of general relativity one can attach an ideal 'boundary at infinity' to spacetime, and this extra term then becomes crucial for understanding the dynamics of the theory (we refer the reader to the Notes for more).

It may seem odd to derive Einstein's equation this way, but there are many spinoffs of this point of view. In general, the initial-value or 'Hamiltonian' approach to classical and quantum field theories is closely intertwined with the variational or Lagrangian approach, and they must be studied together. One immediate bonus for having computed all those variations is that we now know the 'linearized' version of the vacuum Einstein equations. That is, if one considers a metric g solving the vacuum Einstein equation — the simplest example being the Minkowski metric — one can consider a perturbation $g + \epsilon h$ and demand that this new metric also be a solution up to first order in ϵ. Since g was a solution, its Ricci tensor vanishes, and by the formula for $\delta R_{\alpha\beta}$, the Ricci tensor of $g + \epsilon h$ is

$$\frac{\epsilon}{2}\left(g^{\gamma\eta}\nabla_\alpha\nabla_\beta h_{\gamma\eta} + g^{\gamma\eta}\nabla_\gamma\nabla_\eta h_{\alpha\beta} - g^{\gamma\eta}\nabla_\gamma(\nabla_\beta h_{\alpha\eta} + \nabla_\alpha h_{\beta\eta})\right)$$

plus terms of higher order in ϵ. For this to vanish, h must satisfy the **linearized** vacuum Einstein equation

$$g^{\gamma\eta}\nabla_\alpha\nabla_\beta h_{\gamma\eta} + g^{\gamma\eta}\nabla_\gamma\nabla_\eta h_{\alpha\beta} - g^{\gamma\eta}\nabla_\gamma(\nabla_\beta h_{\alpha\eta} + \nabla_\alpha h_{\beta\eta}) = 0.$$

This is what people often use to study the propagation of small ripples in the Minkowski metric, or **gravitational waves**. These have yet not been observed directly, but there is a project underway to detect them. Most people believe in the existence of gravitational waves, in part because general relativity has been confirmed in other ways, and in part because of some beautiful indirect evidence: Russell A. Hulse and Joseph H. Taylor won the Nobel prize in 1993 for careful observations of a binary pulsar that turned out to be gradually spiralling down precisely as one would predict by using general relativity to compute the gravitational radiation emitted.

Some rough insight into quantum gravity can be obtained by considering the linearized Einstein equation on Minkowski space: the quantum theory thereof turns out to be the theory of a spin-2 massless particle, the **graviton**. From the viewpoint of standard quantum field theory it is natural to first study the linearized equations and then incorporate the nonlinear terms as 'interactions'. However, this violates the spirit of general relativity, since it privileges the particular solution being perturbed about (Minkowski space), which plays the role of a 'background geometry'. This violation is punished by intractable infinities — one says the theory is nonrenormalizable — and these days the most interesting work on quantum gravity is 'nonperturbative'.

Exercise 28. *Work out the linearized Einstein equation more explicitly in the case where g is the Minkowski metric. Use a plane wave ansatz to find solutions.*

Exercise 29. *Derive Einstein's equations from the Einstein-Hilbert action when the metric has arbitrary signature. Derive the equations for Yang-Mills fields coupled to gravity from the Lagrangian $R\,\mathrm{vol} + \frac{1}{2}\mathrm{tr}(F \wedge \star F)$ by varying both the metric and the Yang-Mills vector potential A.*

Exercise 30. *Show that one can pull back $(0, s)$ tensors in a manner similar to how one pulls back differential forms. If g is a semi-Riemannian metric on M and $\phi: M \to M$ is a diffeomorphism, show that the Einstein-Hilbert Lagrangian of ϕ^*g equals the pullback of the Einstein-Hilbert Lagrangian of g. Use this to show that if g satisfies Einstein's equation, so does ϕ^*g, so that Einstein's equation is diffeomorphism-invariant.*

The Palatini Action

The Palatini action for general relativity is simply the Einstein-Hilbert action

$$S(g) = \int_M R\,\mathrm{vol}$$

rewritten so that it is not a function of the metric, but instead a function of a connection and a 'frame field'. For us, its main importance will be as a warmup for our discussion of Ashtekar's 'new variables' for general relativity in Chapter 5, but it is certainly beautiful and useful in its own right.

We begin by defining the concept of a 'frame field'. Suppose that M is an oriented n-dimensional manifold diffeomorphic to $\mathbb{R}^n$. Physically, we can think of M as a small open subset of spacetime (since every manifold can be covered with charts diffeomorphic to $\mathbb{R}^n$). Since the tangent bundle of $\mathbb{R}^n$ is trivial, so is TM. A trivialization of TM, recall, is a vector bundle isomorphism

$$e: M \times \mathbb{R}^n \to TM$$

sending each fiber $\{p\} \times \mathbb{R}^n$ of the trivial bundle $M \times \mathbb{R}^n$ to the corresponding tangent space T_pM. A trivialization of TM is also called a **frame field**, since for each p it sends the standard basis of $\mathbb{R}^n$ to a basis of tangent vectors at p, or **frame**:

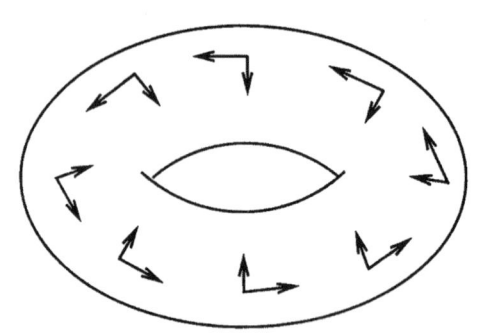

Fig. 1. A frame field on the torus

If M is 3-dimensional, a frame field on M is also called a **triad** or **dreibein**, depending on whether one prefers Greek or German; if M is 4-dimensional, a frame field on it is called a **tetrad** or **vierbein**.

The idea of the Palatini formalism is to do a lot of work on the trivial bundle $M \times \mathbb{R}^n$, which serves as a kind of substitute for the tangent bundle. We can pass back and forth between $M \times \mathbb{R}^n$ and TM by using the frame field e and its inverse,

$$e^{-1} \colon TM \to M \times \mathbb{R}^n.$$

We need to develop some notation to do this efficiently. To keep things specific, let us suppose that we are in the n-dimensional Lorentzian case (other cases work similarly). A section of $M \times \mathbb{R}^n$ is just an $\mathbb{R}^n$-valued function on M, so there is a natural basis of sections $\xi_0, \ldots, \xi_n$ given by

$$\begin{aligned} \xi_0(p) &= (1,0,0,\ldots) \\ \xi_1(p) &= (0,1,0,\ldots) \\ \xi_2(p) &= (0,0,1,\ldots) \end{aligned}$$

and so on, and we can write any section s as

$$s = s^I \xi_I$$

where we use Einstein summation to sum over I. In this game, $\mathbb{R}^n$ is often called the **internal space**. To keep from getting mixed up, we will

The Palatini Action

use upper-case Latin letters $I, J, \ldots$ for **internal indices** associated to the basis of sections ξ_I, and use lower-case Greek letters for **spacetime indices** associated to the coordinate vector fields ∂_μ on a chart.

We can think of the frame field $e: M \times \mathbb{R}^n \to TM$ as defining a map from sections of $M \times \mathbb{R}^n$ to vector fields on M, which we also call e. Applying this map to the sections ξ_I, we get a basis of vector fields $e(\xi_I)$ on M, and in a chart we can write these as

$$e(\xi_I) = e_I^\alpha \partial_\alpha$$

where the components e_I^α are functions on M. In relativity it is typical to abbreviate $e(\xi_I)$ as just e_I, so we will do this. Also, since either the coefficients e_I^α or the vector fields $e_I = e(\xi_I)$ are enough to determine the frame field e, it is common to call either of these things the frame field.

Now the real key to the Palatini formalism is that $M \times \mathbb{R}^n$, as a kind of 'imitation tangent bundle', has one thing the real tangent bundle lacks, namely a canonical inner product. In other words, given two sections s and s' of $M \times \mathbb{R}^n$, we can define their inner product $\eta(s, s')$ by

$$\eta(s, s') = \eta_{IJ} s^I s'^J$$

where η_{IJ} is copied after the Minkowski metric:

$$\eta_{IJ} = \begin{pmatrix} -1 & 0 & 0 & 0 \\ 0 & 1 & 0 & 0 \\ 0 & 0 & \ddots & 0 \\ 0 & 0 & 0 & 1 \end{pmatrix},$$

and is called the **internal metric**. We can raise and lower internal indices with η_{IJ} and its inverse η^{IJ}, just as we raise and lower spacetime indices using a metric. Of course, what we are really doing thereby is mapping $\mathbb{R}^n$ to its dual (or vice versa) using the internal metric.

Now suppose that M has a Lorentzian metric g on it. This means that we can take inner products of vector fields on M by

$$g(v, v') = g_{\alpha\beta} v^\alpha v'^\beta.$$

We say that the frame field e is **orthonormal** if the vector fields e_I are orthonormal, that is,

$$g(e_I, e_J) = \eta_{IJ}.$$

If the frame field is orthonormal, the metric g on M is nicely related to the internal metric η, as follows:

$$g(e(s), e(s')) = \eta(s, s')$$

for any sections s, s' of $M \times \mathbb{R}^n$. To see this, we just compute:

$$\begin{aligned} g(e(s), e(s')) &= g(e(s^I \xi_I), e(s^J \xi_J)) \\ &= s^I s^J g(e_I, e_J) \\ &= \eta_{IJ} s^I s^J \\ &= \eta(s^I \xi_I, s^J \xi_J) \\ &= \eta(s, s'). \end{aligned}$$

Exercise 31. *Conversely, show that if $g(e(s), e(s')) = \eta(s, s')$ for all sections s, s' of $M \times \mathbb{R}^n$, then $g(e_I, e_J) = \eta_{IJ}$.*

In the Palatini formalism we work with orthonormal frame fields rather than metrics on M. If the frame e is orthonormal, the result above implies that the metric on M is given in terms of the inverse frame field by

$$g(v, v') = \eta(e^{-1}v, e^{-1}v').$$

Conversely, since we are assuming M is diffeomorphic to $\mathbb{R}^n$ (which we can always arrange by taking M to be a small open subset of spacetime), one can show that every metric g admits some orthonormal frame field.

The formula above for the metric in terms of an orthonormal frame may look a bit abstract, but for computations one can use the following formulas. Suppose that e is an orthonormal frame field. Then

$$\eta_{IJ} = g(e_I, e_J) = g_{\alpha\beta} e_I^\alpha e_J^\beta.$$

Starting with this, by suitable index gymnastics one can show

$$\delta_J^I = e_\alpha^I e_J^\alpha.$$

Exercise 32. *Prove this identity.*

The Palatini Action

It follows that the inverse frame field is given by the following formula:
$$e^{-1}v = e_\alpha^I v^\alpha \xi_I,$$
since if $v = e(s)$ for some section s of $M \times \mathbb{R}^4$, this formula gives
$$e^{-1}v = e_\alpha^I v^\alpha \xi_I = e_\alpha^I e_J^\alpha s^J \xi_I = \delta_J^I s^J \xi_I = s^I \xi_I = s$$
as it should. From this, we can derive a formula for the metric g in terms of the **coframe field** e_α^I (also known as a **cotriad** or **cotetrad** in dimensions 3 and 4):
$$\begin{aligned} g_{\alpha\beta} &= g(\partial_\alpha, \partial_\beta) \\ &= \eta(e^{-1}\partial_\alpha, e^{-1}\partial_\beta) \\ &= \eta(e_\alpha^I \xi_I, e_\beta^J \xi_J) \\ &= \eta_{IJ} e_\alpha^I e_\beta^J. \end{aligned}$$

Besides the frame field, the other ingredient in the Palatini formalism is a connection on the trivial bundle $M \times \mathbb{R}^n$. By analogy with the definition of a metric-preserving connection, we say a connection D on this bundle is a **Lorentz connection** if
$$v\eta(s, s') = \eta(D_v s, s') + \eta(s, D_v s').$$
This is the same as being an $SO(n, 1)$-connection in the sense of gauge theory. Note that it makes no sense to ask if a connection on $M \times \mathbb{R}^n$ is torsion free! There is thus no 'Levi-Civita connection' on $M \times \mathbb{R}^n$. However, there is a specially nice Lorentz connection, the standard flat connection D^0, given by
$$D_v^0 s = v(s^I)\xi_I.$$
We can write any connection D as $D^0 + A$ for some vector potential A, which is an $\text{End}(\mathbb{R}^n)$-valued 1-form on M:
$$D_v s = (v(s^J) + A_{\mu I}^J v^\mu s^I)\xi_J.$$
We write the curvature of the connection D as F, or using indices,
$$F_{\alpha\beta}^{IJ} = \partial_\alpha A_\beta^{IJ} - \partial_\beta A_\alpha^{IJ} + [A_\alpha, A_\beta]^{IJ}.$$
It is easy to recognize a Lorentz connection by looking at its vector potential:

Exercise 33. *Show that a connection D on $M \times \mathbb{R}^n$ is a Lorentz connection precisely when $A_\mu^{IJ} = -A_\mu^{JI}$, which is just a way of saying that A_μ lives in the Lorentz Lie algebra* $\mathfrak{so}(n,1)$.

Exercise 34. *Show that if A is a Lorentz connection then $F_{\alpha\beta}^{IJ} = -F_{\beta\alpha}^{IJ} = -F_{\alpha\beta}^{JI}$.*

Suppose now that we have both a frame field e and a Lorentz connection D. We can use the frame field to transfer the Lorentz connection from the trivial bundle $M \times \mathbb{R}^n$ to the tangent bundle TM. When we do this, we obtain a connection $\widetilde{\nabla}$ on TM given by

$$\widetilde{\nabla}_\alpha \partial_\beta = \widetilde{\Gamma}^\gamma_{\alpha\beta} \partial_\gamma,$$

where the coefficients $\widetilde{\Gamma}^\gamma_{\alpha\beta}$ are defined by

$$\widetilde{\Gamma}^\gamma_{\alpha\beta} = A^J_{\alpha I} e^I_\beta e^\gamma_J.$$

We will call $\widetilde{\nabla}$ the **imitation Levi-Civita connection** and call the $\widetilde{\Gamma}^\gamma_{\alpha\beta}$ the **imitation Christoffel symbols**. Note that the imitation Christoffel symbols are obtained by converting internal indices in the vector potential A to spacetime indices, using the frame field and coframe field. Similarly, we can define an **imitation Riemann tensor** by

$$\widetilde{R}^\gamma{}_{\alpha\beta}{}^\delta = F_{\alpha\beta}^{IJ} e^\delta_I e^\gamma_J,$$

an **imitation Ricci tensor** by

$$\widetilde{R}_{\alpha\beta} = \widetilde{R}^\gamma{}_{\alpha\gamma\beta},$$

and an **imitation Ricci scalar** by

$$\widetilde{R} = \widetilde{R}^\alpha_\alpha.$$

Exercise 35. *Show that the imitation Riemann tensor is the curvature of the imitation Levi-Civita connection.*

We are now ready to describe the Palatini action! It is basically the Einstein-Hilbert action in disguise, but we emphasize again that, unlike the Einstein-Hilbert action, it is *not* a function of a metric on

The Palatini Action

M. Instead it is a function of a frame field e and a Lorentz connection A. In the Palatini approach, the metric on M is not a fundamental field; instead, it is a function of the frame field given by

$$g_{\alpha\beta} = \eta_{IJ} e^I_\alpha e^J_\beta.$$

The **Palatini action** is given by

$$S(A, e) = \int_M e^\alpha_I e^\beta_J F^{IJ}_{\alpha\beta} \text{ vol}.$$

Here the volume form is given by the usual formula in terms of g, but now g is a function of the frame field!

Now we shall show that the Palatini action gives Einstein's equations. More precisely, if we vary S with respect to both A and e, the equation $\delta S = 0$ will imply that the metric $g_{\alpha\beta} = \eta_{IJ} e^I_\alpha e^J_\beta$ satisfies the vacuum Einstein equation. We begin by computing the variation with respect to the frame field, that is, computing δS assuming $\delta A = 0$. As in the previous section

$$\delta \text{vol} = -\frac{1}{2} g_{\alpha\beta} (\delta g^{\alpha\beta}) \text{vol},$$

but now

$$\delta g^{\alpha\beta} = \delta(\eta^{IJ} e^\alpha_I e^\beta_J) = 2\eta^{IJ} e^\beta_J \delta e^\alpha_I$$

so

$$\delta \text{vol} = -\eta^{IJ} g_{\alpha\beta} e^\beta_J (\delta e^\alpha_I) \text{vol}$$

or expressing g in terms of e and doing some index gymnastics,

$$\delta \text{vol} = -e^K_\gamma (\delta e^\gamma_K) \text{vol}.$$

Exercise 36. *Perform the gymnastics required to derive the above formula.*

This lets us compute the variation of the action as follows:

$$\begin{aligned}\delta S &= \int_M \left((\delta e^\alpha_I) e^\beta_J F^{IJ}_{\alpha\beta} + e^\alpha_I (\delta e^\beta_J) F^{IJ}_{\alpha\beta} - e^K_\gamma (\delta e^\gamma_K) e^\alpha_I e^\beta_J F^{IJ}_{\alpha\beta}\right) \text{vol} \\ &= 2\int_M \left(e^\beta_J F^{IJ}_{\alpha\beta} - \frac{1}{2} e^I_\alpha e^\gamma_K e^\delta_L F^{KL}_{\gamma\delta}\right) (\delta e^\alpha_I) \text{vol}\end{aligned}$$

where we used the result of Exercise 34. Expressing this in terms of the imitation Ricci tensor and scalar, we obtain

$$\delta S = 2 \int_M (\tilde{R}_{\alpha\beta} - \frac{1}{2}\tilde{R}g_{\alpha\beta}) \eta^{IJ} e_J^\beta (\delta e_I^\alpha) \,\text{vol}.$$

Exercise 37. *Check this result.*

It follows that $\delta S = 0$ for an arbitrary variation of the frame field precisely when

$$\tilde{R}_{\alpha\beta} - \frac{1}{2}\tilde{R}g_{\alpha\beta} = 0.$$

This looks a lot like Einstein's equation, and it *is* Einstein's equation when the imitation Riemann tensor is equal to the Riemann tensor of g! By Exercise 35, this will hold when $\tilde{\nabla} = \nabla$.

Next let us vary the Lorentz connection. In other words, let us compute δS assuming $\delta e = 0$. We will show that in this case $\delta S = 0$ precisely when $\tilde{\nabla} = \nabla$. Combined with the results of the previous paragraph, this means that $\delta S = 0$ for all variations of the connection frame field *and* connection precisely when $\tilde{\nabla}$ equals the Levi-Civita connection of g and g satisfies the vacuum Einstein equation.

Note that the Palatini Lagrangian is given by

$$e_I^\alpha e_J^\beta F_{\alpha\beta}^{IJ} \,\text{vol} = \tilde{R}\,\text{vol}.$$

It follows that when $\delta e = 0$, we have

$$\begin{aligned}\delta S &= \int_M (\delta \tilde{R}) \,\text{vol} \\ &= \int_M g^{\alpha\beta} (\delta \tilde{R}_{\alpha\beta}) \,\text{vol}.\end{aligned}$$

Copying a formula from the previous section, we have $\delta \tilde{R}^\lambda{}_{\alpha\sigma\beta} = 2\tilde{\nabla}_{[\alpha} \delta \tilde{\Gamma}^\lambda_{\sigma]\beta}$ hence

$$\delta \tilde{R}_{\alpha\beta} = 2\tilde{\nabla}_{[\alpha} \delta \tilde{\Gamma}^\gamma_{\gamma]\beta}.$$

If we write

$$\tilde{\Gamma}^\gamma_{\alpha\beta} = \Gamma^\gamma_{\alpha\beta} + C^\gamma_{\alpha\beta},$$

we have

$$\delta \tilde{\Gamma}^\gamma_{\alpha\beta} = \delta C^\gamma_{\alpha\beta}$$

The Palatini Action

since $\delta\Gamma^\gamma_{\alpha\beta} = 0$. A computation then gives

$$\frac{1}{2}\delta\widetilde{R} = g^{\alpha\beta}\widetilde{\nabla}_{[\alpha}\,\delta C^\gamma_{\gamma]\beta} =$$

$$g^{\alpha\beta}\nabla_{[\alpha}\delta C^\gamma_{\gamma]\beta} + \frac{1}{2}g^{\alpha\beta}\bigl(-C^\gamma_{\gamma\eta}\delta C^\eta_{\alpha\beta} + C^\eta_{\gamma\alpha}\delta C^\gamma_{\eta\beta} - C^\eta_{\alpha\beta}\delta C^\gamma_{\gamma\eta} + C^\eta_{\gamma\beta}\delta C^\gamma_{\alpha\eta}\bigr)$$

Exercise 38. *Check this result.*

The first term above is a total divergence and contributes nothing to the integral over M. With some work, the remaining terms can be shown to vanish if and only if $C^\gamma_{\alpha\beta} = 0$, in other words, $\widetilde{\nabla} = \nabla$.

Exercise 39. *Do the work necessary to prove the claim above.*

This completes the derivation of Einstein's equation from the Palatini action. We present an alternative, more conceptual derivation in the following exercises:

Exercise 40. *Suppose D is a Lorentz connection and $\widetilde{\nabla}$ the corresponding imitation Levi-Civita connection. Show that $\widetilde{\nabla}$ is metric preserving, and conclude that $\widetilde{\nabla} = \nabla$ if and only if $\widetilde{\nabla}$ is torsion free.*

Exercise 41. *The inverse frame field $e^{-1}\colon TM \to M \times \mathbb{R}^n$ can be thought of as an $\mathbb{R}^n$-valued 1-form. Using the Lorentz connection D to define exterior covariant derivatives of $\mathbb{R}^n$-valued forms, show that $\widetilde{\nabla}$ is torsion free if and only if $d_D e^{-1} = 0$.*

Exercise 42. *Express the Palatini action S in terms of the $\mathbb{R}^n$-valued 1-form e^{-1} and the $\mathrm{End}(\mathbb{R}^n)$-valued 2-form F, the curvature of D. Using the formula $\delta F = d_D \delta A$ and Stokes' theorem, show that when we vary A, $\delta S = 0$ implies $d_D e^{-1} = 0$. As a consequence, if $\delta S = 0$ for both variations in the frame field and variations in the connection, $\widetilde{\nabla} = \nabla$ and $\widetilde{R}_{\alpha\beta} - \frac{1}{2}\widetilde{R}g_{\alpha\beta} = 0$, hence $R_{\alpha\beta} - \frac{1}{2}Rg_{\alpha\beta} = 0$.*

Chapter 4
The ADM Formalism

World-wide instants are not natural cleavage planes of time ... they are imaginary partitions which we find it convenient to adopt .... There is a difference between simplicity and familiarity. A pig may be most familiar to us in the form of rashers, but the unstratified pig is a simpler object to the biologist who wishes to understand how the animal functions. — Sir Arthur Eddington.

While Einstein's equation is very beautiful, it takes a lot of work to extract the physics it contains. In this section we will describe how to think of Einstein's equation as a rule that tells the geometry of space how to evolve as time passes. Thus we will consider the case of a Lorentzian manifold M diffeomorphic to $\mathbb{R} \times S$, where the manifold S represents 'space' and $t \in \mathbb{R}$ represents 'time'. Of course, the particular slicing of spacetime into 'instants of time' is an arbitrary choice, rather than something intrinsic to the world. In other words, if someone simply hands us the spacetime M, there are lots of ways to pick a diffeomorphism

$$\phi \colon M \to \mathbb{R} \times S.$$

These give different ways to define a **time coordinate** τ on M, namely the pullback by ϕ of the standard time coordinate t on $\mathbb{R} \times S$:

$$\tau = \phi^* t.$$

Different people could pick different time coordinates on M this way, so we should make sure that anything we do applies equally well to any one of them. Let us say that a submanifold $\Sigma \subset M$ is a **slice** of M if it equals $\{\tau = \text{constant}\}$ for some time coordinate τ. In the figure below we show a few different slices of a 2-dimensional cylindrical spacetime $\mathbb{R} \times S^1$:

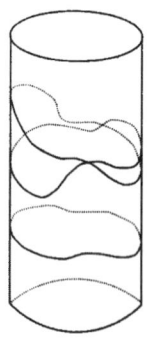

Fig. 1. Slices of the spacetime $\mathbb{R} \times S^1$

In what follows we will concentrate on the case of a 4-dimensional spacetime. In this case Einstein's equation is really 10 different equations, since there are 10 independent components in the Einstein tensor. We will rewrite these equations in terms of the metric on the slice Σ, or '3-metric,' which we write as ^3g, and the 'extrinsic curvature' K of the slice Σ, which describes the curvature of the way it sits in M. As we shall see in the next chapter, the extrinsic curvature can also be thought of as representing the time derivative of the 3-metric. We can think of $(^3g, K)$ as **Cauchy data** for the metric, just as earlier we thought of the vector potential on space and the electric field as Cauchy data for electromagnetism or the Yang-Mills field. We will see that of Einstein's 10 equations, 4 are *constraint* equations that the Cauchy data must satisfy, while 6 are *evolutionary* equations saying how the 3-metric changes with time. This is called the Arnowitt-Deser-Misner, or **ADM**, formulation of Einstein's equation.

Extrinsic Curvature

Let Σ be a slice of the spacetime $\mathbb{R} \times S$. We will assume that Σ is **spacelike**, that is, when we restrict the metric g on M to Σ, we get a

Extrinsic Curvature

Riemannian metric on Σ, meaning that

$$g(v,v) > 0$$

for all nonzero $v \in T_p\Sigma$. We denote this metric on Σ by ^3g, and call it the **3-metric**. In this situation we can find a field of timelike unit vectors n normal to Σ:

$$g(n,n) = -1 \quad \text{and} \quad \forall v \in T_p\Sigma \quad g(n,v) = 0.$$

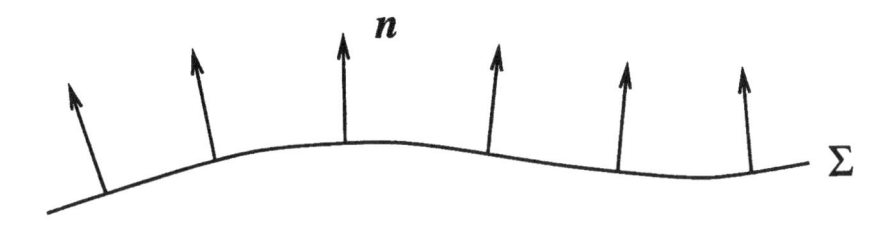

Fig. 2. A field of timelike vectors normal to a spacelike slice

There are actually two choices of a normal vector field n, since we can switch the sign of n. One can think of these two choices as pointing in the 'future' and 'past' directions, as long as one remembers that the physics problem does not know which is the future and which is the past. We may simply pick a choice of n and think of it as pointing towards the future!

Now, given any vector $v \in T_pM$ we can decompose it into a component tangent to Σ and a normal component proportional to n:

$$v = \underbrace{-g(v,n)n}_{\perp} + \underbrace{(v + g(v,n)n)}_{\parallel}$$

The signs here may seem odd, but they are a consequence of the fact that $g(n,n) = -1$. We can check, for example, that the normal component of n is really n with this definition:

$$-g(n,n)n = n,$$

and that the tangent component of any vector field v is is really orthogonal to n:

$$g(v + g(v,n)n, n) = g(v,n) + g(v,n)g(n,n) = 0.$$

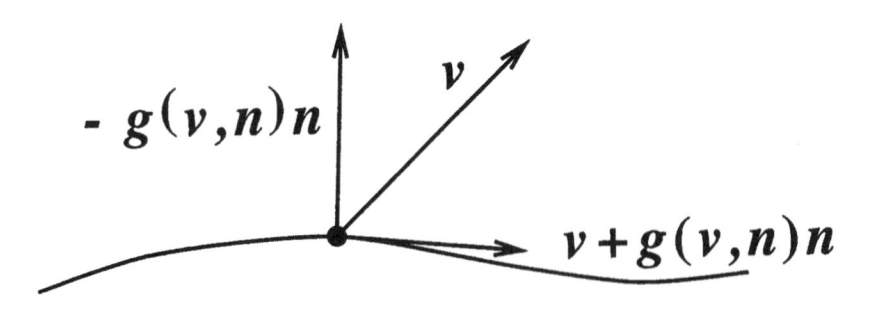

Fig. 3. Decomposing a vector into tangent and normal components

In particular, given any vector fields u, v on Σ, we can split $\nabla_u v$ into normal and tangent parts:

$$\nabla_u v = -g(\nabla_u v, n)n + (\nabla_u v + g(\nabla_u v, n)n).$$

We write the first term as

$$-g(\nabla_u v, n)n = K(u,v)n,$$

where $K(u,v)$ is called the **extrinsic curvature**. This measures how much the surface Σ is curved in the way it sits in M, because it says how much a vector tangent to Σ will fail to be tangent if we parallel translate it a bit using the Levi-Civita connection ∇ on M. We should note that the difference between extrinsic and intrinsic properties is basic to differential geometry. For example, consider a cylinder in $\mathbb{R}^3$. If we restrict the usual Euclidean metric on $\mathbb{R}^3$ to the cylinder we get a *flat* metric on the cylinder — as we can see by simply unrolling it — but the extrinsic curvature is not zero.

We write the second term as

$$^3\nabla_u v = \nabla_u v + g(\nabla_u v, n)n,$$

Extrinsic Curvature

because it turns out to be just the Levi-Civita connection on Σ associated to the metric ^3g. Let us prove this fact! First let us show that it is a connection. As usual, the only nontrivial part is the Leibniz law:

$$^3\nabla_v(fw) = v(f)w + f\,^3\nabla_w,$$

for any $u, v \in \text{Vect}(\Sigma)$ and $f \in C^\infty(\Sigma)$. To prove this we simply use the definition of $^3\nabla$ and the fact that $g(n, w) = 0$:

$$\begin{aligned}
^3\nabla_v(fw) &= \nabla_v(fw) + g(n, \nabla_v(fw))n \\
&= v(f)w + f\nabla_v w + g(n, v(f)w)n + g(n, f\nabla_v w)n \\
&= v(f)w + f\nabla_v w + fg(n, \nabla_v w)n \\
&= v(f)w + f(\nabla_v w + g(n, \nabla_v w)n) \\
&= v(f)w + f\,^3\nabla_v w.
\end{aligned}$$

Next, let us show that $^3\nabla$ is metric preserving. Letting $u, v, w \in \text{Vect}(\Sigma)$, we check:

$$\begin{aligned}
ug(v, w) &= g(\nabla_u v, w) + g(v, \nabla_u w) \\
&= g(K(u,v)n + {}^3\nabla_u v, w) + g(v, K(u,w)n + {}^3\nabla_u w) \\
&= g(^3\nabla_u v, w) + g(v, {}^3\nabla_u w)
\end{aligned}$$

since $g(n, w) = g(v, n) = 0$. Finally, let us show that $^3\nabla$ is torsion-free. This follows from the fact that $K(u, v) = K(v, u)$ for all $u, v \in \text{Vect}(M)$, which we prove in a bit:

$$\begin{aligned}
^3\nabla_u v - {}^3\nabla_v u &= \nabla_u v - K(u,v)n - \nabla_v u + K(v,u)n \\
&= \nabla_u v - \nabla_v u \\
&= [u, v]
\end{aligned}$$

since ∇ itself is torsion free.

The basic properties of the extrinsic curvature are that it is a tensor and that it is symmetric. Recall that by $K(u, v)$ being a tensor we mean that it depends $C^\infty(\Sigma)$-linearly on the vector fields u and v, so that for any u, v we have

$$K(u, v) = K_{ij} u^i v^j$$

in local coordinates, where
$$K_{ij} = K(\partial_i, \partial_j).$$

To see that $K(u,v)$ is $C^\infty(\Sigma)$-linear in u we merely note that
$$K(fu,v) = -g(\nabla_{fu}v, n) = -g(f\nabla_u v, n) = -fg(\nabla_u v, n) = fK(u,v)$$
for all $f \in C^\infty(\Sigma)$. To show that it is $C^\infty(\Sigma)$-linear in v takes only a bit more work:
$$\begin{aligned} K(u, fv) &= -g(\nabla_u fv, n) \\ &= -g(u(f)v + f\nabla_u v, n) \\ &= -fg(\nabla_u v, n) \\ &= -fK(u,v) \end{aligned}$$
using the fact that $g(v,n) = 0$.

To see the symmetry property $K(u,v) = K(v,u)$, it thus suffices to show that $K_{ij} = K_{ji}$, which we do using the fact that ∇ is torsion free:
$$\begin{aligned} K_{ij} - K_{ji} &= K(\partial_i, \partial_j) - K(\partial_j, \partial_i) \\ &= -g(\nabla_i \partial_j, n) + g(\nabla_j \partial_i, n) \\ &= -g(\nabla_i \partial_j - \nabla_j \partial_i, n) \\ &= -g([\partial_i, \partial_j], n) \\ &= 0. \end{aligned}$$

It is worth knowing about another definition of the extrinsic curvature, namely
$$K(u,v) = g(\nabla_u n, v).$$
This agrees with the earlier definition because ∇ is metric preserving:
$$0 = ug(n,v) = g(\nabla_u v, n) + g(n, \nabla_u v).$$

This other definition gives us another way to think about $K(u,v)$: it measures how much the unit normal n rotates in the direction v when we parallel translate it in the direction u.

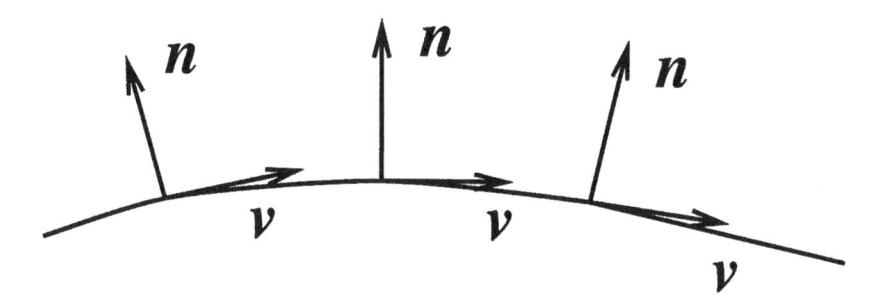

Fig. 4. Rotation of the unit normal n in the v direction

The Gauss-Codazzi Equations

In this section we will show that in 4-dimensional spacetime, 4 of Einstein's equations are constraints on the 3-metric and extrinsic curvature. The remaining 6 describe how the 3-metric changes with time. Of course, there are lots of different ways to split spacetime into space and time, hence lots of different ways to push a given spacelike slice 'forwards in time'. We cannot expect Einstein's equations to tell us how the extrinsic curvature and 3-metric will change with time until we *explicitly say* how we are pushing the spacelike slice forwards in time! This is expressed by the 'lapse' and 'shift', as follows.

Let us pick a particular way of splitting time into space and time, that is, a diffeomorphism

$$\phi: M \to \mathbb{R} \times S.$$

This gives us a time coordinate $\tau = \phi^* t$ on M and thus a particular way of getting slices $\{\tau = s\}$. It also gives us a particular vector field ∂_τ on M, namely the pushforward by ϕ^{-1} of the vector field ∂_t on $\mathbb{R} \times S$. This vector field points 'forwards in time', but it is not necessarily orthogonal to the slices $\{\tau = s\}$! Let us concentrate on a particular slice Σ, say the 'time-zero' slice $\{\tau = 0\}$, and assume it is spacelike. We can split ∂_τ into a component that is normal to Σ and a component that is tangent to Σ:

$$\partial_\tau = -g(\partial_\tau, n)n + (\partial_\tau + g(\partial_\tau, n)n)$$

We will write this for short as

$$\partial_\tau = Nn + \vec{N}$$

where $\vec{N}$ is called the **shift** vector field and N is called the **lapse** function. Thus we have

Lapse : $N = -g(\partial_\tau, n)$
Shift : $\vec{N} = \partial_\tau + g(\partial_\tau, n)n$

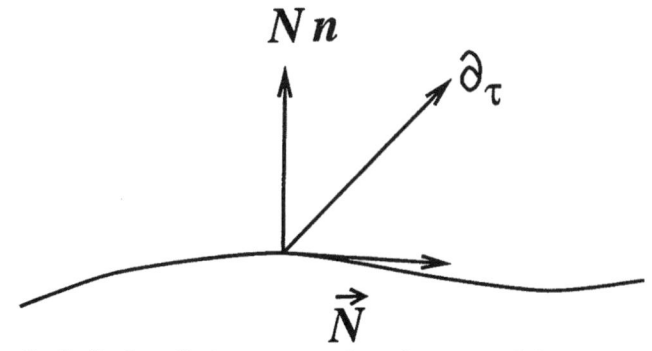

Fig. 5. Splitting ∂_τ into normal and tangential components

We can, of course, solve for the unit normal in terms of the lapse and shift:

$$n = \frac{1}{N}(\partial_\tau - \vec{N}).$$

We shall now show that 4 of Einstein's equations are constraints that the the 3-metric and extrinsic curvature must satisfy, The reason for this is that some components of the Riemann tensor depend only on the extrinsic curvature K and the **intrinsic curvature**, that is, the curvature of ^3g. The formulas that describe this are known as the Gauss-Codazzi equations, which we now derive.

Pick a point p on Σ, and choose local coordinates x^0, x^1, x^2, x^3 in a neighborhood of p in such a way that $x^0 = \tau$, $\partial_0 = \partial_\tau$, and the vector fields $\partial_1, \partial_2, \partial_3$ are tangent to Σ at p.

Exercise 43. *Show that this can be done.*

As we have done now and then, we will use Greek letters as indices ranging from 0 to 3, and Roman letters $i, j, k, \ldots$ as **spacelike indices**

The Gauss-Codazzi Equations

ranging from 1 to 3. We will write the Christoffel symbols of the connection $^3\nabla$ on our slice Σ as $^3\Gamma^i_{jk}$, and write the Riemann tensor of ^3g as $^3R^m{}_{ijk}$.

We will compute the components $R^\alpha{}_{ijk}$ in terms of K_{ij} and $^3R^m{}_{ijk}$. To do this we need to compute

$$R(\partial_i, \partial_j)\partial_k = \nabla_i\nabla_j\partial_k - \nabla_j\nabla_i\partial_k.$$

Let us calculate the first term and then get the second term by switching i and j. The main fact we will use is that $\nabla_u v = K(u,v)n + {}^3\nabla_u v$ for any vector fields u, v on Σ, which implies

$$\nabla_i \partial_j = K_{ij} n + {}^3\Gamma^m_{ij}\partial_m.$$

We will also need to use the fact that $K(u,v) = g(\nabla_u n, v)$, which implies

$$\nabla_i n = K_i^m \partial_m.$$

Using these, the first term becomes:

$$\begin{aligned}
\nabla_i\nabla_j\partial_k &= \nabla_i\left(K_{jk} n + {}^3\Gamma^m_{jk}\partial_m\right) \\
&= K_{jk,i} n + K_{jk}\nabla_i n + {}^3\Gamma^m_{jk,i}\partial_m + {}^3\Gamma^m_{jk}\nabla_i \partial_m \\
&= K_{jk,i} n + K_{jk}K_i^m \partial_m + {}^3\Gamma^m_{jk,i}\partial_m + {}^3\Gamma^m_{jk}(K_{im} n + {}^3\Gamma^\ell_{im}\partial_\ell) \\
&= (K_{jk,i} + {}^3\Gamma^m_{jk}K_{im})n + K_{jk}K_i^m \partial_m + ({}^3\Gamma^m_{jk,i} + {}^3\Gamma^\ell_{jk}{}^3\Gamma^m_{i\ell})\partial_m
\end{aligned}$$

Subtracting the second term, which has i and j switched, we get

$$\begin{aligned}
R(\partial_i, \partial_j)\partial_k &= (K_{jk,i} - K_{ik,j} + {}^3\Gamma^m_{jk}K_{im} - {}^3\Gamma^m_{ik}K_{jm})n + \\
&\quad (K_{jk}K_i^m - K_{ik}K_j^m)\partial_m + \\
&\quad ({}^3\Gamma^m_{jk,i} - {}^3\Gamma^m_{ik,j} + {}^3\Gamma^\ell_{jk}{}^3\Gamma^m_{i\ell} - {}^3\Gamma^\ell_{ik}{}^3\Gamma^m_{j\ell})\partial_m.
\end{aligned}$$

Now note that the first line on the right hand side is just

$$({}^3\nabla_i K_{jk} - {}^3\nabla_j K_{ik})n,$$

while the third is just

$$^3R^m{}_{ijk}\partial_m.$$

Thus we have the **Gauss-Codazzi equations**:

$$R(\partial_i, \partial_j)\partial_k = (^3\nabla_i K_{jk} - {}^3\nabla_j K_{ik})n + (^3R^m_{ijk} + K_{jk}K^m_i - K_{ik}K^m_j)\partial_m.$$

To understand these equations, let us assume that $\partial_0 = n$, that is, that the lapse is 1 and the shift is 0. This will make the formulas a little simpler; we will say at the end of this section what happens in general. In this situation, if we apply the 1-form dx^0 both sides of the Gauss-Codazzi equations we get the **Gauss equation**:

$$R^0{}_{ijk} = {}^3\nabla_i K_{jk} - {}^3\nabla_j K_{ik}.$$

Alternatively, if we apply the 1-form dx^m to both sides, we get the **Codazzi equation**:

$$R^m{}_{ijk} = {}^3R^m{}_{ijk} + K_{jk}K^m_i - K_{ik}K^m_j.$$

Note that the Codazzi equation implies that the intrinsic curvature $^3R^m{}_{ijk}$ equals part of the Riemann tensor of M, namely $R^m{}_{ijk}$, when the extrinsic curvature of Σ vanishes.

Now let us rewrite the 4 of Einstein's equations that involve G^0_α using the Gauss-Codazzi equations. Recall that the Einstein tensor looks like $G_{\mu\nu} = R_{\mu\nu} - \frac{1}{2}g_{\mu\nu}R$, where $R_{\mu\nu} = R^\alpha{}_{\mu\alpha\nu}$ and $R = R^\mu_\mu = R^{\alpha\beta}{}_{\alpha\beta}$. Using the symmetries of the Riemann tensor that we derived in Chapter 1, we have

$$R^\alpha{}_{\mu\alpha\nu} = -R^\alpha{}_{\alpha\mu\nu} = -R_{\mu\nu}{}^\alpha{}_\alpha = R_\mu{}^\alpha{}_{\nu\alpha},$$

or

$$R_{\mu\nu} = R_\mu{}^\alpha{}_{\nu\alpha}.$$

By raising the μ index using the metric, we get

$$R^\mu_\nu = R^{\mu\alpha}{}_{\nu\alpha}.$$

We thus have

$$G^\mu_\nu = R^{\mu\alpha}{}_{\nu\alpha} - \frac{1}{2}\delta^\mu_\nu R^{\alpha\beta}{}_{\alpha\beta}.$$

If we work this out for the case $\mu = \nu = 0$, we get (in 4 dimensions)

$$G^0_0 = -(R^{12}{}_{12} + R^{23}{}_{23} + R^{31}{}_{31}).$$

The Gauss-Codazzi Equations

Exercise 44. *Check this result; note that a lot of terms in the formula for G_0^0 cancel due to the symmetries of the Riemann tensor.*

If we apply the Codazzi equation, we obtain

$$\begin{aligned}-G_0^0 &= {}^{(3)}R^{12}{}_{12} + {}^{(3)}R^{23}{}_{23} + {}^{(3)}R^{31}{}_{31} + \\ &\quad (K_1^2 K_2^1 - K_2^2 K_1^1) + \\ &\quad (K_2^3 K_3^2 - K_3^3 K_2^2) + \\ &\quad (K_3^1 K_1^3 - K_1^1 K_3^3)\end{aligned}$$

Now we claim that the terms involving K are equal to $-\frac{1}{2}((K_i^i)^2 - K_{ij}K^{ij})$, while if we use ${}^3R$ to denote the Ricci scalar curvature of ${}^3g$,

$${}^{(3)}R^{12}{}_{12} + {}^{(3)}R^{23}{}_{23} + {}^{(3)}R^{31}{}_{31} = \frac{1}{2}{}^3R$$

Exercise 45. *Check the above claims.*

Using these facts, we obtain

$$G_0^0 = -\frac{1}{2}({}^3R + (K_i^i)^2 - K_{ij}K^{ij})$$

or, if we think of K as a matrix, simply

$$G_0^0 = -\frac{1}{2}({}^3R + (\operatorname{tr}K)^2 - \operatorname{tr}(K^2)).$$

It follows that Einstein's equation $G_0^0 = 8\pi\kappa T_0^0$ is a constraint relating the extrinsic curvature of any spacelike slice to its scalar curvature!

The three Einstein equations $G_i^0 = 0$ are also constraints. Take the case $i = 1$, for example. Starting with

$$G_\nu^\mu = R^{\mu\alpha}{}_{\nu\alpha} - \frac{1}{2}\delta_\nu^\mu R^{\alpha\beta}{}_{\alpha\beta}.$$

we set $\mu = 0$ and $\nu = 1$; now the second term will not contribute. We obtain

$$G_1^0 = R^{0\alpha}{}_{1\alpha} = R^0{}_{\alpha 1}{}^\alpha = R^0{}_{01}{}^0 + R^0{}_{11}{}^1 + R^0{}_{21}{}^2 + R^0{}_{31}{}^3,$$

where the first two terms vanish due to the symmetries of the Riemann tensor. Therefore
$$G^0_1 = R^0{}_{21}{}^2 + R^0{}_{31}{}^3$$
Using the Gauss equation we get
$$\begin{aligned} G^0_1 &= (^3\nabla_2 K^2_1 - {}^3\nabla_1 K^2_2) + (^3\nabla_3 K^3_1 - {}^3\nabla_1 K^3_3) \\ &= (^3\nabla_1 K^1_1 - {}^3\nabla_1 K^j_j) - (^3\nabla_1 K^1_1 - {}^3\nabla_j K^j_1) \\ &= {}^3\nabla_j K^j_1 - {}^3\nabla_1 K^j_j \end{aligned}$$

This works the same way for all the other G^0_i, so we have
$$G^0_i = {}^3\nabla_j K^j_i - {}^3\nabla_i K^j_j.$$

In other words, the 3 Einstein equations $G^0_i = 8\pi\kappa T^0_i$ are constraints on the extrinsic curvature of any spacelike slice!

It is not much harder to show that similar results hold when we drop the assumption that $\partial_0 = n$. We urge the reader to do this:

Exercise 46. *Show using the Gauss-Codazzi equations that for any choice of lapse and shift,*
$$G_{\mu\nu} n^\mu n^\nu = -\frac{1}{2}(^3R + (\mathrm{tr}K)^2 - \mathrm{tr}(K^2)),$$
and if the vectors $\partial_1, \partial_2, \partial_3$ are tangent to Σ at the point in question,
$$G_{\mu i} n^\mu = {}^3\nabla_j K^j_i - {}^3\nabla_i K^j_j.$$

The 6 remaining Einstein equations $G_{ij} = 0$ are dynamical equations describing the time evolution of ^3g. These are the equations which involve the second time derivatives of the metric. We will say more about them in the next section!

Canonical Quantization

We have touched upon one approach to quantum theory, the path-integral or Lagrangian approach. There is a complementary approach, the 'canonical' or 'Hamiltonian' approach, which is equally important.

Canonical Quantization

In this section, we briefly describe the idea of canonical quantization in the simple case of the quantum mechanics of a particle in $\mathbb{R}^n$, and then sketch how it is applied to quantum gravity. As we shall see, this approach to quantum gravity is closely connected to the ADM formalism.

Consider a classical particle in $\mathbb{R}^n$, as we did in Chapter 4 of Part II. We call $\mathbb{R}^n$ the **configuration space** of the particle, since as time t passes it traces out a path $q(t)$ in this space. Its path satisfies the Euler-Lagrange equations

$$\frac{\partial L}{\partial q^i} = \frac{d}{dt}\frac{\partial L}{\partial \dot{q}^i},$$

where the Lagrangian $L(q, \dot{q})$ is a function of the position and velocity. In the simplest case of a particle with mass m in a potential V,

$$L = \frac{1}{2}m\dot{q}^2 - V(q)$$

and the Euler-Lagrange equations give

$$m\ddot{q} = -\nabla V(q),$$

which is just $F = ma$.

There is a general recipe for starting with a Lagrangian and obtaining a formula for the Hamiltonian, or energy. First, we start with the Lagrangian and define the **momentum p_i conjugate** to the position coordinate q^i by

$$p_i = \frac{\partial L}{\partial \dot{q}^i}.$$

In the case of a particle in a potential, this gives the usual formula for momentum, $p = m\dot{q}$. In general, if we are able to solve for the velocity $\dot{q}$ in terms of the position and momentum we can define the **Hamiltonian** by

$$H(p, q) = p \cdot \dot{q} - L(q, \dot{q}).$$

For the particle in a potential, this equals $p^2/2m + V(q)$, as expected. In what follows, we assume we can solve for $\dot{q}$ in this way. However, this will *not* hold in general relativity, because that theory has constraints.

In general relativity, the 3-metric plays the role of 'position', and the time derivative of the 3-metric can be expressed in terms of its conjugate momentum *together with* the lapse and shift.

The position and momentum taken together give a point (q, p) in the **phase space** $\mathbb{R}^{2n}$. In a more general approach to classical mechanics, the configuration space can be any manifold M, and the velocity is a tangent vector to M, while the momentum is a cotangent vector. (This is why we write the components p_i with subscripts.) The phase space is then the cotangent bundle of the configuration space, and the state of a classical system is represented by a point in the phase space. The idea of the Hamiltonian approach is to convert the Euler-Lagrange equations into equations describing how the state evolves in time. To do this in the case of a particle in $\mathbb{R}^n$, we simply compute dH in two different ways: on the one hand,

$$dH = \frac{\partial H}{\partial p_i} dp_i + \frac{\partial H}{\partial q^i} dq^i,$$

but on the other hand,

$$\begin{aligned} dH &= \dot{p}_i d\dot{q}^i + \dot{q}^i dp_i - \frac{\partial L}{\partial q^i} dq^i - \frac{\partial L}{\partial \dot{q}^i} d\dot{q}^i \\ &= \dot{p}_i d\dot{q}^i + \dot{q}^i dp_i - \frac{d}{dt}\frac{\partial L}{\partial \dot{q}^i} dq^i - p_i d\dot{q}^i \\ &= \dot{q}^i dp_i - \dot{p}_i dq^i, \end{aligned}$$

where we have used the definition of momentum and the Euler-Lagrange equations. Equating these two formulas for dH, we get **Hamilton's equations**:

$$\dot{q}^i = \frac{\partial H}{\partial p_i}, \qquad \dot{p}_i = -\frac{\partial H}{\partial q^i}.$$

An elegant way to think of Hamilton's equations involves the formalism of 'Poisson brackets'. Note that the momentum coordinates p_i, the position coordinates q^i, and the Hamiltonian can all be thought of as functions on phase space. Indeed, in classical mechanics any **observable** may be regarded as a function on phase space. One can define a Lie bracket on the algebra of functions on phase space, the so-called

Canonical Quantization   427

Poisson bracket, by

$$\{f,g\} = \frac{\partial f}{\partial p_i}\frac{\partial g}{\partial q^i} - \frac{\partial f}{\partial q^i}\frac{\partial g}{\partial p_i},$$

using the Einstein summation convention as usual.

Exercise 47. *Check that $\{\cdot,\cdot\}$ satisfies the Lie algebra axioms as well as the Leibniz law $\{f,gh\} = \{f,g\}h + g\{f,h\}$.*

In terms of Poisson brackets, Hamilton's equations become

$$\dot{q}^i = \{H, q^i\}, \qquad \dot{p}_i = \{H, p_i\}.$$

Note that in going from the Lagrangian to the Hamiltonian approach, we have exchanged n 2nd-order differential equations for $2n$ first-order equations. More generally, if f is any observable, we have

$$\begin{aligned}\frac{d}{dt}f(p,q) &= \frac{\partial f}{\partial q^i}\dot{q}^i + \frac{\partial f}{\partial p_i}\dot{p}_i \\ &= \frac{\partial H}{\partial p_i}\frac{\partial f}{\partial q^i} - \frac{\partial H}{\partial q^i}\frac{\partial f}{\partial p_i} \\ &= \{H, f\},\end{aligned}$$

or $\dot{f} = \{H, f\}$ for short. In other words, the rate of change of an observable is determined by its Poisson bracket with the Hamiltonian. We say that the Hamiltonian **generates** time evolution.

Turning to the Hamiltonian approach in *quantum* mechanics, recall from Chapter 6 of Part I that the Hilbert space of states for a quantum particle on $\mathbb{R}^n$ is given by $L^2(\mathbb{R}^n)$. To quantize the particle, the idea is to replace observables that are functions on phase space with observables that are self-adjoint *operators* on $L^2(\mathbb{R}^n)$, in such a way that Poisson brackets go over to *commutators*. In other words, to each important function f on phase space we would like to associate a self-adjoint operator $\hat{f}$ on $L^2(\mathbb{R}^n)$, in such a way that if

$$\{f,g\} = k$$

then

$$[\hat{f},\hat{g}] = -i\hat{k}.$$

The factor of i is required for the operator $\hat{k}$ to be self-adjoint; also, there is really a factor of $\hbar$ in the right hand side of the last equation, but we will use units in which it equals 1. Unfortunately, it is typical in quantization that many problems and subtleties arise. In particular, it is impossible to assign operators to *all* observables in such a way that the above relation holds. Thus one should think of this prescription as an ideal to strive towards, rather than a simple recipe to follow. If one succeeds in assigning operators to observables in a satisfactory way, including the Hamiltonian, one then describes the time evolution of observables by setting

$$\hat{f}_t = e^{it\hat{H}} \hat{f} e^{-it\hat{H}},$$

so that, in analogy with classical mechanics, we have

$$\frac{d}{dt}\hat{f}_t = i[\hat{H}, \hat{f}_t].$$

Again we say that time evolution is 'generated' by the Hamiltonian.

A simple example of this quantization procedure is the **free particle**, by which we mean a particle on $\mathbb{R}^n$ in the potential $V = 0$. The most fundamental observables are the momentum, position, and the Poisson brackets of these are given by

$$\{p_j, q^k\} = \delta^k_j, \qquad \{p_j, p_k\} = \{q^j, q^k\} = 0.$$

If we define operators on $L^2(\mathbb{R}^n)$ by

$$(\hat{q}^j \psi)(x) = x^j \psi(x),$$
$$(\hat{p}_j \psi)(x) = -i\partial_j \psi(x),$$

we will obtain analogous relations for the commutators — the so-called **canonical commutation relations**:

$$[\hat{p}_j, \hat{q}^k] = -i\delta^k_j, \qquad [\hat{p}_j, \hat{p}_k] = [\hat{q}_j, \hat{q}_k] = 0.$$

Since classically we have
$$H = p^2/2m,$$
we define the Hamiltonian for the quantum free particle by
$$\hat{H} = \hat{p}^2/2m,$$
and use this to define the dynamics of the theory.

Canonical Quantization

Exercise 48. *Compute the commutator of $\hat{H}$ with $\hat{p}^j$ and $\hat{q}_j$, and compare it with the Poisson brackets of H with p^j and q_j.*

Suppose now that we attempt to apply this recipe to general relativity! This is quite complicated, so to make the essential ideas stand out clearly we will skip all the long calculations and just present the main results, all the time working by analogy with the case of a particle in $\mathbb{R}^n$. There are many subtleties and problematic issues that we will ignore, so for more details, we urge the reader to the references in the Notes.

To keep life simple, we will only consider the vacuum Einstein equation. We assume that our spacetime M is diffeomorphic to $\mathbb{R} \times S$, where S is a 3-dimensional manifold, and we fix a spacelike slice Σ. The analog of the 'position' q in this case is the 3-metric, and the analog of the configuration space $\mathbb{R}^n$ is the space $\mathrm{Met}(\Sigma)$ of all Riemannian metrics on Σ. We say that $\mathrm{Met}(\Sigma)$ is the **configuration space** for gravity. More flamboyantly, $\mathrm{Met}(\Sigma)$ is also known as **superspace**. To emphasize the analogy with the position of a particle, it is typical to write the 3-metric as q_{ij} rather than ${}^3 g_{ij}$, and we will do this. We will also write simply q for the determinant of q_{ij}.

To define the momentum conjugate to the 3-metric we will write the Einstein-Hilbert Lagrangian in terms of q_{ij} and $\dot{q}_{ij}$. Of course, to define time derivatives we need to proceed as in the previous section by fixing a diffeomorphism between M and $\mathbb{R} \times S$, thus obtaining a time coordinate τ and vector field ∂_τ. We take $\Sigma = \{\tau = 0\}$, and work in local coordinates such that $\partial_0 = \partial_\tau$, and such that $\partial_1, \partial_2, \partial_3$ are tangent to Σ. If we do this, it turns out that $\dot{q}_{ij}$ is closely related to the extrinsic curvature! Namely, one can show that

$$K_{ij} = \frac{1}{2} \, N^{-1} \, (\dot{q}_{ij} - {}^3\nabla_i N_j - {}^3\nabla_j N_i).$$

The Lagrangian is given by $R \, (-\det g)^{1/2} \, d^4x$, but we will factor out the form d^4x and instead work with the function

$$\mathcal{L} = R \sqrt{-\det g}.$$

In terms of the 3-metric and lapse function this is given by

$$\mathcal{L} = q^{1/2} \, N R,$$

and if one expresses R in terms of the 3-metric and extrinsic curvature, discarding terms that give total divergences (since these would integrate out to zero, at least when Σ is compact), one obtains

$$\mathcal{L} = q^{1/2} N ({}^3R + \text{tr}(K^2) - (\text{tr}K)^2).$$

From this version of the Lagrangian and the relation between K_{ij} and $\dot{q}_{ij}$ one can obtain the momentum conjugate to q_{ij}:

$$p^{ij} = \frac{\partial \mathcal{L}}{\partial \dot{q}_{ij}},$$

and it works out to be

$$p^{ij} = q^{1/2}(K^{ij} - \text{tr}(K)q^{ij}).$$

Then one can work out the Hamiltonian — or more precisely, Hamiltonian density — by the formula

$$\mathcal{H}(p^{ij}, q_{ij}) = p_{ij}\dot{q}^{ij} - \mathcal{L}.$$

The integral of this quantity over Σ is the Hamiltonian for general relativity

$$H = \int_\Sigma \mathcal{H}\, d^3x,$$

in the case of a compact space S. (In the noncompact case one cannot throw out total divergences as we have been doing.) If one computes $\mathcal{H}$, discarding total divergences again, one obtains

$$\mathcal{H} = q^{1/2}(NC + N^i C_i)$$

where

$$C = -{}^3R + q^{-1}(\text{tr}(p^2) - \frac{1}{2}\text{tr}(p)^2)$$

and

$$C_i = -2\,{}^3\nabla^j(q^{-1/2}\, p_{ij}).$$

The fact that the Hamiltonian involves terms proportional to the lapse and shift should not be surprising, since the role of the Hamiltonian is to generate time evolution, and in general relativity we need

Canonical Quantization

to specify the lapse and shift to know the meaning of time evolution! There is, however, something much more surprising to be seen here. If we express the quantities C and C_i in terms of the extrinsic curvature using our formula for p^{ij}, we find that

$$C = -2G_{\mu\nu}n^\mu n^\nu$$

and

$$C_i = -2G_{\mu i}n^\mu.$$

Exercise 49. *Check these equations using Exercise 46.*

This implies that the Hamiltonian density for general relativity must *vanish* by the vacuum Einstein equation! In other words, Einstein's equation imply

$$H = 0.$$

This fact seems rather puzzling at first. A theory with Hamiltonian equal to zero might seem to be completely trivial, yet the dynamics of general relativity is very interesting. How can this be?

The key is that the equations

$$C = C_i = 0$$

are precisely the 4 Einstein equations that are *constraints* on the initial data. We said the configuration space of general relativity is Met(Σ). It is natural then to expect that the phase space is the space of all pairs (q_{ij}, p^{ij}), or the cotangent bundle $T^*\text{Met}(\Sigma)$. However, not all points of this phase space represent allowed states! The Einstein equations that are constraints must be satisfied, and this restriction picks out a subspace of the phase space called the **physical phase space**:

$$X = \{C = C_i = 0\} \subset T^*\text{Met}(\Sigma).$$

The Hamiltonian vanishes on this subspace. However, as we shall see, Hamilton's equations still give nontrivial dynamics.

To formulate Hamilton's equations, we can formally define the Poisson bracket of two functions on phase space by

$$\{f,g\} = \int_\Sigma \left\{ \frac{\partial f}{\partial p^{ij}(x)} \frac{\partial g}{\partial q_{ij}(x)} - \frac{\partial f}{\partial q_{ij}(x)} \frac{\partial g}{\partial p^{ij}(x)} \right\} q^{1/2} d^3x.$$

Here the derivatives on the right are called **functional derivatives**. These can be a bit confusing at first, but they are really just another way of thinking about the concept of 'variation' we have been using. Suppose, for example, that f is a function on $\text{Met}(\Sigma)$. Then we write

$$\frac{\partial f}{\partial q_{ij}(x)}$$

for the function on $\text{Met}(\Sigma)$, if it exists, such that

$$\int_\Sigma h_{ij}(x) \frac{\partial f}{\partial q_{ij}(x)} q^{1/2} d^3x = \frac{d}{ds} f(q+sh)\Big|_{s=0}.$$

for every symmetric $(0,2)$ tensor field h. If we wrote h as δg, we would also write the right-hand side of this equation as δf. The case of functions on $T^*(\text{Met}(\Sigma))$ is similar. We can use the formula for the Poisson brackets to compute the brackets of the p_{ij} and q^{ij}, obtaining formulas analogous to those for the particle in $\mathbb{R}^n$:

$$\{p^{ij}(x), q_{kl}(y)\} = (\delta^i_k \delta^j_l + \delta^i_l \delta^j_k)\ \delta^{(3)}(x-y)$$
$$\{p^{ij}(x), p^{kl}(y)\} = 0$$
$$\{q_{ij}(x), q_{kl}(y)\} = 0.$$

We can also obtain the evolutionary part of Einstein's equations by this means. These are really just the equations $G_{ij} = 0$ in disguise, which are equations for the second time derivative of the 3-metric, but rewritten so as to give twice as many first-order equations. In brief, they are just

$$\dot{q}^{ij} = \{H, q^{ij}\}, \qquad \dot{p}_{ij} = \{H, p_{ij}\}.$$

However, one can also work out the brackets explicitly, obtaining the rather terrifying equations

$$\dot{q}_{ij} = 2q^{-1/2} N(p_{ij} - \frac{1}{2} p^k_k q_{ij}) + 2\,^3\nabla_{[i}\, N_{j]},$$

$$\dot{p}^{ij} = -Nq^{1/2}(^3R^{ij} - \frac{1}{2}\,^3R\, q^{ij}) + \frac{1}{2} N q^{-1/2} q^{ij}(p_{ab}p^{ab} - \frac{1}{2}(p^a_a)^2)$$
$$- 2Nq^{-1/2}(p^{ia}\, p^j_a - \frac{1}{2} p^a_a p^{ij}) + q^{1/2}(\nabla^i \nabla^j N - q^{ij}\,^3\nabla^a\,^3\nabla_a N)$$
$$+ q^{1/2}\nabla_a(q^{-1/2} N^a p^{ij}) - 2p^{a[i}\,^3\nabla_a N^{j]}$$

Canonical Quantization

which we present solely to impress the reader. The point is that, even on the physical phase space X where $H = 0$, the time evolution given by Hamilton's equations is nontrivial.

Recall that the lapse and shift measure how much time evolution pushes the slice Σ in the *normal* direction and the *tangent* direction, respectively. In particular, if we set the shift equal to zero, the Hamiltonian for general relativity is equal to

$$C(N) = \int_\Sigma NC\, q^{1/2} d^3x,$$

and it generates time evolution in a manner that corresponds to pushing Σ forwards in the normal direction. On the other hand, if we set the lapse equal to zero, the Hamiltonian becomes

$$C(\vec{N}) = \int_\Sigma N^i C_i\, q^{1/2} d^3x,$$

which generates a funny sort of 'time evolution' that pushes Σ in a direction tangent to itself. More precisely, this quantity generates transformations of X corresponding to the flow on Σ generated by $\vec{N}$. This flow is a 1-parameter family of diffeomorphisms of Σ. For this reason, $C(\vec{N})$ or C_i is called the **diffeomorphism constraint**, while $C(N)$ or C is called the **Hamiltonian constraint**. It is actually no coincidence that C and C_i play a dual role as both constraints and terms in the Hamiltonian. This is, in fact, a crucial special feature of field theories with no fixed background structures!

It is interesting to see the Poisson brackets of the Hamiltonian and diffeomorphism constraints. The formulas are nicer if we use the integrated versions $C(N)$ and $C(\vec{N})$; one obtains:

$$\begin{aligned}
\{C(\vec{N}), C(\vec{N}')\} &= C([\vec{N}, \vec{N}']) \\
\{C(\vec{N}), C(N)\} &= C(\vec{N} N') \\
\{C(N), C(N')\} &= C((N\partial^i N' - N'\partial^i N)\partial_i).
\end{aligned}$$

where $\vec{N} N'$ is simply the derivative of the function N' in the direction $\vec{N}$, and $(N\partial^i N' - N'\partial^i N)\partial_i$ is the result of converting the 1-form $NdN' - N'dN$ into a vector field by raising indices. The relations above are known as the **Dirac algebra**. Note that the constraints are closed under taking Poisson brackets; that is, the bracket of any two constraints is again a constraint.

Now let us see what happens if we try to *quantize* gravity using the Hamiltonian approach. At this point, we warn the reader that quantum gravity is still poorly understood, and that we will be sketching a program that people *hoped* would lead to a theory of quantum gravity, but which is ridden with difficulties. First we need to define a Hilbert space for the theory. The natural choice is to use L^2 of the configuration space. However, Met(Σ) is infinite-dimensional, so it is unclear what a square-integrable function on this space would be! This is a common problem when doing canonical quantization of a field theory, and there are occasionally ways around it, so rather than giving up we will pretend we know how to make sense of $L^2(\text{Met}(\Sigma))$. Next, we need to find operators corresponding to 3-metric and its conjugate momentum. By analogy with the case of the particle in $\mathbb{R}^n$, we take the operator corresponding to the 3-metric to be

$$(\hat{q}_{ij}(x)\psi)(q) = g_{ij}(x)\,\psi(q),$$

where $g \in \text{Met}(\Sigma)$ is a 3-metric and x is any point of Σ. Similarly, we define the momentum operator to be the functional derivative

$$(\hat{p}^{ij}(x)\psi)(q) = -i\frac{\partial}{\partial q_{ij}(x)}\psi(q).$$

These operators satisfy the canonical commutation relations

$$[\hat{p}^{ij}(x), \hat{q}_{kl}(y)] = -i(\delta_k^i\,\delta_l^j + \delta_l^i\,\delta_k^j)\,\delta^{(3)}(x,y)$$
$$[\hat{p}^{ij}(x), \hat{p}^{kl}(y)] = 0$$
$$[\hat{q}_{ij}(x), \hat{q}_{kl}(y)] = 0$$

Next let us try to quantize the Hamiltonian. For this, we take the formulas for the Hamiltonian and diffeomorphism constraints and replace q_{ij} and p^{ij} wherever they appear with the operators $\hat{q}_{ij}$ and $\hat{p}^{ij}$, to obtain quantum versions $\hat{C}$ and $\hat{C}_i$ of the constraints as operators on $L^2(\text{Met}(\Sigma))$. To do this, we must confront **operator ordering problems**: since $\hat{q}_{ij}$ and $\hat{p}^{ij}$ do not commute, different ways of writing down the classical formulas for the constraints yield different operators!

Canonical Quantization

A 'good' operator ordering would make the quantum constraints satisfy commutation relations analogous to the classical ones:

$$[\hat{C}(\vec{N}), \hat{C}(\vec{N}')] = -i\hat{C}([\vec{N}, \vec{N}'])$$
$$[\hat{C}(\vec{N}), \hat{C}(N')] = -i\hat{C}(\vec{N} N')$$
$$[\hat{C}(N), \hat{C}(N')] = -i\hat{C}((N\partial^i N' - N'\partial^i N)\partial_i).$$

Unfortunately, this seems well-nigh impossible to achieve. Particularly problematic is the fact that the constraints involve $q^{1/2}$, the square root of the determinant of the metric. Operator-ordering problems are notoriously tricky when one deals with operators that are not polynomials in the basic position and momentum operators.

Suppose, nonetheless, that we succeeded in obtaining operators $\hat{C}$ and $\hat{C}_i$ that we were happy with. We could then write down the Hamiltonian for the quantum theory as

$$\hat{H} = \int_\Sigma (N\hat{C} + N^i \hat{C}_i) q^{1/2} d^3 x.$$

Classically, the Hamiltonian vanishes on the physical phase space X because of the 4 Einstein equations that serve as constraints. How are we to deal with these constraints in the quantum theory? The subject of constraints in quantum theory is a profoundly vexed and complicated one, but there is an approach due to Dirac which goes as follows (in a very simplified form). We say that a vector $\psi \in L^2(\text{Met}(\Sigma))$ is a **physical** state if it satisfies the constraints in their quantum form:

$$\hat{C}(N)\psi = \hat{C}(\vec{N})\psi = 0$$

for all $N, \vec{N}$. Alternatively, we may require that the **Wheeler-DeWitt equation**

$$\hat{H}\psi = 0,$$

hold for all choices of lapse and shift.

At this point, the program of canonical quantization runs into difficulties too severe to ignore. There are many subtle technical issues we cannot go into, but we should note three basic problems that are absolutely devastating. First, nobody has ever found any solutions of the Wheeler-DeWitt equation posed in this form! Certain expressions can be written down that appear to be solutions at a formal level, but it is very difficult to make sense of them.

Second, if one were to find physical states, they would span some vector space, the **physical state space**,

$$\mathcal{H}_{phys} = \{\psi \colon \forall N, \vec{N} \ \hat{C}(N)\psi = \hat{C}(\vec{N})\psi = 0\}.$$

There is, however, no reason to expect the physically relevant inner product on $\mathcal{H}_{phys}$ to agree with the inner product in $L^2(\mathrm{Met}(\Sigma))$ (which, as we noted, is not easy to define in the first place). This is called the **inner product problem**. It appears that one should determine the correct inner product by requiring that observables come out to be self-adjoint operators. Unfortunately, this leads us into the jaws of the third problem! The Hamiltonian vanishes on $\mathcal{H}_{phys}$, so any operator A on $\mathcal{H}_{phys}$ automatically commutes with the Hamiltonian. It follows that such operators correspond to observables that do not change with time:

$$\frac{d}{dt} A_t = i[\hat{H}, A_t] = 0.$$

Where has the dynamics of the theory gone? It turns out that the states in $\mathcal{H}_{phys}$ do not describes states of quantum gravity *at a particular time* the way the pair $({}^3g, K)$ does in classical gravity, rather, they describe states *for all time*, or more precisely, just that information about the state that is invariant under all spacetime diffeomorphisms. This is the famous **problem of time** in quantum gravity. We are not used to doing physics in a manifestly diffeomorphism-invariant way, so we do not know any candidate observables that should be represented as operators on $\mathcal{H}_{phys}$.

These problems held up progress in canonical quantum gravity for many years. Indeed, the inner product problem and the problem of time are just as frustrating as ever, so there may well be something fundamentally misguided about the whole project. However, there is new hope in canonical quantum gravity these days, because some solutions of the Wheeler-DeWitt equation *have* been found in terms of the 'new variables' for general relativity, and also in terms of the 'loop representation'. In the next chapter, we briefly sketch how this works. As we shall see, knot theory makes an interesting appearance here!

Chapter 5
The New Variables

Electromagnetism is, as we have seen, a gauge field. That gravitation is a gauge field is universally accepted, although exactly how it is a gauge field is a matter still to be clarified. — C. N. Yang

As we saw in the last chapter, the dynamics of general relativity is generated by constraints. That is, the 6 Einstein equations describing time evolution can be obtained by calculating the Poisson brackets of the 3-metric and its conjugate momentum with a Hamiltonian that is a linear combination of constraints: quantities that must vanish on the physical phase space by the other 4 Einstein equations. While there are profound conceptual problems associated with quantizing gravity (and, we may hope, unifying it with the other forces), many of the *technical* problems in canonical quantum gravity revolve around these constraints. In the early 1980s, Abhay Ashtekar and others developed 'new variables' to describe general relativity, in terms of which the constraints radically simplify. This eases the otherwise intractable factor-ordering problems one runs into when one tries to turn the constraints into operators in the process of canonical quantization. These new variables also bring the mathematical structure of general relativity much closer to that of Yang-Mills theory. As a result, one can apply techniques from gauge theory to quantum gravity. In particular, as we shall see, one can use Chern-Simons theory to obtain a solution of the Wheeler-DeWitt equation,

$$\hat{H}\psi = 0,$$

in the case of quantum gravity with a nonzero cosmological constant. The physical significance — if any — of this 'Chern-Simons state' is still under investigation, as are its mathematical properties, but at least now we have a solution to study!

We have alluded already to the 'loop representation' of a gauge theory, in which the key observables are Wilson loops. Using the new variables, Lee Smolin and Carlo Rovelli were able to devise a loop representation of quantum gravity in the late 1980s. In this representation, states of quantum gravity correspond to isotopy invariants of 'generalized links', that is, collections of loops (not necessarily embedded) in space. This has led to the discovery of intriguing relationships between quantum gravity and knot theory. The Chern-Simons state, for example, corresponds to the Kauffman bracket link invariant. It appears that these ideas can be used to construct many more solutions of the Wheeler-DeWitt equation, as well. It seems, therefore, that gauge fields, knots and gravity are different facets of a single subject! Research on these topics is extremely active, and the reader will need to study the references in the notes, as well as the stream of new papers on the subject, in order to learn the state of the art. By the same token, the reader should take with a grain of salt anything we write about quantum gravity (as opposed to, for example, the new variables in classical general relativity), since today's conventional wisdom could easily be overthrown tomorrow.

As Yang noted in the quotation above, while general relativity is clearly a gauge theory in some sense, the precise relation between general relativity and other gauge theories is not a simple matter. Certainly the notions of *connection* and *curvature* are crucial in general relativity, but in the original Einstein-Hilbert formulation they are derived from a more basic entity: the *metric*. For many years people have tried to invent formulations of general relativity in which the connection plays a more fundamental role and the metric is de-emphasized.

In the Palatini formalism, for example, the metric is a secondary concept, the basic fields being a Lorentz connection on the 'imitation tangent bundle' $M \times \mathbb{R}^n$ and a frame field $e: M \times \mathbb{R}^n \to TM$. It is interesting to see what happens when one attempts to canonically quantize gravity using the Palatini formalism. Since there are more variables one expects more constraints. Indeed, one finds that in ad-

The New Variables

dition to the Hamiltonian and diffeomorphism constraints, there is a Gauss law constraint analogous to that in electromagnetism and Yang-Mills theory. The form of these constraints is much simpler than in the Einstein-Hilbert approach we discussed in the last chapter. In particular, one can write the constraints so that they are polynomials (and spatial derivatives thereof) in terms of fields satisfying the canonical commutation relations. Unfortunately, the constraints are not closed under Poisson brackets. This complicates the quantization of the theory in such a way that the Palatini formalism is little better than the Einstein-Hilbert one for the purposes of quantum gravity.

The 'new variables' can be thought of as a modification of the Palatini formalism that avoids this problem. The main idea is to take advantage of the special features of 4-dimensional spacetime and work with the 'self-dual part' of the Lorentz connection. Let us explain this notion! In what follows we will speak as physicists and not worry about the difference between a connection and its vector potential.

As with Maxwell's equations, using self-duality in gravity when the metric is Lorentzian requires working with complex-valued fields. Thus we define the **complexified tangent bundle** of M, written $\mathbb{C}TM$, to be the vector bundle whose fiber at each point $p \in M$ is the vector space $\mathbb{C} \otimes T_p M$ consisting of complex linear combinations of tangent vectors. There is also an 'imitation' complexified tangent bundle, namely the trivial bundle $M \times \mathbb{C}^4$. A **complex frame field** is then a vector bundle isomorphism $e: M \times \mathbb{C}^4 \to \mathbb{C}TM$.

We define the internal metric η on $M \times \mathbb{C}^4$ by the same formula as for $M \times \mathbb{R}^4$. This allows us to raise and lower internal indices. A connection A on $M \times \mathbb{C}^4$ is an $\text{End}(\mathbb{C}^4)$-valued 1-form on M. Its components are written $A_{\alpha J}^I$, where α is a spacetime index and I, J are internal indices. Alternatively, we can raise an index and think of the connection as having components A_α^{IJ}. We say that A is a **Lorentz connection** if $A_\alpha^{IJ} = -A_\alpha^{JI}$. Because of this antisymmetry property, we can think of a Lorentz connection as a $\Lambda^2 \mathbb{C}^4$-valued 1-form. Recall that the Hodge star operator maps 2-forms to 2-forms in 4 dimensions, which is the basis of duality symmetry. There is an analogous **internal Hodge star operator** mapping $\Lambda^2 \mathbb{C}^4$ to itself: denoting it by $*$, it is

given by
$$*T^{IJ} = \frac{1}{2}\epsilon^{IJ}{}_{KL}T^{KL}$$
for any quantity with two antisymmetric raised internal indices, by analogy with the formula for the usual Hodge star operator (see Exercise 69 in Chapter 5 of Part I). In particular, we can define the internal Hodge dual of a Lorentz connection by
$$(*A)^{IJ}_\alpha = \frac{1}{2}\epsilon^{IJ}{}_{KL}A^{KL}_\alpha,$$
and we can write any Lorentz connection A as a sum of self-dual and anti-self-dual parts:
$$A = {}^+A + {}^-A, \qquad *{}^\pm A = \pm i\, {}^\pm A.$$
Explicitly, we have
$$^\pm A = (A \mp i * A)/2.$$

In the self-dual formulation of general relativity, one of the two basic fields is a **self-dual** Lorentz connection, that is, a Lorentz connection ^+A on $M \times \mathbb{C}^4$ with
$$*{}^+A = i\, {}^+A.$$
The other basic field is a complex frame field $e\colon M \times \mathbb{C}^4 \to \mathbb{C}TM$. The action in the self-dual formulation is built using the curvature of the self-dual Lorentz connection, which is written ^+F and given by
$$^+F^{IJ}_{\alpha\beta} = \partial_\alpha\, {}^+A^{IJ}_\beta - \partial_\beta\, {}^+A^{IJ}_\alpha + [{}^+A_\alpha, {}^+A_\beta]^{IJ}.$$
As in the Palatini formalism, one can use the frame field to define a metric g on M by
$$g_{\alpha\beta} = \eta_{IJ}e^I_\alpha e^J_\beta,$$
where the coefficients e^I_α are defined using the inverse frame field:
$$e^{-1}\partial_\alpha = e^I_\alpha \xi_I.$$
However, because the frame field is complex, the metric g is now *complex*. The **self-dual action** is given by
$$S_{SD}({}^+A, e) = \int_M e^\alpha_I\, e^\beta_J\, {}^+F^{IJ}_{\alpha\beta}\, \mathrm{vol},$$

The New Variables

where, as in the Palatini formalism, the volume form is given by

$$\text{vol} = \sqrt{-\det g}\, d^4x,$$

g being defined in terms of e as above.

Before discussing the equations one gets from the self-dual action, we should say a bit more about self-duality. One can define the internal Hodge dual of the curvature of a connection on $M \times \mathbb{C}^4$ as follows:

$$(*F)^{IJ}_{\alpha\beta} = \frac{1}{2}\epsilon^{IJ}{}_{KL} F^{KL}_{\alpha\beta},$$

and one says the curvature is **self-dual** if

$$*F = iF.$$

It turns out, quite pleasantly, that the curvature of a self-dual Lorentz connection is self-dual! One can check this fact directly with a computation:

Exercise 50. *Check by a computation in local coordinates that the curvature of a self-dual Lorentz connection on $M \times \mathbb{C}^4$ is self-dual.*

However, this fact is so important the we should explain the deep underlying reason for it. Given a real Lie algebra $\mathfrak{g}$ — one which is a vector space over the *real* numbers — we can make the vector space $\mathfrak{g} \otimes \mathbb{C}$ into a Lie algebra by defining, for any $x, y \in \mathfrak{g}$ and $\alpha, \beta \in \mathbb{C}$,

$$[x \otimes \alpha, y \otimes \beta] = [x, y] \otimes \alpha\beta.$$

The Lie algebra $\mathfrak{g} \otimes \mathbb{C}$ is called the **complexification** of $\mathfrak{g}$. Now, every complex Lie algebra can be thought of as a real Lie algebra if we ignore our ability to multiply its elements by imaginary numbers. If $\mathfrak{g}$ came from a complex Lie algebra in this manner, it turns out that $\mathfrak{g} \otimes \mathbb{C}$ is isomorphic to the direct sum of two copies of $\mathfrak{g}$, which we can think of as 'right-handed' and 'left-handed' copies:

Exercise 51. *Show that the complexification of a real Lie algebra $\mathfrak{g}$ is a complex Lie algebra. If $\mathfrak{g}$ comes from a complex Lie algebra as described above, show that*

$$\mathfrak{g}_\pm = \{x \otimes 1 \pm ix \otimes i \colon x \in \mathfrak{g}\}$$

are Lie subalgebras of $\mathfrak{g} \otimes \mathbb{C}$ that are isomorphic as Lie algebras to $\mathfrak{g}$, and that $\mathfrak{g} \otimes \mathbb{C}$ is the direct sum of the Lie algebras $\mathfrak{g}_\pm$.

Now, a Lorentz connection on $M \times \mathbb{C}^4$ is basically just an $\mathfrak{so}(3,1) \otimes \mathbb{C}$-valued 1-form. However, in Exercise 31 of Part II, the hard-working reader showed that $SL(2,\mathbb{C})$ was a double cover of $SO_0(3,1)$. This implies that Lie algebras $\mathfrak{so}(3,1)$ and $\mathfrak{sl}(2,\mathbb{C})$ are isomorphic. It follows that $\mathfrak{so}(3,1) \otimes \mathbb{C}$ is isomorphic to $\mathfrak{sl}(2,\mathbb{C}) \otimes \mathbb{C}$. As a consequence, $\mathfrak{so}(3,1) \otimes \mathbb{C}$ is a direct sum of two Lie subalgebras isomorphic to $\mathfrak{sl}(2,\mathbb{C})$, called the 'self-dual' and 'anti-self-dual' parts. The self-dual part of a Lorentz connection is simply the part having components in the self-dual Lie subalgebra. Since a Lie subalgebra is closed under taking brackets, if the connection ^+A is self-dual, then so is its curvature ^+F. From this point of view, it is the existence of the double cover $\rho: SL(2,\mathbb{C}) \to SO_0(3,1)$ that makes self-duality so useful in 4-dimensional gravity! The splitting of $\Lambda^2 \mathbb{C}^4$ into self-dual and anti-self-dual parts corresponds to the splitting of $\mathfrak{so}(3,1) \otimes \mathbb{C}$ into two Lie subalgebras isomorphic to $\mathfrak{sl}(2,\mathbb{C})$.

Now let us see what happens when one computes the variation of the self-dual action and demands that

$$\delta S_{SD} = 0.$$

Since the action is similar to the Palatini action, the computation proceeds in a similar manner. One obtains two equations. First, by varying the self-dual connection, one obtains an equation saying that ^+A is the self-dual part of a Lorentz connection A on $M \times \mathbb{C}^4$ for which the corresponding imitation Christoffel symbols

$$\tilde{\Gamma}^{\gamma}_{\alpha\beta} = A^J_{\alpha I} e^I_{\beta} e^{\gamma}_J$$

equal the Christoffel symbols of the metric g. This implies that the self-dual part of the Riemann tensor of g,

$$^+R^{\alpha}{}_{\beta\gamma}{}^{\delta} = \frac{1}{2}(R^{\alpha}{}_{\beta\gamma}{}^{\delta} - \frac{i}{2}\epsilon^{\alpha\delta}{}_{\mu\nu}R^{\mu}{}_{\beta\gamma}{}^{\nu}),$$

is related to ^+F by the frame field as follows:

$$^+R^{\alpha}{}_{\beta\gamma}{}^{\delta} = {}^+F^{IJ}_{\beta\gamma} e^{\alpha}_I e^{\delta}_J.$$

Second, by varying the frame field, one obtains a self-dual analog of Einstein's equation:

$$^+R_{\alpha\beta} - \frac{1}{2}g_{\alpha\beta} {}^+R = 0$$

The New Variables 443

where
$$^+R_{\alpha\beta} = {}^+R^\lambda{}_{\alpha\lambda\beta}, \qquad ^+R = {}^+R^\alpha_\alpha.$$

However, using symmetries of the Riemann tensor, this analog of Einstein's equation is equivalent to the vacuum Einstein equation!

Exercise 52. *Check the computations above and show that $^+R_{\alpha\beta} - \frac{1}{2}{}^+Rg_{\alpha\beta} = 0$ implies the vacuum Einstein equation $R_{\alpha\beta} = 0$.*

Note, however, that we have recovered the vacuum Einstein equation for *complex* metrics on the spacetime M. To obtain ordinary general relativity, we need to impose **reality conditions** on the complex frame field that make g real-valued. Dealing with the reality conditions requires extra work, both in the classical and in the quantum theory. There are a number of strategies for handling them — indeed, this is a somewhat controversial aspect of the theory — but we will not go into these here. Instead, let us sketch the Hamiltonian formalism that goes along with the self-dual action.

Let Σ be a spacelike slice of the manifold $\mathbb{R} \times S$, and for simplicity let us work in coordinates such that ∂_0 is normal to Σ, while ∂_i is tangent to Σ for the spacelike indices. Given a self-dual Lorentz connection $^+A^{IJ}_\alpha$ on $M \times \mathbb{C}^4$, we can restrict it to a connection A^{IJ}_i on $\Sigma \times \mathbb{C}^4$. (It is customary to leave out the plus sign here.) This connection A^{IJ}_i still satisfies
$$A^{IJ}_i = -A^{JI}_i, \qquad *A = iA,$$
so we will still call it a self-dual Lorentz connection. Since $\mathfrak{sl}(2,\mathbb{C})$ has a basis in terms of Pauli matrices, we can also write this self-dual Lorentz connection as $-\frac{i}{2}A^a_i \sigma_a$, where we use letters such as a, b, c, to denote indices running from 1 to 3, associated to the Pauli matrices. We raise and lower these indices using the Kronecker deltas δ_{ab} and δ^{ab}.

The configuration space for the 'new variables' formulation of general relativity is the space $\mathcal{A}$ of all self-dual Lorentz connections on $\Sigma \times \mathbb{C}^4$. In other words, the field playing the role analogous to the 'position' in classical mechanics is now the self-dual Lorentz connection A^a_i on space, rather than the 3-metric. The momentum conjugate to A^a_i turns out to be
$$\tilde{E}^i_a = q^{1/2} e^i_a$$

where q is the determinant of the 3-metric, and we have the Poisson bracket relations:

$$\{\tilde{E}_a^i(x), A_j^b(y)\} = -i\delta_a^b \delta_j^i \delta^{(3)}(x,y)$$
$$\{\tilde{E}_a^i(x), \tilde{E}_b^j(y)\} = 0$$
$$\{A_i^a(x), A_j^b(y)\} = 0$$

In terms of the new variables, the Hamiltonian and diffeomorphism constraints are given by

$$\tilde{C} = \epsilon^{abc} \tilde{E}_a^i \tilde{E}_b^j F_{ijc}$$
$$C_j = \tilde{E}_a^k F_{jk}^a,$$

where F_{ij}^a, an $\mathfrak{so}(3) \otimes \mathbb{C}$-valued 2-form, is the curvature of the connection A on Σ. We write the Hamiltonian constraint with a tilde because it is **densitized**, that is, it equals $q^{1/2}$ times that of the one in the previous chapter. Also, we have suppressed some constant factors which clutter things up. There is also a Gauss law constraint given by

$$G_a = D_i \tilde{E}_a^i,$$

where D is the connection corresponding to the vector potential A.

While we will not go into the details, it is absolutely crucial that these constraints are closed under taking Poisson brackets. It is also important to note the relationship between general relativity and Yang-Mills theory that follows from the above facts. The field $\tilde{E}_a^i$ plays the part of the electric field in Yang-Mills theory, while F_{ij}^a plays the part of the magnetic field. The Gauss law $G_a = 0$ for gravity in the new variables format is identical to that for the Yang-Mills equations, but the Hamiltonian and diffeomorphism constraints are new, as is the fact that time evolution is generated by constraints.

Now let us turn to the quantization of gravity in the 'new variables' formalism. In the quantum theory we would expect states to be vectors in the space $L^2(\mathcal{A})$ of all square-integrable functions on $\mathcal{A}$. Of course, until we have a working theory of the 'Lebesgue measure' $\mathcal{D}A$ on $\mathcal{A}$, this L^2 space is purely formal. Thus we will ignore the condition of square-integrability for now and think of states as arbitrary functions ψ on

The New Variables

$\mathcal{A}$. To quantize, we replace the classical 'position' and 'momentum' variables A_i^a and $\tilde{E}_a^i$ by the operators

$$(\hat{A}_i^a(x)\psi)(A) = A_i^a(x)\,\psi(A)$$

and

$$(\hat{E}_a^i(x)\psi)(A) = \frac{\partial}{\partial A_i^a(x)}\psi(A),$$

which have commutation relations

$$\begin{aligned} [\hat{E}_a^i(x), \hat{A}_j^b(y)] &= \delta_a^b \delta_j^i \, \delta^{(3)}(x,y) \\ [\hat{E}_a^i(x), \hat{E}_b^j(y)] &= 0 \\ [\hat{A}_i^a(x), \hat{A}_j^b(y)] &= 0 \end{aligned}$$

analogous to the classical Poisson bracket relations. With these operators in hand we can then make the Hamiltonian, diffeomorphism and Gauss law constraints into operators. There are a number of operator orderings to choose from, but a convenient choice is the following:

$$\begin{aligned} \hat{C} &= \epsilon^{abc} \hat{E}_a^i \hat{E}_b^j \hat{F}_{ijc} \\ \hat{C}_j &= \hat{E}_a^k \hat{F}_{jk}^a \\ \hat{G}_a &= \hat{D}_i \hat{E}_a^i, \end{aligned}$$

where

$$(\hat{F}_{jk}^a(x)\psi)(A) = F_{jk}^a(x)\,\psi(A).$$

While there is some controversy about this, owing in part to the formal nature of mathematics involved, it seems that these operators satisfy commutation relations analogous to the Poisson brackets of the classical constraints. As in the previous chapter, we define the **physical state space** $\mathcal{H}_{phys}$ to be the space of functions ψ on $\mathcal{A}$ that satisfy the constraints in quantum form. In other words:

$$\mathcal{H}_{phys} = \{\psi \colon \hat{C}\psi = \hat{C}_j\psi = \hat{G}_a\psi = 0\}.$$

The problem, then, is to find functions ψ in $\mathcal{H}_{phys}$.

What do the constraint equations mean? The diffeomorphism constraint generates flows on the slice Σ, so

$$\hat{C}_j\psi = 0$$

is really saying that $\psi(A) = \psi(A')$ whenever A' is obtained from A by applying a diffeomorphism that is connected to the identity by a flow. Similarly, the Gauss law constraint turns out to generate gauge transformations, so that

$$\hat{G}_a \psi = 0$$

says that $\psi(A) = \psi(A')$ whenever A' is obtained from A by a small gauge transformation. How about the Hamiltonian constraint? This is the one that really encodes the *4-dimensional* diffeomorphism invariance of general relativity, and all the dynamics of the theory lurk within it. Precisely for this reason, it is difficult to find explicit functions ψ on $\mathcal{A}$ for which

$$\hat{C}\psi = 0$$

holds.

Now, finally, we come to the fascinating relationship between Chern-Simons theory and quantum gravity! The point is that Chern-Simons theory gives rise to a solution of all three constraint equations, *if* we work with a version of quantum gravity in which the cosmological constant, Λ, is nonzero. If the cosmological constant Λ is nonzero, the Gauss law and diffeomorphism constraints are unchanged, but the Hamiltonian constraint becomes

$$\hat{C} = \epsilon^{abc} \hat{E}^i_a \hat{E}^j_b \hat{F}_{ijc} - \frac{\Lambda}{6} \epsilon_{ijk} \epsilon^{abc} \hat{E}^i_a \hat{E}^j_b \hat{E}^k_c.$$

We define the **Chern-Simons state** Ψ_{CS} to be the following function on $\mathcal{A}$:

$$\Psi_{CS}(A) = e^{-\frac{2}{\Lambda} S_{CS}(A)}$$

where as in Part II,

$$S_{CS(A)} = \int_\Sigma \operatorname{tr}(A \wedge dA + \frac{2}{3} A \wedge A \wedge A).$$

We claim that Ψ_{CS} lies in the physical state space for quantum gravity with cosmological constant:

$$\hat{C}_j \Psi_{CS} = \hat{G}_a \Psi_{CS} = \hat{C} \Psi_{CS} = 0.$$

The New Variables

Two of these three equations are obvious! Since the Gauss law constraint generates gauge transformations, and the Chern-Simons action is invariant under small gauge transformations, we have

$$\hat{G}_a \Psi_{CS} = 0.$$

Since the diffeomorphism constraint generates diffeomorphisms of Σ, and the Chern-Simons action is preserved by diffeomorphisms that are connected to the identity, we also have

$$\hat{C}_j \Psi_{CS} = 0.$$

The fact that $\hat{C}\Psi_{CS} = 0$ is not so obvious. Why should the Chern-Simons state, a creature of 3 dimensions, satisfy an equation that expresses 4-dimensional diffeomorphism invariance? The ultimate explanation for this puzzle appears to be the relation between the Chern-Simons class and the 2nd Chern class, which lives in 4 dimensions. We will not go into this, however; instead, we simply offer a direct argument to show that $\hat{C}\Psi_{CS} = 0$.

For this, recall that in Chapter 4 of Part II we showed that

$$\delta S_{CS} = 2 \int_\Sigma \text{tr}(F \wedge \delta A).$$

In coordinate notation we have $F = \frac{1}{2} F_{ij} dx^i \wedge dx^j$, or putting in the internal indices,

$$F = -\frac{i}{4} F_{ij}^c \sigma_c dx^i \wedge dx^j.$$

Similarly, we have

$$\delta A = -\frac{i}{2} \delta A_k^d \sigma_d dx^k.$$

Since $\text{tr}(\sigma_c \sigma_d) = 2\delta_{cd}$, it follows that in index notation

$$\delta S_{CS} = -\frac{1}{2} \int_\Sigma \epsilon^{ijk} F_{ijc} \delta A_k^c \, d^3x.$$

In functional derivative notation, the above equation is written

$$\frac{\partial}{\partial A_k^c(x)} S_{CS}(A) = -\frac{1}{2} \epsilon^{ijk} F_{ijc}.$$

It follows that

$$\frac{\partial}{\partial A_k^c}\Psi_{CS}(A) = \frac{\partial}{\partial A_k^c}e^{-\frac{6}{\Lambda}S_{CS}(A)}$$
$$= \frac{3}{\Lambda}\epsilon^{ijk}F_{ijc}e^{-\frac{6}{\Lambda}S_{CS}(A)}$$
$$= \frac{3}{\Lambda}\epsilon^{ijk}F_{ijc}\Psi_{CS}(A).$$

As a consequence,

$$\epsilon_{ijk}\frac{\partial}{\partial A_k^c}\Psi_{CS}(A) = \frac{6}{\Lambda}F_{ijc}\Psi_{CS}(A),$$

so

$$\hat{C}\Psi_{CS} = \epsilon^{abc}\frac{\partial}{\partial A_i^a}\frac{\partial}{\partial A_j^b}(F_{ijc} - \frac{\Lambda}{6}\epsilon_{ijk}\frac{\partial}{\partial A_k^c})\Psi_{CS} = 0.$$

What is the physical meaning of the Chern-Simons state? This is still unclear, but a clue is provided by the work of Kodama, who first wrote down this state. Kodama's work indicates that the Chern-Simons state is a quantized version of 'anti-deSitter space', a simple solution of the vacuum Einstein equations with nonzero cosmological constant. However, the extent to which the Chern-Simons state is physically realistic is still controversial. In particular, to better understand the dynamics of the Chern-Simons state, more work on the problem of time in quantum gravity is needed.

Recall from Chapter 5 of Part II that — at least formally — every 'measure' on $\mathcal{A}$ that is invariant under diffeomorphisms connected to the identity gives rise to a link invariant. Thus, in terms of the new variables, states of quantum gravity should yield invariants of links, as well as collections of loops that are not necessarily embedded. In particular, the 'measure'

$$\Psi_{CS}(A)\mathcal{D}A$$

corresponds to the Kauffman bracket link invariant! This hints at a relationship between knot theory and quantum gravity that we are currently only beginning to fathom. The 'loop representation' of quantum gravity proposed by Rovelli and Smolin exploits this relationship by expressing the Hamiltonian constraint as a constraint that an invariant of

The New Variables

generalized links must satisfy in order to come from a state of quantum gravity. At a formal level, this allows them to construct *many* solutions of the Wheeler-DeWitt equation starting from the simplest invariants of links, those which equal 1 on a given isotopy class of link and 0 on the rest. On the other hand, starting with the Chern-Simons state, Brügmann, Gambini, Pullin and others have been constructing solutions of the Wheeler-DeWitt equation from link invariants such as the coefficients of the Alexander-Conway polynomial. As with the Chern-Simons state, the physical significance of all these states is a matter of debate, and the mathematics involved has not been made rigorous. Nonetheless, a tantalizing picture has emerged in which links serve as *flux tubes of area* in the theory of quantum gravity, the area of a surface being given by the number of such tubes that pierce it times $\ell_P^2/2$, where ℓ_P is the Planck length. If something like this were the case, the relation between gauge fields, knots, and gravity would be a truly profound one.

Unfortunately, we must leave at this point, just when things are getting really interesting! The reader can continue to follow the story in some of the references provided in the Notes. We conclude with one more exercise, as a challenge, and one more quotation, for inspiration.

Exercise 53. *Construct a theory of physics reconciling gravity and quantum theory. (Hint: you may have to develop new mathematical tools.) Design and conduct experiments to test the theory.*

The way in which Faraday made use of his ideas of lines of force in coordinating the phenomena of magneto-electric induction shew him to have been in reality a mathematician of a very high order — one from whom the mathematicians of the future may derive valuable and fertile methods. For the advance of the exact sciences depends upon the discovery and development of appropriate and exact ideas, by means of which we may form a mental representation of the facts, sufficiently general, on the one hand, to stand for any particular case, and sufficiently exact, on the other, to warrant the deductions we may draw from them by the application of mathematical reasoning. From the straight line of Euclid to the lines of force of Faraday this has been the character of the ideas by which science has been advanced, and by the free use of dynamical as well as geometrical ideas we may hope for a further advance. The use of mathematical calculations is to compare the results of the application of these ideas with our measurements of the quantities concerned in our experiments... . We are probably ignorant even of the name of the science which will be developed out of the materials we are now collecting... .

James Clerk Maxwell

Notes to Part III

1. Semi-Riemannian Geometry

The quote by William Clifford is from the abstract of his paper for the the Cambridge Philosophical Society, 'On the space theory of matter', written in 1876, and reprinted in Volume 1 of *The World of Mathematics* by James Newman, Simon and Schuster, New York, 1956. The quote by Einstein and Grossman is from their 1913 paper, 'Entwerf einer verallgemeinerten Relativitätstheorie und einer Theorie der Gravitation', in *Zeit. Math. Phys.* **62**, 225-261, translated in Misner, Thorne and Wheeler's *Gravitation* (see below).

Most of the books on differential geometry listed in the notes to Chapter 2 of Part I discuss Riemannian or semi-Riemannian geometry; in particular, Choquet-Bruhat, DeWitt-Morette and Dillard-Bleick's book is a good overview. For a good book precisely on this one subject, we recommend *Semi-Riemannian geometry: with Applications to Relativity* by Barrett O'Neill, Academic Press, New York, 1983.

2. Einstein's Equation

The quote by Einstein and Grossman is from the 1913 paper cited above, while the quote by Einstein is from his 1916 paper 'Hamiltonsches Princip and allgemeine Relativitätstheorie', *Sitzungberichte der Preussichen Akad. Wissenschaften*, which was reprinted in translation in *The Principle of Relativity*, cited in the notes for Chapter 2 of Volume I.

A gentle introduction to the ideas of general relativity is Robert M. Wald's *Space, Time, and Gravity: the Theory of the Big Bang and Black Holes*, U. of Chicago Press, Chicago, 1977. To dig in a bit deeper but still not get overwhelmed by the details, try Hans Adolph Buchdahl's *Seventeen Simple Lectures on General Relativity Theory*, Wiley, New York, 1981, or *Essential Relativity; Special, General, and Cosmological*, by Wolfgang Rindler, Springer-Verlag, New York, 1979.

Eventually, however, one must get thoroughly immersed in the wonderful complexities of relativity, and for this there are two more books that stand out in our opinion. Splendidly original and eccentric, *Gravitation* by Charles W. Misner, Kip S. Thorne and John Archibald Wheeler, published by W. H. Freeman, San Francisco, 1973, was one of the first thorough treatments of general relativity to embrace modern coordinate-free notation. It has been suggested that the title is an allusion to the book's large mass: a wide variety of topics are covered in detail, for a grand total of 1279 pages. While some love and others hate this book, everybody interested in general relativity, especially quantum gravity, should have a copy. A more compact and up-to-date treatment is Robert M. Wald's *General Relativity*, U. of Chicago Press, Chicago, 1984, which is especially notable for good introductions to the ADM formalism, the singularity theorems and Hawking radiation. Wald gives a good description of the philosophy of abstract index notation.

It is worth noting that while energy and momentum are locally conserved, meaning that $\nabla^\mu T_{\mu\nu} = 0$, conservation of 'total' energy or momentum — meaning the existence of quantities given by integrals over space that do not change with the passage of time — is more problematic. As we note in Chapter 4, one obvious candidate for the total energy, namely the Hamiltonian for general relativity, vanishes thanks to Einstein's equations in the case of a universe with compact spacelike slices. See the books above for more on this interesting subject.

3. Lagrangians for General Relativity

The quote due to Maupertuis is from his 1746 paper 'Recherche des loix du mouvement', *Acad. R. Sci. Berlin*, and was taken in translation from Misner, Thorne and Wheeler's *Gravitation*, cited in the notes to Chapter 2.

Our version of the Palatini formalism is a relatively modern one; in the original one, the metric and a connection on the tangent bundle were taken as independent fields. A good introduction to the Einstein-Hilbert and Palatini actions appears in *Gravitation*, cited above; the treatment in Wald's *General Relativity* (also cited above) is also worth reading, as is that in Ashtekar's *Lectures on Non-perturbative Canonical Gravity*, cited in the notes to Chapter 5. Peter Peldan's paper 'Actions for gravity, with generalizations: a review', to appear in *Class. Quant. Grav.* sometime in 1994, presents a thorough discussion of various Lagrangians for general relativity.

4. The ADM Formalism

The quote from Eddington is from *The Nature of the Physical World*, chapter III, 'Time'. By the way, 'rashers' are essentially the same as bacon. Wald's *General Relativity* has a good introduction to the ADM formalism and the initial-value problem for general relativity, that is, the problem of constructing solutions from Cauchy data. *Gravitation*, while less detailed, is also worth reading for physical insight. Another good source for this subject and many other aspects of relativity, especially the singularity theorems, is *The Large Scale Structure of Space-Time*, by S. W. Hawking and G. F. R. Ellis, Cambridge U. Press, Cambridge, 1973.

The original paper by R. Arnowitt, S. Deser and C. S. Misner is still very much worth reading; it is 'The dynamics of general relativity', in *Gravitation: an Introduction to Current Research*, ed. Louis Witten, Wiley, New York, 1962, pp. 227-265. Since then there has been a lot of rigorous mathematical work on the ADM formalism, some of which is reviewed in A. Fischer and J. E. Marsden's paper 'The initial value problem and the dynamical formulation of general relativity', in *General Relativity, an Einstein Centenary Survey* ed. S. W. Hawking and W. Israel, Cambridge U. Press, Cambridge, 1979, pp. 138-211.

For the Hamiltonian approach to classical mechanics and classical field theory, see any of the texts on classical mechanics cited in the notes to Chapter 4 of Part II. Most of the texts on quantum mechanics and quantum field theory cited in the notes to Chapter 6 of Part I and Chapter 3 of part II discuss canonical quantization. In addition to a general knowledge of these subjects, work in gauge theory and quantum gravity requires an especially good understanding of the role played by constraints. Constraints are mentioned in passing in most of the quantum field theory books cited above, but for a deeper study of them, the place to start is still P. A. M. Dirac's *Lectures on quantum mechanics*, Yeshiva University, New York, 1964. For more modern ideas, try Marc Henneaux and Claudio Teitelboim's *Quantization of Gauge Systems*, Princeton U. Press, New Jersey, 1992, and the references therein.

A good place to start reading about canonical quantum gravity is the original series of papers by Bryce S. DeWitt, 'Quantum theory of gravity, I-III', *Phys. Rev.* **160** (1967), 1113-1148, **162** (1967) 1195-1239, 1239-1256. Discussions of canonical quantum gravity and other approaches to quantum gravity can also be found in *Quantum Gravity: An Oxford Symposium*, eds. Chris J. Isham, Roger Penrose, and Dennis W. Sciama, Oxford U. Press,

Oxford, 1975, and *Quantum Gravity 2: A Second Oxford Symposium*, with the same editors, Oxford U. Press, Oxford, 1981. For modern (circa 1990) work on canonical quantum gravity, the best place to start is with the two books by Ashtekar cited below.

5. The New Variables

Ashtekar and collaborators have already written a couple of excellent books summarizing the state of the art concerning canonical quantum gravity, particularly the new variables and the loop representation. In a sense, our book should be regarded as preparation for these. The first, *New Perspectives in Canonical Gravity*, lecture notes by Abhay Ashtekar and invited contributors, was published in 1988 by Bibliopolis, Napoli, Italy. It is available through the American Institute of Physics. Errata have been published as Syracuse University preprint by Joseph D. Romano and Ranjeet S. Tate, but are also available from the Center for Gravitational Physics and Geometry at Pennsylvania State University. The second, Abhay Ashtekar's *Lectures on Non-perturbative Canonical Gravity*, prepared in collaboration with Ranjeet Tate, was published in 1991 by World Scientific, Singapore.

The loop representation of quantum gravity was initiated in Lee Smolin and Carlo Rovelli's paper 'Loop representation for quantum general relativity', *Nucl. Phys.* **B331** (1990), 80-152, and this is still a good place to begin the serious study of it. A very good review article on the loop representation is 'Recent developments in nonperturbative quantum gravity' by Lee Smolin, in *Quantum Gravity and Cosmology: Proceedings of the XXIIth GIFT International Seminar on Theoretical Physics*, ed. Juan Perez-Mercader *et al*, World Scientific, Singapore 1992. There are more nice review articles in *Conceptual Problems of Quantum Gravity*, edited by Abhay Ashtekar and John Stachel, Birkhauser, Boston, 1991; this book also treats broader issues such as the problem of time. For a review of recent work that emphasizes the relationship to knot theory, try *Knots and Quantum Gravity*, ed. John Baez, Oxford U. Press, Oxford, 1994, which is the proceedings of a conference held in 1993. The volume *Knots, Topology and Quantum Field Theories*, cited in the notes to Chapter 5 of Part II, also has some articles on knots and quantum gravity.

H. Kodama's paper on the Chern-Simons state is 'Holomorphic wavefunction of the universe,' *Phys. Rev.* **D42** (1990), 2548-2565. For more on this state, see Jorge Pullin's review paper 'Knot theory and quantum gravity in loop space: a primer', to appear in *Proceedings of the Vth Mexican*

Notes to Part III 455

School of Particles and Fields, ed. J. L. Lucio, World Scientific, Singapore, and the references therein.

The quotation from Maxwell is from his article 'Faraday', and it appears in his *Scientific Papers*, cited in the notes to Chapter 5 of Part II. We found it in Yang's *Selected Papers*, as cited in the notes to Chapter 1 of Part II.

Index

abelian, 167
acceleration, 268
action, 136, 166, 267, 269, 398
action principle, 136, 267, 269
additive link invariant, 335
ADM formalism, 414, 425
Alexander-Conway polynomial, 327, 328
algebra, 25
    commutative, 25
    graded commutative, 61
    supercommutative, 61
alternating link diagram, 300
ambient isotopy, 297, 344
amphicheiral knot, 299, 304
angular momentum, 197
anti-deSitter space, 448
anti-self-duality, 97, 440, 442
antisymmetrization, 383
arc-connectedness, 105
arclength, 76
atlas, 19

baryon, 343
base space, 200
basepoint, 110
basis of sections, 208
beta-decay, 216
Bianchi identity, 253, 255, 257, 390

bilinear map, 74
    nondegenerate, 75
    symmetric, 74
blackboard framing, 306, 307
Bohm-Aharonov effect, 139
Borromean rings, 296
boson, 180
boundary, 116
bundle, 200
    locally trivial, 202, 205
    trivial, 202

cable, 325
canonical commutation relations, 428, 434
canonical quantization, 424
Cauchy data, 128, 129, 264, 414
charge density, 7
chart, 18, 115
    oriented, 85
Chern class, 281
Chern form, 280
Chern-Simons action, 287
Chern-Simons form, 285
Chern-Simons state, 438, 446
Chern-Simons theory, 287
Christoffel symbols, 374
    imitation, 408
closed form, 104
closed set, 16

cocycle, 179
    essential, 180
cocycle condition, 179, 213
Codazzi equation, 422
coframe field, 407
cohomologous forms, 123
cohomology class, 123
    integral, 282
color, 168, 219
commutativity, 37
commutator, 35, 186, 427
compactness, 21
components, 27, 54, 295, 367
configuration space, 425, 426, 429
connected component, 105
connected Lorentz group, 165
connected sum, 299
connected to the identity, 288
connectedness, 105
connection, 200, 223
    flat, 227, 244
    Lorentz, 407, 439, 443
    modulo gauge transformations, 229, 342, 349
    self-dual, 440, 443
constraint, 129
continuity, 17
continuity equation, 95, 388
contractible loop, 112
contraction, 368
contravariance, 32, 47
coordinate 1-forms, 50
coordinate transformation
    active, 52, 53
    passive, 52
coordinate vector fields, 50
core, 322

cosmological constant, 393, 438, 446
cotangent bundle, 210
cotangent vector, 44, 45, 47, 77
cotetrad, 407
cotriad, 407
covariance, 32
covariant derivative, 223, 234, 376, 379
covariant exterior derivative, 249
cover, 18, 181, 196
curl, 65
current, 93, 131, 261
current density, 7
curvature, 243
    self-dual, 441
curvature 2-form, 249
curve, 29

degree, 57
densitized tensor field, 444
deRham cohomology, 123
diagram, 298
diffeomorphism constraint, 433
differential, 41, 42, 63
differential form, 59
    E-valued, 249
    p-form, 60
    1-form, 40–44
    closed, 104
    End(E)-valued, 248
    exact, 104
Dirac algebra, 433
direct product, 168
direct sum, 58, 168, 169, 198
direct sum vector bundle, 210
disjoint union topology, 21

Index

disk D^n, 116
distance, 379
divergence, 65
divergence-free tensor, 388
double cover, 177, 178
doublet, 216
dreibein, 404
dual basis, 54, 209
dual form, 88
dual linear map, 47
dual vector bundle, 209
dual vector space, 46
duality, 9, 149

E-valued differential form, 249
Einstein summation convention, 24, 79
Einstein tensor, 382
Einstein's equation, 370, 387, 392, 393
    linearized, 402
    vacuum, 392
Einstein-Hilbert action, 398
electric field, 7, 104
    gravitational, 444
    Yang-Mills, 262, 264
electromagnetic field, 3, 12
electron, 5, 216
electroweak force, 167
embedding, 121, 297
empty link, 295
End(E)-valued 1-form, 225
End(E)-valued differential form, 248
endomorphism, 220
endomorphism bundle, 221
energy, 389

energy density, 81, 388
energy-momentum, 99
Euler-Lagrange equations, 271, 425
evolutionary equation, 129
exact form, 104
expected value, 134
exponential, 184
exponential map, 189
exterior algebra, 56, 59
exterior algebra bundle, 211
exterior covariant derivative, 250
exterior derivative, 41, 42, 48, 63–65
exterior product, 56
extrinsic curvature, 416

fermion, 180
fiber, 201, 206
fiberwise linearity, 206
figure-eight knot, 296
flavor, 219
flow, 35
flux, 114, 133
flux tube, 316, 449
force, 268
form bundle, 210
frame, 404
frame field, 404
    complex, 439
framed link, 306
framed Reidemeister moves, 308
framing, 306, 346
free particle, 428
frequency of a plane wave, 100
functional derivative, 432

G-bundle, 214

G-connection, 228
gauge boson, 180
gauge choice, 127
gauge equivalence, 229
gauge field, 5
gauge freedom, 126
gauge group, 214
gauge invariance, 206, 222, 241
gauge transformation, 126, 222
    large, 289
    small, 288, 446
gauge-invariant, 262
Gauss equation, 422
Gauss integral, 315
Gauss law, 129, 264, 439, 444
Gauss-Codazzi equations, 422
general linear group, 162, 166
geodesic, 379
gluon, 5
graded commutator, 258
graded cyclic property, 276
graded Jacobi identity, 258
gradient, 39, 40, 65
grand unified theory, 168
gravitational constant, 370
gravitational waves, 394, 402
graviton, 402
group, 162

hadron, 150, 216, 219, 343
Hamilton's equations, 426
Hamiltonian, 268, 425, 426, 430
    generating time evolution, 427, 428
Hamiltonian constraint, 433
Higgs boson, 6
Hodge star operator, 87, 88, 261

internal, 439
holonomy, 238, 247
homomorphism, 165, 191
homotopy, 107
Hopf link, 295, 305

identity, 162
identity component, 165, 177
identity loop, 239
indices
    abstract, 370
    dummy, 369
    internal, 225, 246, 405
    raising and lowering, 76, 79, 369
    spacelike, 420
    spacetime, 405
induced topology, 20
inner product, 80
inner product problem, 436
instanton, 98, 279
integral curve, 34
internal metric, 405
internal space, 404
intersection number, 320
intrinsic curvature, 420
invariant subspace, 170
inverse, 162
inverse path, 239
isomorphism, 166, 191, 203
isospin, 216
isotopic, 302
isotopy class, 299
isotopy invariant, 299

Jacobi identity, 38, 191
Jones polynomial, 340

Index

Kauffman bracket, 335
kinetic energy, 268
knot, 293
Kronecker delta, 43

Lagrangian, 135, 269, 341, 425
lapse function, 420
left circularly polarized wave, 101
left multiplication, 192
left-handed basis, 84
left-handed crossing, 310
Leibniz law, 26
length, 379
lepton, 5
level, 345
Levi-Civita connection, 372
    imitation, 408
Levi-Civita symbol, 91
Lie algebra, 183, 191
    semisimple, 198
Lie bracket, 35, 186
Lie group, 165
Lie subalgebra, 192
lightlike vector, 74
line bundle, 206
linearity over C^∞, 40
link, 294, 344
linking number, 310, 314
living in $\mathfrak{g}$, 215, 222
living in G, 214
local coordinates, 50
local trivialization, 205
locally constant function, 124
locally finite linear combination, 59
loop, 110
    contractible, 110, 111

loop number, 331
loop representation, 438
loop value, 335
Lorentz force law, 7
Lorentz group, 163
Lorentz transformation, 10
Lorentzian metric, 75

magnetic charge, 150
magnetic field, 7, 69, 131
    gravitational, 444
    Yang-Mills, 262, 264
magnetic flux, 133, 135
magnetic monopole, 10, 150
manifold, 18
    Lorentzian, 75
    orientable, 84
    oriented, 85
    Riemannian, 75
    semi-Riemannian, 75
    smooth, 20
    topological, 20
    with boundary, 115
map, 32
mass, 268
matrix groups, 162, 163, 165
matrix units, 221
Maxwell's equations, 3, 7, 69, 72, 92, 93
    static, 69, 70, 94
    vacuum, 8, 96
meson, 343
metric, 73, 75
metric-preserving connection, 372
Minkowski metric, 75
module, 27
Möbius strip, 84, 202, 203, 212

momentum, 99, 388, 389
    conjugate, 425
monopole, 150
morphism, 203, 221
multi-index, 65
multiplet, 219
multiplicative link invariant, 336
muon, 5, 217

naturality, 48, 67
negative orientation, 84
neighborhood, 16
neutrino, 5, 216, 217
neutron, 5, 216
new variables, 437
Newton's law, 271
nondegeneracy, 80
nonorientability, 84
nucleon, 216, 217
nugatory crossing, 300, 301
null vector, 74, 75

observable, 134, 426
one-form, *see* differential form
open set, 16, 17
operator ordering problems, 434, 445
orientation, 82, 84
orientation-preserving map, 85
oriented link, 305
orthogonal group, 163
orthonormal basis, 75

p-form, *see* differential form
Palatini action, 408, 409
Palatini formalism, 404–406
pancake proof, 318
parallel translation, 234

parallel transport, 231, 234, 379
parity, 61
partition function, 333, 342
partition of unity, 118
path, 104
path integral, 135, 137, 341
path-ordered exponential, 236
path-ordered product, 236
paths
    composable, 238
    product of, 238
Pauli matrices, 172
Perko pair, 300
phase, 134, 166
phase space, 426
    physical, 431
photon, 5
physical state, 435
physical state space, 436, 445
pion, 217
Planck length, 6, 449
Planck's constant, 135, 152
plane wave, 99
    frequency, 100
Poincaré group, 164
Poisson bracket, 427
position, 268
positive orientation, 84
potential energy, 268
prime knot, 299
problem of time, 436
product, 162
product topology, 21
projection map, 200
proper time, 76
proton, 5, 216
pullback, 31, 32, 47, 53, 61

Index

pushforward, 32, 33

quantum chromodynamics, 219
quantum group, 350
quark, 5, 219, 220
quaternions, 173

rapidity, 10
reality conditions, 443
regular isotopy, 309
Reidemeister move, 301–303, 308, 336–338
representation
    contragredient, 175
    dual, 175
    equivalent, 168
    fundamental, 169
    irreducible, 170
    of a group, 166
    of a Lie algebra, 197
    projective, 179
    trivial, 175
restriction, 204
Ricci flat metric, 392
Ricci scalar, 382
    imitation, 408
Ricci tensor, 382
    imitation, 408
Riemann curvature tensor, 380
Riemann tensor
    imitation, 408
Riemannian metric, 75
right-handed basis, 84
right-handed crossing, 310

scalar curvature, 382
scalar field, 366
scalar potential, 104

Schur's lemma, 171
section, 200, 207, 208
    over a set, 205
Seifert surface, 319
self-dual action, 440
self-duality, 97, 279, 440–442
self-linking number, 311
semi-Riemannian manifold, 75
semi-Riemannian metric, 74
shift vector field, 420
sign, 310
signature, 75
simply connected, 108
site, 333
skein relations, 312
    for Alexander-Conway polynomial, 327
    for Jones polynomial, 340
    for linking number, 314
    for writhe, 313
skew-adjoint, 187, 189
slice, 414
smearing, 346
smooth function, 19, 115, 116
smooth manifold, 20
smooth map, 32
smoothness, 7
solenoid, 132
spacelike slice, 414
spacelike vector, 74, 75
spacetime, 72
    splitting, 72
spacetime interval, 73, 74
special linear group, 163
special orthogonal group, 163
special unitary group, 164
specific heat, 333

sphere S^n, 15, 16, 20
spin-down state, 198, 216
spin-up state, 197, 216
standard fiber, 202, 205, 214
standard flat connection, 227
standard model, 5
standard orientation, 84
star operator, 58
state, 330
state sum, 331
state vector, 134
static spacetime, 76
Stokes' theorem, 119
strand, 302
stress-energy tensor, 387
strong force, 167
subgroup, 162
submanifold, 120
    with boundary, 121
subrepresentation, 170
superconductivity, 317
superfluidity, 317
superspace, 429
symmetrization, 383

tangent bundle, 201
    complexified, 439
tangent space, 28, 116
tangent vector, 27, 28, 30, 77
    to a curve, 29
tau particle, 5
temporal gauge, 127, 231
tensor, 366
tensor field, 366
tensor product, 169, 170
tensor product vector bundle, 210
tetrad, 404

three-metric, 415
time coordinate, 413
time-reversal, 164
timelike vector, 74, 75
topological space, 16
topology, 16
torsion-free connection, 372
torus T^n, 141, 142
total divergence, 401
total space, 200
trace, 172, 272
traceless matrix, 172
transition amplitude, 134
transition function, 18, 19, 212
transition probability, 134
transversality, 319
trefoil knot, 294, 314
triad, 404
triplet, 217

unitary group, 164
unitary matrix, 164
unitary representation, 178
unknot, 294, 312

vacuum expectation value, 342
variation, 270, 271, 274, 432
vector bundle
    complex, 206
    real, 205
vector bundle morphism, 206
vector field, 24, 25
    integrable, 35
    left-invariant, 192
vector fields
    basis of, 27
    linearly independent, 27

spanning, 27
vector potential, 104, 126, 224, 225
velocity, 268, 426
vierbein, 404
volume element, 83
volume form, 83, 85
    associated to a metric, 86
vortex lines, 292, 317
vorticity, 292

W boson, 5
wavefunction, 136
wedge product, 56
Wheeler-DeWitt equation, 435, 437
Whitehead link, 311
Whitney trick, 308
Wilson loop, 242, 342, 343, 438
winding number, 111
wormhole, 141
writhe, 311, 312, 314

Yang-Mills action, 274, 279
Yang-Mills equation, 261, 262, 277, 278
Yang-Mills Lagrangian, 273

Z boson, 5

SERIES ON KNOTS AND EVERYTHING

Editor-in-charge: Louis H. Kauffman *(Univ. of Illinois, Chicago)*

The Series on Knots and Everything: is a book series polarized around the theory of knots. Volume 1 in the series is Louis H Kauffman's Knots and Physics.

One purpose of this series is to continue the exploration of many of the themes indicated in Volume 1. These themes reach out beyond knot theory into physics, mathematics, logic, linguistics, philosophy, biology and practical experience. All of these outreaches have relations with knot theory when knot theory is regarded as a pivot or meeting place for apparently separate ideas. Knots act as such a pivotal place. We do not fully understand why this is so. The series represents stages in the exploration of this nexus.

Details of the titles in this series to date give a picture of the enterprise.

Published*:

- Vol. 1: Knots and Physics (3rd Edition)
  by L. H. Kauffman
- Vol. 2: How Surfaces Intersect in Space — An Introduction to Topology (2nd Edition)
  by J. S. Carter
- Vol. 3: Quantum Topology
  edited by L. H. Kauffman & R. A. Baadhio
- Vol. 4: Gauge Fields, Knots and Gravity
  by J. Baez & J. P. Muniain
- Vol. 5: Gems, Computers and Attractors for 3-Manifolds
  by S. Lins
- Vol. 6: Knots and Applications
  edited by L. H. Kauffman
- Vol. 7: Random Knotting and Linking
  edited by K. C. Millett & D. W. Sumners
- Vol. 8: Symmetric Bends: How to Join Two Lengths of Cord
  by R. E. Miles
- Vol. 9: Combinatorial Physics
  by T. Bastin & C. W. Kilmister
- Vol. 10: Nonstandard Logics and Nonstandard Metrics in Physics
  by W. M. Honig
- Vol. 11: History and Science of Knots
  edited by J. C. Turner & P. van de Griend

*The complete list of the published volumes in the series can also be found at
http://www.worldscientific.com/series/skae

Vol. 12: Relativistic Reality: A Modern View
edited by J. D. Edmonds, Jr.

Vol. 13: Entropic Spacetime Theory
by J. Armel

Vol. 14: Diamond — A Paradox Logic
by N. S. Hellerstein

Vol. 15: Lectures at KNOTS '96
by S. Suzuki

Vol. 16: Delta — A Paradox Logic
by N. S. Hellerstein

Vol. 17: Hypercomplex Iterations — Distance Estimation and Higher Dimensional Fractals
by Y. Dang, L. H. Kauffman & D. Sandin

Vol. 18: The Self-Evolving Cosmos: A Phenomenological Approach to Nature's Unity-in-Diversity
by S. M. Rosen

Vol. 19: Ideal Knots
by A. Stasiak, V. Katritch & L. H. Kauffman

Vol. 20: The Mystery of Knots — Computer Programming for Knot Tabulation
by C. N. Aneziris

Vol. 21: LINKNOT: Knot Theory by Computer
by S. Jablan & R. Sazdanovic

Vol. 22: The Mathematics of Harmony — From Euclid to Contemporary Mathematics and Computer Science
by A. Stakhov (assisted by S. Olsen)

Vol. 23: Diamond: A Paradox Logic (2nd Edition)
by N. S. Hellerstein

Vol. 24: Knots in HELLAS '98 — Proceedings of the International Conference on Knot Theory and Its Ramifications
edited by C. McA Gordon, V. F. R. Jones, L. Kauffman, S. Lambropoulou & J. H. Przytycki

Vol. 25: Connections — The Geometric Bridge between Art and Science (2nd Edition)
by J. Kappraff

Vol. 26: Functorial Knot Theory — Categories of Tangles, Coherence, Categorical Deformations, and Topological Invariants
by David N. Yetter

Vol. 27: Bit-String Physics: A Finite and Discrete Approach to Natural Philosophy
by H. Pierre Noyes; edited by J. C. van den Berg

Vol. 28: Beyond Measure: A Guided Tour Through Nature, Myth, and Number
by J. Kappraff

Vol. 29: Quantum Invariants — A Study of Knots, 3-Manifolds, and Their Sets
by T. Ohtsuki

Vol. 30: Symmetry, Ornament and Modularity
by S. V. Jablan

Vol. 31: Mindsteps to the Cosmos
by G. S. Hawkins

Vol. 32: Algebraic Invariants of Links
by J. A. Hillman

Vol. 33: Energy of Knots and Conformal Geometry
by J. O'Hara

Vol. 34: Woods Hole Mathematics — Perspectives in Mathematics and Physics
edited by N. Tongring & R. C. Penner

Vol. 35: BIOS — A Study of Creation
by H. Sabelli

Vol. 36: Physical and Numerical Models in Knot Theory
edited by J. A. Calvo et al.

Vol. 37: Geometry, Language, and Strategy
by G. H. Thomas

Vol. 38: Current Developments in Mathematical Biology
edited by K. Mahdavi, R. Culshaw & J. Boucher

Vol. 39: Topological Library
Part 1: Cobordisms and Their Applications
edited by S. P. Novikov & I. A. Taimanov

Vol. 40: Intelligence of Low Dimensional Topology 2006
edited by J. Scott Carter et al.

Vol. 41: Zero to Infinity: The Fountations of Physics
by P. Rowlands

Vol. 42: The Origin of Discrete Particles
by T. Bastin & C. Kilmister

Vol. 43: The Holographic Anthropic Multiverse
by R. L. Amoroso & E. A. Ranscher

Vol. 44: Topological Library
Part 2: Characteristic Classes and Smooth Structures on Manifolds
edited by S. P. Novikov & I. A. Taimanov

Vol. 45: Orbiting the Moons of Pluto
Complex Solutions to the Einstein, Maxwell, Schrödinger and Dirac Equations
by E. A. Rauscher & R. L. Amoroso

Vol. 46: Introductory Lectures on Knot Theory
edited by L. H. Kauffman, S. Lambropoulou, S. Jablan & J. H. Przytycki

Vol. 47: Introduction to the Anisotropic Geometrodynamics
by S. Siparov

Vol. 48: An Excursion in Diagrammatic Algebra: Turning a Sphere from Red to Blue
by J. S. Carter

Vol. 49: Hopf Algebras
by D. E. Radford

Vol. 50: Topological Library
Part 3: Spectral Sequences in Topology
edited by S. P. Novikov & I. A. Taimanov

Vol. 51: Virtual Knots: The State of the Art
by V. O. Manturov & D. P. Ilyutko

Vol. 52: Algebraic Invariants of Links (2nd Edition)
by J. Hillman

Vol. 53: Knots and Physics (4th Edition)
by L. H. Kauffman

Vol. 54: Scientific Essays in Honor of H Pierre Noyes on the Occasion of His 90th Birthday
edited by J. C. Amson & L. H. Kauffman

Vol. 55: Knots, Braids and Möbius Strips
by J. Avrin

Vol. 56: New Ideas in Low Dimensional Topology
edited by L. H. Kauffman & V. O. Manturov